Nitric Oxide Protocols

METHODS IN MOLECULAR BIOLOGY™

John M. Walker, SERIES EDITOR

100. **Nitric Oxide Protocols**, edited by *M. A. Titheradge, 1998*
99. **Human Cytokines and Cytokine Receptors**, edited by *Reno Debets, 1998*
98. **DNA Profiling Protocols**, edited by *James M. Thomson, 1997*
97. **Molecular Embryology:** *Methods and Protocols,* edited by *Paul T. Sharpe, 1998*
96. **Adhesion Proteins Protocols**, edited by *Elisabetta Dejana, 1998*
95. **DNA Topology and DNA Topoisomerases:** *II. Enzymology and Topoisomerase Targetted Drugs,* edited by *Mary-Ann Bjornsti, 1998*
94. **DNA Topology and DNA Topoisomerases:** *I. DNA Topology and Enzyme Purification,* edited by *Mary-Ann Bjornsti, 1998*
93. **Protein Phosphatase Protocols**, edited by *John W. Ludlow, 1998*
92. **PCR in Bioanalysis**, edited by *Stephen Meltzer, 1997*
91. **Flow Cytometry Protocols**, edited by *Mark J. Jaroszeski, 1998*
90. **Drug–DNA Interactions:** *Methods, Case Studies, and Protocols,* edited by *Keith R. Fox, 1997*
89. **Retinoid Protocols**, edited by *Christopher Redfern, 1997*
88. **Protein Targeting Protocols**, edited by *Roger A. Clegg, 1997*
87. **Combinatorial Peptide Library Protocols**, edited by *Shmuel Cabilly, 1997*
86. **RNA Isolation and Characterization Protocols**, edited by *Ralph Rapley, 1997*
85. **Differential Display Methods and Protocols**, edited by *Peng Liang and Arthur B. Pardee, 1997*
84. **Transmembrane Signaling Protocols**, edited by *Dafna Bar-Sagi, 1997*
83. **Receptor Signal Transduction Protocols,** edited by *R. A. J. Challiss, 1997*
82. ***Arabidopsis* Protocols,** edited by *José M Martinez-Zapater and Julio Salinas, 1998*
81. **Plant Virology Protocols,** edited by *Gary D. Foster, 1998*
80. **Immunochemical Protocols,** *SECOND EDITION,* edited by *John Pound, 1998*
79. **Polyamine Protocols,** edited by *David M. L. Morgan, 1998*
78. **Antibacterial Peptide Protocols,** edited by *William M. Shafer, 1997*
77. **Protein Synthesis:** *Methods and Protocols,* edited by *Robin Martin, 1998*
76. **Glycoanalysis Protocols,** edited by *Elizabeth F. Hounsel, 1998*
75. **Basic Cell Culture Protocols,** edited by *Jeffrey W. Pollard and John M. Walker, 1997*
74. **Ribozyme Protocols,** edited by *Philip C. Turner, 1997*
73. **Neuropeptide Protocols,** edited by *G. Brent Irvine and Carvell H. Williams, 1997*
72. **Neurotransmitter Methods,** edited by *Richard C. Rayne, 1997*
71. **PRINS and *In Situ* PCR Protocols,** edited by *John R. Gosden, 1997*
70. **Sequence Data Analysis Guidebook,** edited by *Simon R. Swindell, 1997*
69. **cDNA Library Protocols,** edited by *Ian G. Cowell and Caroline A. Austin, 1997*
68. **Gene Isolation and Mapping Protocols,** edited by *Jacqueline Boultwood, 1997*
67. **PCR Cloning Protocols:** *From Molecular Cloning to Genetic Engineering,* edited by *Bruce A. White, 1996*
66. **Epitope Mapping Protocols,** edited by *Glenn E. Morris, 1996*
65. **PCR Sequencing Protocols,** edited by *Ralph Rapley, 1996*
64. **Protein Sequencing Protocols,** edited by *Bryan J. Smith, 1996*
63. **Recombinant Proteins:** *Detection and Isolation Protocols,* edited by *Rocky S. Tuan, 1996*
62. **Recombinant Gene Expression Protocols,** edited by *Rocky S. Tuan, 1996*
61. **Protein and Peptide Analysis by Mass Spectrometry,** edited by *John R. Chapman, 1996*
60. **Protein NMR Protocols,** edited by *David G. Reid, 1996*
59. **Protein Purification Protocols,** edited by *Shawn Doonan, 1996*
58. **Basic DNA and RNA Protocols,** edited by *Adrian J. Harwood, 1996*
57. **In Vitro Mutagenesis Protocols,** edited by *Michael K. Trower, 1996*
56. **Crystallographic Methods and Protocols,** edited by *Christopher Jones, Barbara Mulloy, and Mark Sanderson, 1996*
55. **Plant Cell Electroporation and Electrofusion Protocols,** edited by *Jac A. Nickoloff, 1995*
54. **YAC Protocols,** edited by *David Markie, 1995*
53. **Yeast Protocols:** *Methods in Cell and Molecular Biology,* edited by *Ivor H. Evans, 1996*
52. **Capillary Electrophoresis:** *Principles, Instrumentation, and Applications,* edited by *Kevin D. Altria, 1996*
51. **Antibody Engineering Protocols,** edited by *Sudhir Paul, 1995*
50. **Species Diagnostics Protocols:** *PCR and Other Nucleic Acid Methods,* edited by *Justin P. Clapp, 1996*
49. **Plant Gene Transfer and Expression Protocols,** edited by *Heddwyn Jones, 1995*
48. **Animal Cell Electroporation and Electrofusion Protocols,** edited by *Jac A. Nickoloff, 1995*
47. **Electroporation Protocols for Microorganisms,** edited by *Jac A. Nickoloff, 1995*
46. **Diagnostic Bacteriology Protocols,** edited by *Jenny Howard and David M. Whitcombe, 1995*
45. **Monoclonal Antibody Protocols,** edited by *William C. Davis, 1995*
44. ***Agrobacterium* Protocols,** edited by *Kevan M. A. Gartland and Michael R. Davey, 1995*
43. **In Vitro Toxicity Testing Protocols,** edited by *Sheila O'Hare and Chris K. Atterwill, 1995*
42. **ELISA:** *Theory and Practice*, by *John R. Crowther, 1995*
41. **Signal Transduction Protocols,** edited by *David A. Kendall and Stephen J. Hill, 1995*
40. **Protein Stability and Folding:** *Theory and Practice,* edited by *Bret A. Shirley, 1995*
39. **Baculovirus Expression Protocols,** edited by *Christopher D. Richardson, 1995*
38. **Cryopreservation and Freeze-Drying Protocols,** edited by *John G. Day and Mark R. McLellan, 1995*
37. **In Vitro Transcription and Translation Protocols,** edited by *Martin J. Tymms, 1995*

METHODS IN MOLECULAR BIOLOGY™

Nitric Oxide Protocols

Edited by

Michael A. Titheradge

University of Sussex, Brighton, UK

Humana Press Totowa, New Jersey

999 Riverview Drive, Suite 208
Totowa, New Jersey 07512

This publication is printed on acid-free paper. ∞
ANSI Z39.48-1984 (American Standards Institute)
Permanence of Paper for Printed Library Materials.

Cover design by Patricia Cleary; from Fig. 1 of "Methods for the Study of NO-Induced Apoptosis in Cultured Cells," by Anne C. Loweth and Noel G. Morgan.

For additional copies, pricing for bulk purchases, and/or information about other Humana titles, contact Humana at the above address or at any of the following numbers: Tel.: 973-256-1699; Fax: 973-256-8341; E-mail: humana@mindspring.com; or visit our Website: http://humanapress.com

Printed in the United States of America. 10 9 8 7 6 5 4 3 2 1

Preface

It is now recognized that nitric oxide (NO) plays an essential role in many biological systems, both as an inter- and intracellular signaling mechanism in such diverse areas as the vascular system, the immune system, and neural communication. In addition, overproduction of NO has also been implicated as a crucial factor in many pathological situations, particularly in inflammation, diabetes, stroke, neurodegeneration, and sepsis. The involvement of NO in physiological and pathological situations is now far-ranging, and interest in the biochemistry, physiology, and pharmacology of NO is still expanding rapidly. Therefore, it seems an appropriate time to produce a book containing protocols relevant to all workers in the nitric oxide field.

In *Nitric Oxide Protocols* I have attempted to gather together chapters from all areas of nitric oxide research and to provide detailed methods covering a very wide variety of techniques, including cloning, expression, and purification of the different NO synthase isoforms; quantitation of the rate of transcription and translation of NO synthases using RT-PCR, Northern, and Western blotting; quantitation of NO production itself both in vivo and in vitro; and direct measurement of the activity of NO synthase. In addition, a number of issues—such as the use NO donors, peroxynitrite, and NO gas to mimic endogenous NO production—have been included, together with chapters on the use of inhibitors of NO synthase and the measurement of nitrotyrosine residues in proteins,on DNA damage, and on apoptosis caused by NO production. Two chapters have also been included on the assay of GTP cyclohydrolase I activity and the measurement of biopterin, and since these are particularly relevant to NO production, on the use of inhibitors of biopterin biosynthesis. Throughout the book, the aim has been to highlight the merits of each assay or procedure and compare them with other available methods, and where possible a number of different alternative procedures are described. Potential problems and common errors encountered with each of these protocols have been highlighted, and in many cases the Introduction and Notes sections of related chapters provide a valuable source of information. The reader is therefore encouraged to browse through related chapters to extract the maximum benefit.

This book could not have been produced without the help and cooperation of all those authors who kindly contributed their chapters, and I should like to thank them for their efforts. In particular I should like to thank Richard Knowles both for his chapters and also advice as to the content of the book. I should also like to thank the series editor, John Walker, for his rapid response and constant advice and encouragement throughout the preparation of the book.

Michael A. Titheradge

Contents

Contributors

STEPHEN L. ARCHER • *Cardiovascular Section (111C), Minneapolis VA Medical Center, Minneapolis, MN*
JOSEPH BECKMAN • *Department of Anesthesiology, University of Alabama at Birmingham, AL*
TIMOTHY R. BILLIAR • *Department of Surgery, University of Pittsburgh School of Medicine, Presbyterian University Hospital, Pittsburgh, PA*
LISARDO BOSCÁ • *Facultad de Farmacia, Instituto de Bioquímica CSIC-UCM, Universidad Complutense, Madrid, Spain*
VICTORIA CATTELL • *Department of Histopathology, Imperial College School of Medicine at St. Mary's, London, UK*
IAN G. CHARLES • *The Cruciform Project, The Rayne Institute, London, UK*
ANN CHUBB • *The Cruciform Project, The Rayne Institute, London, UK*
H. TERENCE COOK • *Department of Histopathology, Imperial College School of Medicine at St. Mary's, London, UK*
JOHN P. CROW • *Department of Anesthesiology, University of Alabama at Birmingham, AL*
JAMES C. CUNNINGHAM • *Department of Pharmacy, University of Brighton, Brighton, UK*
VICTOR M. DARLEY-USMAR • *Molecular and Cellular Division, Department of Pathology, University of Alabama at Birmingham,AL*
JOHN DAWSON • *Glaxo-Wellcome Research and Development, Stevenage, UK*
WOLFGANG DESCH • *Institut für Mathematik, Karl-Franzens-Universität Graz, Graz, Austria*
NEALE FOXWELL • *The Cruciform Project, The Rayne Institute, London, UK*
SARAH E. GARNER • *Institute for Environmental Medicine, University of Pennsylvania, Philadelphia, PA*
EDWARD P. GARVEY • *Glaxo Wellcome, Research Triangle Park, NC*
ANDREW J. GOW • *Institute for Environmental Medicine, University of Pennsylvania, Philadelphia, PA*
IRENE C. GREEN • *Department of Biochemistry and Molecular Genetics, University of Sussex, Brighton, UK*
MICHAEL H. L. GREEN • *MRC Cell Mutation Unit, University of Sussex, Brighton, UK*

Kazuyuki Hatakeyama • *Department of Surgery, University of Pittsburgh, Pittsburgh, PA*
Sunna Hauschild • *Department of Immunobiology, Institute of Zoology, University of Leipzig, Germany*
Benjamin Hemmens • *Institut für Pharmakologie und Toxikologie, Karl-Franzens-Universität Graz, Graz, Austria*
Neil Hogg • *Biophysics Research Institute, Medical College of Wisconsin, Milwaukee, WI*
Harry Ischiropoulos • *Institute for Environmental Medicine, University of Pennsylvania, Philadelphia, PA*
Mark L. Johnson • *Department of Surgery, Presbyterian University Hospital, University of Pittsburgh School of Medicine, Pittsburgh, PA*
B. Kalyanaraman • *Biophysics Research Institute, Medical College of Wisconsin, Milwaukee, WI*
Peter Klat • *Institut für Pharmakologie und Toxikologie, Technische Universität München, Germany*
Richard G. Knowles • *Enzyme Pharmacology Group, Glaxo-Wellcome Research, Stevenage, UK*
Walter R. Kukovetz • *Institut für Pharmakologie und Toxikologie, Karl-Franzens-Universität Graz, Graz, Austria*
Suzanne Laychock • *Department of Pharmacology, SUNY at Buffalo School of Medicine, Buffalo, NY*
Ignacio Lizasoain • *Departamento de Farmacologica, Facultad Medicina, Universidad Complutense de Madrid, Spain*
Jillian E. Lowe • *MRC Cell Mutation Unit, University of Sussex, Brighton, UK*
Anne Loweth • *Department of Biological Sciences, University of Keele, Staffs, UK*
Paloma Martín-Sanz • *Instituto de Bioquímica CSIC-UCM, Facultad de Farmacia, Universidad Complutense, Madrid, Spain.*
Bernd Mayer • *Institut für Pharmakologie und Toxikologie, Karl-Franzens-Universität Graz, Austria*
Stuart Malcom • *Biokinetics Laboratory, Temple University, Philadelphia, PA*
Molly McClelland • *Institute for Environmental Medicine, University of Pennsylvania, Philadelphia, PA*
Evangelos D. Michelakis • *Cardiovascular Section (111C), Minneapolis VA Medical Center, Minneapolis, MN*
Noel G. Morgan • *Department of Biological Sciences, University of Keele, Staffs, UK*
María A. Moro • *Departamento de Farmacologica, Facultad Medicina, Universidad Complutense de Madrid, Spain*

KAREN MOSLEY • *Department of Histopathology, Imperial College School of Medicine at St. Mary's, London, UK*
S. TSUYOSHI OHNISHI • *Philadelphia Biomedical Research Institute, King of Prussia, PA*
RICHARD C. RAYNE • *Department of Biology, Birkbeck College, University of London, UK*
DARYL D. REES • *Centre for Clinical Pharmacology, University College London, The Rayne Institute, London, UK*
NORBERT REILING • *The Picower Institute for Medical Research, Manhasset, NY*
R. J. RUSSELL • *Discovery Biology II, Pfizer Ltd, Sandwich, UK*
MARK SALTER • *Lead Discovery Group, Glaxo-Wellcome Research, Stevenage, UK*
RICHARD A. SHAPIRO • *Department of Surgery, University of Pittsburgh School of Medicine, Presbyterian University Hospital, Pittsburgh, PA*
KURT SCHMIDT • *Institut für Pharmakologie und Toxikologie, Karl-Franzens-Universität Graz, Graz, Austria*
FIONA S. SMITH • *School of Biological Sciences, University of Sussex, Brighton, UK*
NATHAN SPEAR • *Department of Anesthesiology, University of Alabama at Birmingham, AL*
G. MICHAEL TAYLOR • *Department of Respiratory Medicine, Imperial College School of Medicine at St. Mary's, London, UK*
STEVEN THOMAS • *Department of Biochemistry and Molecular Genetics, University of Sussex, Brighton, UK*
MICHAEL A. TITHERADGE • *School of Biological Sciences, University of Sussex, Brighton, UK*
JAYNE TULLETT • *MRC Toxicology Unit, University of Leicester, Leicester, UK*
ARTUR J. ULMER • *Department of Immunology and Cell Biology, Forschungszentrum Borstel, Germany*
C. ROGER WHITE • *Vascular Biology and Hypertension Program, Department of Medicine, University of Alabama at Birmingham, AL*

Barry W. Allen

1

Enzymology of Nitric Oxide Synthases

Benjamin Hemmens and Bernd Mayer

1. Introduction

1.1. A Multifaceted Enzyme Family

Biologically produced nitric oxide (NO) originates from oxygen and L-arginine in the reaction catalyzed by NO synthase (NOS) enzymes (NOS, EC 1.14.13.39), first reported in the late 1980s. It was soon discovered that these enzymes have different biochemical properties depending on the tissue from which they are isolated. This suggested that there are special regulatory mechanisms that match the supply of NO exactly to its function in each tissue where it is produced. Subsequent research has confirmed that indeed these enzymes are subject to sophisticated and complex regulation. It is indispensable to understand this regulation better, both to understand the physiology of NO, and to pick targets for new therapeutic strategies.

NOS enzymes are regulated partly at the level of gene expression. However, the enzymes themselves possess intricate control properties, and it is these, and their basis in the protein structure of the enzymes, that are the concern of the following discussion. At a first glance, NOS may seem to be a colorful patchwork of familiar motifs borrowed from other enzymes: for example, a P450 heme, a pteridine cofactor, flavin adenine dinucleotide (FAD), and flavin mononucleotide (FMN)-containing reductase domains, and a calmodulin (CaM)-dependent activation. On closer inspection, though, the participation of each of these elements in NOS has unique aspects that could not have been guessed from other enzymes. Thus, NOS contains important lessons of general enzymological interest.

We will concentrate here on more recent developments: comprehensive reviews of earlier work are already available (e.g., **refs. *1–3***).

From: *Methods in Molecular Biology, Vol. 100. Nitric Oxide Protocols*
Edited by: Michael A. Titheradge © Humana Press Inc., Totowa, NJ

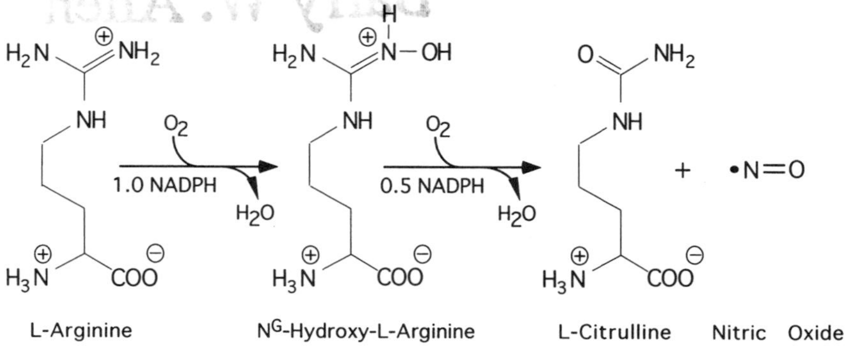

L-Arginine — N^G-Hydroxy-L-Arginine — L-Citrulline — Nitric Oxide

Fig. 1. The NOS reaction.

1.2. Three NOS Isoenzymes

Three different isoenzymes of NOS have been identified that catalyze the oxidation of L-arginine to L-citrulline and NO: neuronal (nNOS), endothelial (eNOS), and inducible NOS (iNOS). Their distinctive properties, tissue distribution, subcellular localization, and the terminology used in the literature to describe the isoenzymes are summarized in **Table 1**.

1.3. Reactions and Assays

The overall reaction catalyzed by NOS is shown in **Fig. 1**. NO synthesis is dependent on reducing equivalents derived from reduced nicotinamide adenine dinucleotide phosphate (NADPH) *(4)*, and requires molecular oxygen as a substrate *(5,6)*. The conversion of L-arginine to L-citrulline with production of NO takes place in two steps with N^G-hydroxy-L-arginine (NOHLA) as an intermediate *(7,8)*. Two moles of O_2 (Abu-Soud, H. M., Presta, A., Baek, K. J., and Stuehr, D. J. unpublished data) and 1.5 mol of NADPH *(6,7)* are consumed per mole of formed products.

The reaction mechanism has been discussed comprehensively in earlier reviews *(1,9,10)*. Both reaction steps are catalyzed by heme iron *(11)*. The initial hydroxylation of L-arginine is formally equivalent to a typical cytochrome P450 monoxygenase reaction. The second step involving 1-electron oxidation of the intermediate NOHLA is unusual but not unique among cytochrome P450-catalyzed reactions. Conventional P450s were shown to catalyze oxidation of NOHLA and related guanidines to NO and the corresponding urea derivative *(12)*. Turnover rates are low, but NO is known to inactivate P450s through high-affinity binding to ferrous heme, resulting in formation of rather stable inhibitory nitrosyl–heme complexes.

Table 1
The Isoenzymes of NOS

NOS isoenzyme	Alternative description	Human chromosome	Molecular mass, kDa	Distinctive properties	Subcellular localization	Tissue expression
Neuronal	Type I nNOS ncNOS bNOS	12	160	Ca^{2+}-dependent constitutively expressed	Binds to specific proteins via an N-terminal PDZ domain	Neuronal cells Skeletal muscle
Endothelial	Type III eNOS ecNOS	7	134	Ca^{2+}-dependent constitutively expressed	Targets to the Golgi and to calveoli via N-terminal myristoylation and palmitoylation	Endothelial cells Epithelial cells Cardiomyocytes
Inducible	Type II iNOS macNOS	17	130	Ca^{2+}-independent induced by inflammatory stimuli (cytokines, LPS)	Soluble?	Macrophages Hepatocytes Astrocytes Smooth muscle cells (and many more)

Formation of NO by purified NOS enzymes has never been directly detected using the NO electrode. There are two usual assays for the full NOS reaction. One follows the conversion of tritiated L-arg to L-citrulline, whereas the other is based on the reaction of NO (or its breakdown products) with hemoglobin, which can be measured photometrically *(13)*.

In the following discussion, we also refer to two other assays that are used to follow part-reactions catalyzed by NOS. One is the NADPH:cytochrome-c-reductase reaction, in which the reduction of cytochrome c is measured at 550 nm; the other is the uncoupled NADPH-oxidase reaction, which results in the formation of H_2O_2.

2. Recent Highlights

2.1. Dual Role of (6R)-5,6,7,8-tetrahydro-L-biopterin

All NOS enzymes contain a (6R)-5,6,7,8-tetrahydro-L-biopterin (BH_4) cofactor. The role of BH_4 in NOS is quite unlike anything known from other enzymes in which it is a cofactor. It has been known for some time that BH_4 is an allosteric activator of NOS: we discuss new developments in our understanding of this effect. However, it is also becoming clear that beyond its allosteric role, BH_4 is absolutely necessary for catalytic turnover of the enzyme, and that its redox properties are important.

2.2. Uncoupling of NADPH Oxidation

In vitro, all NOS enzymes can be persuaded to catalyze NADPH oxidation, leading to the production of superoxide instead of NO. nNOS seems likely to catalyze this reaction under some conditions in vivo. New findings suggest that the intracellular concentrations of L-arginine and BH_4 regulate the nature of the products of the nNOS reaction, rather than the rates of enzyme activity. NO and superoxide may be generated simultaneously and combine to peroxynitrite, a potent cytotoxin exerting a number of actions, including nitration of tyrosine residues in proteins *(14)*.

2.3. Regulation of eNOS

The endothelial isoenzyme appears to exist in vivo mainly bound to membranes, and has been located in the Golgi apparatus and caveolae of endothelial cells *(15–18)*. In an as yet only poorly understood signal-transduction process, receptor agonists appear to modulate both phosphorylation *(19)* and fatty acylation *(20)* of eNOS, resulting in translocation of the protein and subtle changes of activity or other biochemical properties of the enzyme. The sites at which fatty-acyl moieties are attached are also in a special N-terminal domain.

2.4. Receptor Coupling of nNOS

The neuronal isoenzyme is expressed in distinct neurons throughout the central nervous system *(21)*, as well as in nitrergic (nonadrenergic, noncholinergic) terminals in the periphery innervating smooth muscle and many other tissues *(22)*. In the brain, NOS is coupled to the *N*-methyl-D-aspartate (NMDA) subtype of glutamate receptors. Receptor activation results in the influx of Ca^{2+}-ions and activation of NOS through binding of the Ca^{2+}/CaM complex. The enzyme was first isolated as a soluble protein *(23–25)*, but later studies indicated that it exists at least partially in a membrane-associated form. New results suggest a mechanism for this coupling involving an isoenzyme-specific N-terminal domain.

2.5. Regulation of iNOS

The inducible isoenzyme is usually not present in cells under physiological conditions, but is expressed within several hours after stimulation of cells by inflammatory signals such as cytokines or endotoxin. Most of these stimuli exert their effects at the transcriptional level, although some may also affect mRNA translation and/or stability *(26)*. The signal transduction cascades involved in iNOS induction are complex and largely tissue-specific, but activation of the nuclear factor NFκB appears to be a key signal for induction of the iNOS promoter in murine fibroblasts and several other cell types *(27)*.

One of the most important distinctive biochemical features of iNOS is its ability to bind CaM at unusually low Ca^{2+} concentrations (≤30 n*M*) *(28)*. Because $[Ca^{2+}]_i$ is normally well above this level, iNOS is physiologically Ca^{2+}-independent and not regulated by signals inducing transient increases in $[Ca^{2+}]_i$, rendering this isoenzyme a high-output pathway for the long-lasting release of NO that is required for immune defense in infectious disease. We discuss important advances in understanding the CaM and Ca^{2+} activation of NOS enzymes.

3. Cloning and Expression

Each of the three isoenzymes has been cloned from a variety of tissues and species, and the sequences allow us to be fairly confident that there are no further major subtypes of NOS still to be discovered. **Table 2A** summarizes the versions of the enzyme that have been cloned. Of particular interest are the recent sequences from Drosophila *(29)*, which is essentially similar to mammalian nNOS, and chicken macrophage iNOS *(30)*. Both of these sequences are highly conserved compared to the mammalian enzymes.

Table 2A
Cloning of NOS Isoenzymes

Isoenzyme	Species	Tissue	Reference
nNOS	Rat	Brain	*(31,87,88)*
	Mouse	Brain	*(89)*
	Human	Brain	*(90)*
	Drosophila	(genomic)	*(29)*
eNOS	Bovine	Aortic endothelium	*(91–93)*
	Human	Endothelium	*(94,95)*
iNOS	Mouse	Macrophage	*(96–98)*
	Rat	Liver	*(99)*
	Rat	Astroglia	*(100)*
	Rat	Vascular smooth muscle	*(101)*
	Rat	Islets of Langerhans	*(102)*
	Human	Hepatocytes	*(103)*
	Human	DLD1 colorectal carcinoma	*(104)*
	Human	Articular chondrocyte	*(105,106)*
	Human	Foreskin fibroblast HSF42	*(107)*
	Bovine	Macrophage	*(108)*
	Chicken	Macrophage	*(30)*

The last few years have seen the development of efficient expression systems for recombinant NOS enzymes, resulting in an increased availability of pure enzymes and a rapid expansion of functional and structural studies. Until recently, successful overexpression of these enzymes was achieved either in stably transfected mammalian cells *(31)* or baculovirus-infected insect cells *(37)*. More recently, the use of a special expression vector, pCW_{ori}, and the coexpression of chaperone proteins has allowed useful levels of expression in *Escherichia coli*. *(32)*. The expression of the inducible isoenzyme has presented special difficulties. First, it appears to require coexpression of CaM in order to fold correctly and incorporate cofactors *(32)*. Second, because it is permanently active, unlike the other isoenzymes, it can cause NO-mediated killing of the host cells. These problems can be overcome by settling for a lower expression level and by including NOS inhibitors in the culture medium. These expression systems are summarized in **Table 2B**.

4. NOS is a Multidomain Enzyme

4.1. Overview of Domain Structure

The first complete primary structure of a NOS enzyme was deduced from the cDNA sequence by Snyder and colleagues *(31)*. It was immediately clear

Table 2B
Expression of Recombinant NOS Isoenzymes

Isoenzyme	Species	Expressed in	References
nNOS	Rat	293 kidney cells	***(31,109)***
	Human	COS cells	***(90)***
	Human	Baculovirus	***(110)***
	Rat	Baculovirus	***(37,87,111,112)***
	Rat	*E. coli*	***(64,113,114)***
	Rat	Yeast	***(88)***
eNOS	Bovine	COS cells	***(91–93)***
	Human	3T3 cells	***(94)***
	Bovine	Baculovirus	***(56,115)***
	Human	Baculovirus	***(40,110,116,117)***
	Bovine	*E. coli*	***(118)***
	Human	*E. coli*	***(61)***
iNOS	Human	CHO cells	***(105,119)***
	Human	293 kidney cells	***(103)***
	Rat	293 kidney cells	***(102)***
	Rat liver	COS cells	***(99)***
	Human	Baculovirus	***(110)***
	Mouse	Baculovirus	***(120,121)***
	Mouse	*E. coli*	***(32,122)***
	Mouse	Yeast	***(39)***

that the C-terminal half of the enzyme was composed largely of structural units already familiar from other enzymes, and the approximate locations of CaM, flavin, and NADPH-binding sites were assigned (**Fig. 2**). The N-terminal half of the protein, by contrast, displayed very little homology to known proteins.

This impression of an enzyme of two halves was reinforced by the discovery of a site in the center of the polypeptide hypersensitive to proteolysis, allowing the isolation of two fragments which retained ligand binding and catalytic activities corresponding to parts of the overall NOS reaction ***(33,34)***. The C-terminal fragment catalyzed the NADPH-dependent reduction of cytochrome c, whereas the N-terminal fragment retained an absorption spectrum characteristic of bound hem, and after reduction, formed a CO complex absorbing at 450 nm. As a result, the C-terminal half of the enzyme is often called the reductase domain, whereas the N-terminal half is called the oxygenase or P450 domain.

In the light of general experience of autonomously folding domains in proteins, each of these regions is in itself still a large structure and is probably composed of smaller units, each of which can attain its correct tertiary fold

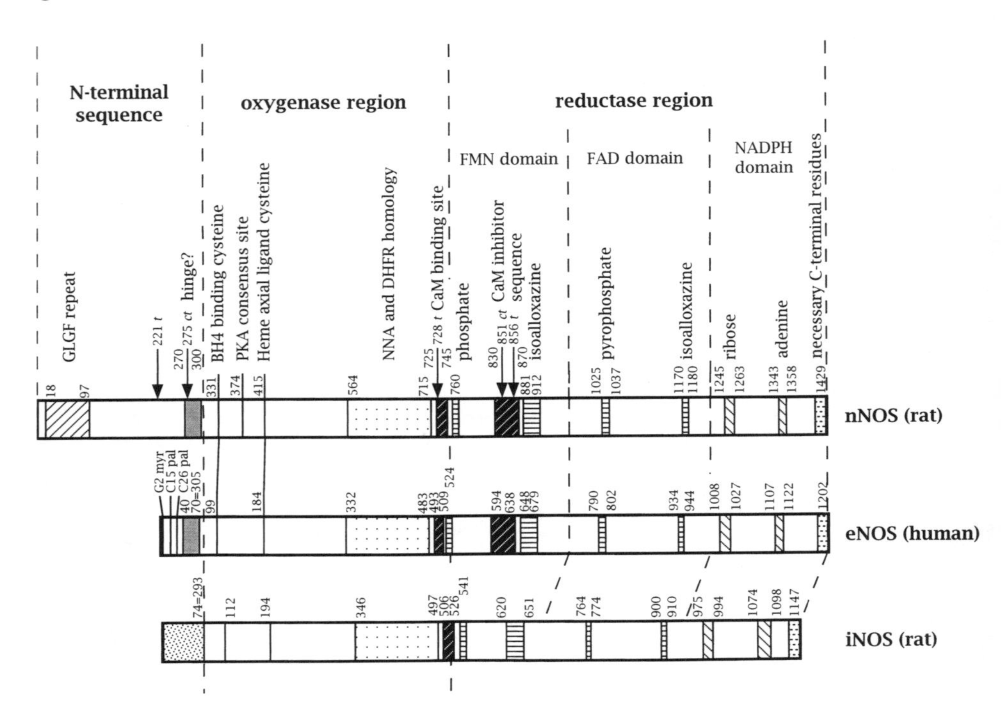

Fig. 2. Domain structure of NOS isoenzymes. The numbering of amino-acid residues is taken from the sequences in the SwissProt database P29476 (rat nNOS), P29474 (human eNOS), and Q06518 (rat iNOS). The discs-large homology (PDZ domain) was identified by Hendriks et al. ***(45)***. Trypsin (*t*) and chymotrypsin (*ct*) cleavage sites and the putative contact site for the phosphate of FMN, are from ***(33,34)***. Cysteine 99 in eNOS ***(40)*** and cysteine 331 in nNOS (Klösch, B., Schmidt, K., and Mayer, B.; unpublished) were shown by site-directed mutagenesis to be necessary for normal BH_4 binding ***(41)***. Mutation of the residues C415 in nNOS ***(37,64)***, C184 in eNOS ***(38)***, and C194 in iNOS ***(39)*** caused loss of hem binding. The DHFR homology region, which as an isolated fragment was able to bind NNA, was identified by Nishimura et al. ***(42)***. The cAMP-dependent protein kinase consensus site, the CaM binding sequence, and the putative contact sites for the isoalloxazine ring of FMN, and for FAD and NADPH were identified for nNOS by Bredt et al. ***(31)***. The CaM binding site was confirmed experimentally by Voherr et al. ***(80)***, Zhang and Vogel ***(81)***, and Venema et al. ***(83)***. The function of the camodulin inhibitory sequence was shown experimentally by Salerno et al. ***(85)***. Deletion of 23 residues from the C-terminus of iNOS completely abolished enzyme activity ***(120)***.

independently. In the reductase region, it is easy to guess the location of these units because of the repetition of classical nucleotide-binding motifs. Such guesses are now being assessed experimentally. In the oxygenase region, our orientation has just recently reached the level that we are fairly confident of the order of the heme, BH_4, and L-arg binding sites. The structural units composing the oxygenase region have not yet been defined.

Of course, the total physiological function of NOS is more than the sum of its basic structural components, and the most fascinating aspect of the enzyme is exactly this integration and adaptation of these different elements to fulfill the unique combination of constraints imposed first by the chemistry of producing NO from L-arg, and second by the physiological need to control the rate and location of this reaction.

4.2. Oxygenase Region is a Unique P450 Structure

The heme-binding site of NOS has been shown—by purification of fragments generated both by limited proteolysis and by expression of partial-cDNA constructs—to be in the N-terminal half of the protein. NOS enzymes have spectral properties typical of cytochromes P450, indicating that the heme has a similar coordination structure: yet there is no convincing sequence homology between the heme-containing domains of P450 enzymes and NOS.

Experiments with heme-deficient enzyme indicate that the activation of molecular oxygen and the consequent reactions take place at the heme-containing active site. The oxygenase region also contains the binding sites for L-arg and BH_4, as well as the necessary determinants for formation of the native enzyme homodimer. This region is therefore the scene of the unique catalytic chemistry which results in NO formation, and the lack of strong homology to other proteins suggests that this chemistry requires an equally unique protein structure. Whether the existing sequence similarity *(35)* is the product of divergent or convergent evolution is a matter of conjecture *(36)*. At present, the basic structural characterization of the oxygenase region is in a relatively early stage, consisting mainly of some mutagenesis studies.

In all three isoenzymes, mutation of a certain cysteine residue (C415 in nNOS, C184 in eNOS, and C194 in iNOS) resulted in the loss of heme binding *(37–39)*. This residue is conserved in all known NOS sequences. It is proposed that the side chain of this residue provides the thiol proximal ligand of the heme iron which is expected from spectroscopic experiments.

In eNOS, another cysteine (C99) was mutated to alanine, which resulted in a large loss of affinity of BH_4 binding, as measured by enzyme activation. High concentrations of BH_4 were able to restore enzyme activity *(40)*. This residue is also conserved in all known NOS sequences, and preliminary results indicate that the equivalent mutation in nNOS (C331A) has a similar effect (Klösch, B., Schmidt, K., and Mayer, B., unpublished results). Another two mutations in iNOS, G450A, and A453I were found to abolish BH_4 binding *(41)*. Both of these mutations are within a sequence that shows some similarity (but not conclusive homology) to the BH_4-binding domain of aromatic amino-acid hydroxylases and the folate-binding domain of dihydrofolate reductase (DHFR). A fragment corresponding to the DHFR-homology region (residues 564–715 of rat nNOS) did not, however, bind BH_4 autonomously *(42)*. Between the first of the three mutations mentioned and the other two lies a stretch of over 300 residues, posing the question of how the folded structure of the enzyme brings these residues into structural communication. Perhaps an ancestral pteridine-binding domain has become dependent on noncovalent contacts with the heme-binding domain, thus creating the unique BH_4-dependent heme chemistry.

The location of the L-arg binding site is also known approximately from the observation that the DHFR-homology fragment just mentioned (residues 564–715 of rat nNOS) did bind the L-arg analog N^G-nitro-L-arginine *(42)*.

4.3. Reductase Region

In contrast to the N-terminal half of the NOS polypeptide, the C-terminal half is easily recognizable as an assembly of FAD- , FMN-, and NADPH-binding domains from its sequence similarity to other enzymes. The individual domains have many homologs, whereas the complete reductase assembly containing both FMN and FAD and catalyzing electron transfer from NADPH to a specific heme-containing subunit or domain, is found in two well-studied examples apart from NOS. NADPH:cytochrome P450 reductase associates noncovalently with its target heme-containing proteins *(36)*. Cytochrome P450 BM-3 from *Bacillus megaterium* represents an analogous structure to the NOS enzymes, insofar as the reductase and heme-containing domains are the C-terminal and N-terminal halves of a single polypeptide *(43)*. Results on the NOS reductase confirm its overall similarities to these homologs; but equally, the knowledge of the homologs serves to point up the unique properties introduced by the integration of the domains into the NOS protein. The requirement for activation by Ca^{2+}/CaM is the most striking innovation compared to other reductases: impressive progress has been made in the last year in this area, and we describe the new developments in detail in a separate section below.

It has been assumed by analogy with cytochrome P450 reductase that electron transport in NOS proceeds from NADPH via FAD to FMN to the heme. However, there have not yet been detailed studies of the mechanism. Recent kinetic studies on cytochrome P450 BM-3 revealed a mechanism that differed in important respects from that of cytochrome P450 reductase *(44)*. The particular combination of redox states of flavins populated during the reaction may be tailored to the requirements of the oxygenase and may well differ from other enzymes.

5. Isoenzyme-Specific N-Terminal Domains Localize NOS Within the Cell

The sequences of the NOS isoenzymes are quite unrelated between their N-termini and a position equivalent to residue 300 of rat nNOS. The evidence so far indicates that these N-terminal sequences fulfill isoenzyme-specific regulatory functions. There is positive evidence that these sequences function in the targeting of NOS to a particular intracellular location, because nNOS and iNOS tolerated substantial deletions in this region without significant loss of activity. **Figure 2** provides an overview of these sequences.

5.1. nNOS Contains a PDZ Domain

The N-terminal sequence of nNOS is longer than those of the other isoenzymes (300 residues). It was noted by Hendriks *(45)* that between residues 18 and 97 the sequence contains a motif present in several other proteins and variously called "discs-large homology," "GLGF repeat," or "PDZ domain." All the known proteins containing this motif are localized to cell–cell junctions *(45)*. Bredt and colleagues reported that nNOS associates via this motif with the dystrophin complex in skeletal muscle, and that this association is responsible for the localization of nNOS at the sarcolemmal membrane *(46)*. A study from the same group confirmed that the PDZ motif in nNOS associates with PDZ motifs in other proteins, for example α1-syntrophin, a component of the dystrophin complex *(47)*. In the same report, another association, with another PDZ motif, was discovered using the yeast two-hybrid system. In this case, the partner was PSD-95, a brain protein that specifically binds to certain ion channels, including the NMDA receptor. A peptide identical to the C-terminus of the NMDA receptor-type 2B competitively displaced nNOS from PSD-95, suggesting that the receptor and the enzyme bind to the same motif of PSD-95 *(47)*. Stimulation of NMDA receptors is known to result in NO-mediated activation of soluble guanylyl cyclase, and it is proposed that these interactions may be part of the functional linkage between the receptor and NOS.

5.2. eNOS is Myristoylated and Palmitoylated

The N-terminal sequence of eNOS is about 70 residues long. Like the N-terminal sequence of nNOS, it functions in localizing the enzyme to membranes, but by a different mechanism. The glycine residue at position 2 becomes myristoylated, resulting in membrane association *(48,49)*. Further, attachment of palmitic acid has been discovered at residues C15 and C26 *(17,50)*. The exact contribution of each of these modifications, and whether they are regulated in response to extracellular signals, has been the subject of some debate. However, it is clear that the enzyme with both palmitic and myristic acid attached is exclusively membrane-bound, whereas mutants in which both modifications are blocked are located entirely in the cytosol.

The physiological regulation of eNOS is rendered even more complex by recent reports showing that chemical and physical signals such as oxygen *(51)*, chronic exercise *(52)*, or shear stress *(53)* can induce long- or short-term changes in NO release through transcriptional control of eNOS expression or direct effects on enzyme activity, respectively. The mechanism by which shear stress or flow lead to an increased release of NO is especially intriguing, as this process appears to involve a Ca^{2+}-independent stimulation of eNOS, possibly mediated by tyrosine phosphorylation and/or increased intracellular pH *(53)*. The molecular basis of this Ca^{2+}-independent stimulation of eNOS is not yet understood. It is conceivable that shear stress activates a signaling cascade that changes the biochemical properties or the microenvironment of eNOS, such that the protein binds CaM at low $[Ca^{2+}]_i$, a phenomenon known for iNOS (*see* below). Future studies should clarify whether the Ca^{2+}-independent activation of eNOS is associated with a translocation of the enzyme to a distinct cellular compartment as described for other stimuli-increasing endothelial-NO release. NO release from endothelial cells could also be increased through increased expression of superoxide dysmutase (SOD) induced by shear stress *(54)*. However, this response is too slow to account for the initial phase of the increased NO release.

5.3. iNOS

iNOS enzymes also have a characteristic N-terminal sequence of about 70 residues, which is conserved in the recently published sequence of the chicken enzyme *(30)*. This conservation suggests that it does have a function, but there is no positive indication what the function is. Deletion of most of it did not affect the enzyme activity.

6. Assembly of the Oxygenase Active Site

6.1. Heme is Necessary for Dimer Formation

The purification of heme-deficient versions of all three isoenzymes has been achieved by expressing the recombinant enzymes under conditions of reduced heme availability—for example, by omitting hemin chloride from baculovirus-infected cell cultures. Although the native enzyme had been predominantly dimeric, the heme-free enzyme was monomeric and could be isolated by gel filtration *(55–58)*. These heme-free monomers could not catalyze L-citrulline formation. They were found to contain no BH_4, and could bind neither BH_4 nor the L-arg analog N^G-nitro-L-arginine. In other respects they were normal: their secondary structure, Ca^{2+}/CaM-sensitive cytochrome-c reductase activity, and flavin content were indistinguishable from the native enzyme. Incubation of the heme-free monomers with hemin chloride resulted in the formation of dimers which could bind BH_4 and N^G-nitro-L-arginine.

A current question is whether the enzyme must bind one mole of heme per mole of subunit in order to form the dimer: two recent observations suggest that under certain conditions a dimer can form where only one of the two subunits contains a heme group. Xie et al. co-expressed wild-type NOS with mutants which could not bind heme, but observed formation of mixed dimers between the two forms: the heterodimers appeared to be enzymatically inactive ***(59)***. In a gel-filtration study of heme-deficient eNOS, we noticed that the proportion of the protein appearing in the dimer peak, when compared to the heme content, suggested the presence of less than two heme groups per dimer ***(56)***.

These results do not exclude differences of detail between the isoenzymes: however, the basic conclusions are clear: The presence of heme is an absolute requirement for assembly of the enzyme dimer, and for the formation of the BH_4 and L-arg binding sites. This agrees with the suggestion made above on the basis of the site-directed mutants—that there is significant structural communication between different parts of the oxygenase region.

6.2. Conversion of the Dimer to a Hyperstable Form

Recently, we discovered an unusual property of NOS. When NOS enzymes are submitted to the normal pretreatment for sodium dodecylsulfate-polyacrylamide gel electrophoresis (SDS-PAGE), involving heating to 100°C in the presence of 2% SDS and 5% 2-mercaptoethanol, the protein migrates normally with the mobility expected for the monomer. However, if the pig-brain enzyme was mixed with the same concentrations of SDS and 2-mercaptoethanol but not heated, a second band appeared on the gels which was shown to be the NOS dimer ***(60)***. As incubation with this concentration of SDS, without heating, would dissociate most oligomeric enzymes, it is remarkable that the NOS enzyme as isolated is partly resistant to this treatment.

It should be noted that the enzyme which behaved in this way was fully dimeric as determined by gel filtration in a nondenaturing buffer, suggesting that some of the dimers present were in a special SDS-resistant state, whereas others were not. The source of this extra stability was illuminated by the finding that preincubation with BH_4 or L-arg increased the proportion of SDS-resistant dimer present, whereas preincubation with both conferred almost complete resistance to dissociation in SDS up to a temperature of about 40°C. These observations were made with pig-brain nNOS: it is becoming apparent that although the other forms of the enzyme also form SDS-resistant dimers, the contribution of added BH_4 and L-arg to this behavior is somewhat dependent on the isoenzyme and its source. Dimers of recombinant-rat nNOS obtained from a baculovirus system were converted less efficiently to the SDS-resistant form than was the pig-brain enzyme ***(58)***, and the SDS-resistant dimer of eNOS was predominant even without pretreatment with L-arg or BH_4 ***(56,61)***.

For both eNOS and nNOS, heme was only detected in the SDS-resistant dimers, and not in the monomers, after electrophoresis ***(56,58)***. This would agree with heme retention or loss being the decisive factor for survival or dissociation of the dimer and the stabilizing effect of BH_4 or L-arg resulting from the inhibition of heme loss ***(40,62,63)***.

6.3. Spin-State Equilibrium and Enzyme Activity

nNOS-containing heme, but not BH_4, has been shown to exist in a dimeric form, but does not catalyze citrulline formation ***(63)***. The absorption spectrum of the BH_4-free enzyme is quite different to that of the native enzyme (which normally, as isolated, contains about 1 mol of BH_4 per mole of dimer), with a maximum at 418 nm instead of 398 nm. By analogy with cytochrome P450, this difference reflects a shift in the spin state of the heme iron, which is low-spin in the BH_4-free enzyme but predominantly high-spin in the native enzyme.

Incubation of pteridine-free nNOS with BH_4 or L-arg was sufficient to shift the spectrum to the high-spin form. The rate of conversion of the spectrum was independent of the BH_4 or L-arg concentration, suggesting that some intrinsic structural transition of the enzyme was a precondition for BH_4 or L-arg binding. This was confirmed by examining the binding kinetics of BH_4. When BH_4 alone was added to the pteridine-free enzyme, the rate of BH_4 binding was equal to the rate of the spin transition, but when BH_4 was added to enzyme which had first been incubated with L-arg and was therefore already in the high-spin form, BH_4 binding was very much faster.

Binding of BH_4 to the BH_4-free enzyme was highly anticooperative, with a difference of more than 1,000-fold between the two affinities ***(63)***. The shift from low- to high-spin was already complete when one BH_4 molecule was bound per dimer: this is the state in which the enzyme is usually isolated.

The implication of this behavior for the function of the enzyme became clear when it was found that the restoration of the high-spin heme exactly correlated with the regain of normal enzyme activity. This contrasts with other P450 enzymes, which generally exist in a low-spin state in the absence of substrate. These results can be combined in a model in which the enzyme exists in an equilibrium between the two spin states, L-arg and BH_4 only bind to the high-spin form and only the high-spin form with bound BH_4 has NOS activity ***(62–64)***.

6.4. The Uncoupled NADPH-Oxidase Reaction

Oxidation of L-arg to L-citrulline with production of NO is not the only reaction which can be catalyzed at the oxygenase-active site. Under certain conditions, the reductive activation of oxygen—a normal step in the NOS reaction—can be followed by the release of superoxide from the enzyme instead of its use for L-citrulline and NO formation *(6,65–67)*. This provides a sink for reducing equivalents from the reductase region and allows the enzyme to work as an NADPH oxidase in the absence of L-arginine.

This uncoupled reaction is inversely dependent on the amount of bound BH_4 *(63)*. The BH_4-free nNOS had high NADPH-oxidase activity in the absence of L-arg, and even in the presence of saturating concentrations of L-arg, no L-citrulline was produced. Titrating the enzyme with BH_4 progressively shifted the path of the reaction from H_2O_2 production to L-citrulline production. In the presence of fully saturating BH_4 and L-arg, the rate of NADPH oxidation was equivalent to the rate of L-citrulline production.

In the BH_4-saturated nNOS, the rate of NADPH oxidation was not affected by N^G-methyl-L-arginine. Thus, the mere binding of a ligand at the L-arg site does not affect the rate of oxygen activation: rather, the true substrate L-arg, when bound here, is optimally positioned to scavenge the activated-oxygen species.

The uncoupled reaction has not yet been studied in so much detail in iNOS and eNOS: however, the existing results indicate that compared to nNOS, the other two isoenzymes exhibit much slower rates of NADPH oxidation in the absence of L-arg *(56,68)*. However, in iNOS and eNOS, the uncoupled reaction can be stimulated by N^G-methyl-L-arginine, which therefore seems able to enhance the rate of turnover of the oxygen-activation reaction. In the absence of L-arg or N^G-methyl-L-arginine, the heme can be reduced, but the reaction does not proceed to the release of superoxide.

If in vivo nNOS becomes unsaturated with BH_4 or L-arg, the enzyme could catalyze formation of significant amounts of superoxide, which would react with NO to form peroxynitrite *(14)*. Though formation of peroxynitrite in cells could be attenuated by SOD, high concentrations of SOD are required to outcompete the peroxynitrite reaction and render NOS-catalyzed superoxide formation detectable *(69)*. At least some cells appear not to contain enough SOD to prevent peroxnitrite formation. A recent report demonstrated peroxynitrite-mediated injury and NOS-catalyzed protein nitration in L-arginine-depleted cells transfected with nNOS *(70)*.

Uncoupling of the nNOS reaction may be important in the ongoing debate on the role of NO in neurotoxicty and/or protection of neurons from ischaemia/reperfusion injury involving overstimulation of NMDA receptors. Work on the susceptibility to experimental stroke of nNOS-knockout mice suggested that nNOS contributes to excitotoxicity, whereas eNOS is protective *(71)*. Protection by eNOS was attributed to the positive effects of endothelial-derived NO on cerebral blood flow, but it is conceivable that the distinct physiological effects of the two isoenzymes could be a consequence of their different biochemical properties. It might be expected that nNOS generates substantial amounts of oxygen radicals and peroxynitrite in conditions of reduced L-arginine or pteridine availability, whereas eNOS, which lacks uncoupled NADPH-oxidase activity, may only slow down its rates of NO production under these conditions, but not switch to formation of peroxynitrite. The finding that nNOS is the sole NOS isoenzyme that does not downregulate NADPH oxidation at low concentrations of L-arginine or BH_4 suggests that this feature of the enzyme is not simply an undesired "leaking" of the catalytic cycle, but could be physiologically important for brain function or development.

6.5. Redox Activity of BH_4

Although the heme is already almost completely in the high-spin form in a dimer containing one molecule of BH_4, the activity of this enzyme can still be doubled by adding more BH_4 to saturate the second, low-affinity site. In what way is the presence of a BH_4 molecule in the oxygenase region of each monomer necessary for NO synthesis? In the aromatic amino-acid hydroxylases, BH_4 functions as a redox shuttle, being oxidized in the course of the reaction, dissociating, and being regenerated by dihydropteridine reductase. In contrast, it does not undergo net oxidation in a complete turnover of the NOS reaction. However, this does not exclude that it is transiently oxidized in the course of the NOS reaction, and indeed, results are accumulating that point in this direction.

The evidence is mainly based on results obtained with BH_4 antagonists of NOS. Unlike other pteridine-dependent enzymes, the NOS family of proteins is highly specific for BH_4 and exhibits only low affinity for most synthetic derivatives serving as cofactors of aromatic amino-acid hydroxylases *(72,73)*. Among the few exceptions, dihydrobiopterin (H_2biopterin) and the recently characterized 4-amino analog of BH_4 (*see* **Fig. 3** for structure) exhibit reasonably high affinity for the pteridine site of NOS *(74,75)*. With the amino function at position C4, 4-amino-BH_4 resembles methotrexate, an inhibitor of dihydropteridine and dihydrofolate reductases. However, while methotrexate does not bind to NOS owing to its bulky side chain at C6, 4-amino-BH_4 was found to bind with 5- to 10-fold higher affinity than the natural cofactor to nNOS *(75)* and iNOS (Mayer, B., Stuehr, D. J., and Werner, E. R., unpublished results), making this drug a useful probe for a possible involvement of pteri-

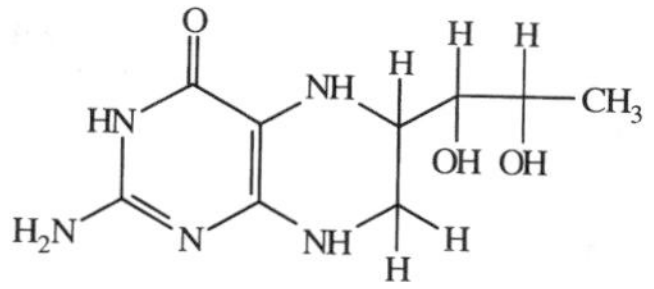

H$_4$Biopterin

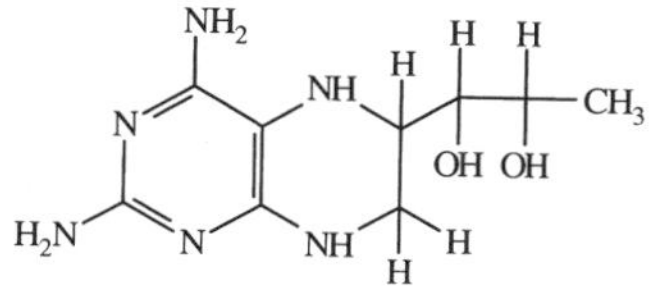

4-Amino-H$_4$Biopterin

Fig. 3. Structures of BH_4 and 4-amino-BH_4. **(Top)** The natural cofactor of NO synthases is referred to in the text as BH_4; its correct description is 5,6,7,8-tetrahydro-L-*erythro*-biopterin or 5,6,7,8-tetrahydro-6-(L-erythro-1,2-dihydroxypropyl)pterin. **(Bottom)** The compound referred to in the text as 4-amino-BH_4 is 2,4-diamino-5,6,7,8-tetrahydro-6-(L-erythro-1,2-dihydroxypropyl)pteridine.

dine redox-cycling in NOS catalysis. In fact, 4-amino-BH_4 turned out to be inactive as a NOS cofactor and potently antagonized enzyme stimulation by BH_4. Because the amino analog exerted the same allosteric effects, i.e., spin transition of the heme and stabilization of dimers as the natural cofactor (Mayer, B., Gorren, A. C. F., and Werner, E. R., unpublished results), the change in protein conformation induced by pteridine binding appears to be necessary but not sufficient to support NOS activity.

These data stongly suggest that BH_4 has a redox-function in NOS catalysis that remains to be uncovered. Conceivably, this function of the pteridine is related to the fact that NOS generates a product that would potently inhibit other P450s through formation of stable, catalytically inactive nitrosyl-heme complexes. Indeed, it was reported that NOS self-inactivates via formation of an inhibitory nitrosyl-heme complex, but dissociation of NO is apparently fast enough to leave about 10% of the enzyme in an active state during turnover *(76)*. Although BH_4 has no apparent effect on the stability of the nitrosyl-heme complexes formed with exogenous NO *(77)*, it is conceivable that a redox-interaction of the pteridine with an intermediate of the NOS reaction may be requisite for dissociation and release of NO. This issue remains open for future research.

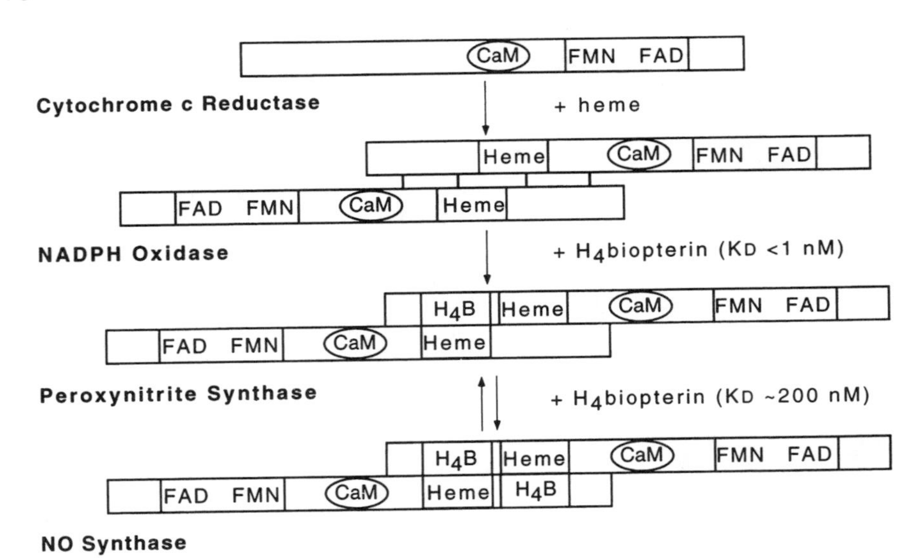

Fig. 4. Model for structural transitions of the oxygenase region.

6.6. Structural Transitions of the Oxygenase: A Model

At least for nNOS, our present knowledge on subunit assembly can be summarized schematically, as shown in **Fig. 4**. We can distinguish at least four different species of the enzyme, which all exhibit different structural properties and catalytic activities. In the presence of heme but absence of BH_4, monomers containing an intact reductase domain assemble to highly unstable dimers that contain mainly low-spin heme and, if activated by Ca^{2+}/CaM, exhibit NADPH-oxidase activity leading to generation of superoxide. Binding of BH_4 to the pteridine-free dimers is anti-cooperative and takes place in two steps *(63)*. The first pteridine molecule binds with a K_D of approx 10^{-11} M to the small fraction of the enzyme containing high-spin heme, and thereby shifts the equilibrium fully to high-spin. This half-pteridine saturated nNOS species i stable dimer and can be regarded as peroxynitrite synthase because it is partially uncoupled and produces equal amounts of NO and superoxide in the presence of Ca^{2+}/CaM (*see* **Subheading 6.4.**). At BH_4 concentrations of 1–10 μM, nNOS becomes pteridine saturated resulting in full coupling of NADPH oxidation to NO synthesis and, hence, complete inhibition of superoxide generation. It should be mentioned that the latter enzyme species cannot be isolated because of rapid dissociation of the second pteridine ligand. Autoxidation of free BH_4 results in nonenzymatic production of superoxide *(69)* and, therefore, has precluded positive identification of NO as the reaction product. (For a more comprehensive discussion of this rather intriguing issue, *see* **ref.** *73*.)

7. Activation by Calmodulin and Ca^{2+}

7.1. All NOS Isoenzymes Need Calmodulin

All three NOS isoenzymes bind CaM. In functional studies of purified enzymes, a difference between iNOS and the other two isoenzymes was immediately apparent: whereas CaM binds to eNOS and nNOS only in the presence of micromolar concentrations of free Ca^{2+}, iNOS appears to carry a permanently bound molecule of CaM, which does not dissociate even at very low Ca^{2+} concentrations *(28)*.

Without bound Ca^{2+}/CaM, eNOS and nNOS do not catalyze citrulline formation and have barely detectable NADPH:cytochrome-c reductase activity. The addition of Ca^{2+}/CaM stongly activates the NADPH:cytochrome-c reductase activity *(68,78)*. This activity, and its activation by Ca^{2+}/CaM is not affected by the absence or presence of heme and therefore seems to involve primarily the reductase region of the enzyme. A stopped-flow study showed that the reduction of the flavins upon addition of NADPH was much faster in the presence of Ca^{2+}/CaM than in their absence *(79)*.

7.2. Calmodulin Recognition of its Target Sequence in nNOS and eNOS

The CaM binding site was clearly recognizable in the first NOS sequence between residues 725–745 *(31)*. Since then some detailed studies have been made that confirmed this identification for nNOS *(80,81)* and extended it to the corresponding sequences in the other isoenzymes *(82,83)*. Vorherr et al. *(80)* demonstrated Ca^{2+}-dependent binding of a synthetic peptide identical to residues 725–754 of nNOS to CaM with a dissociation constant of about 2 n*M*. Zhang and Vogel *(81)*, using a similar nNOS peptide, obtained a similar estimate of the dissociation constant and found that the peptide adopted an α-helical conformation when bound to CaM. The free peptide did not have a stable α-helical structure, but became α-helical in the presence of trifluoroethanol. An NMR study of this complex *(84)* found structural features common to complexes of Ca^{2+}/CaM with other target proteins such as myosin light-chain kinase and adenylate cyclase: the peptide bound with a C to N orientation to the N- and C-terminal domains of CaM and the contacts with the peptide appeared to be made via the usual methionine-rich hydrophobic patches.

A similar dissociation constant of 1 n*M* was obtained with the whole nNOS enzyme by means of quenching of tryptophan fluorescence that resulted from Ca^{2+}/CaM binding *(33)*. Further, in the absence of Ca^{2+}/CaM, the N-peptide bond of A728 was hypersensitive to trypsin proteolysis. Binding of Ca^{2+}/CaM prevented the proteolysis. This agrees well with the results of Zhang and Vogel

(81), suggesting that in the absence of Ca^{2+}/CaM, the binding site is exposed on the surface of the protein and is rather flexible, but adopts a more determined structure in the complex with Ca^{2+}/CaM.

Recently, Salerno et al. *(85)* reported a new aspect of CaM binding to the eNOS and nNOS *(85)*. On the C-terminal side of the CaM-recognition sequence, between the homology regions for the contacts with the phosphate and isoalloxazine ring of FMN, is an approx 45-residue insert which interrupts the homology to other FMN-containing domains. A homology-based model showed that this insert would emerge from the structure on the side facing towards the CaM-recognition site, and might sterically hinder CaM binding. Indeed peptides based on the sequence of the insert were able to inhibit eNOS and nNOS and displace Ca^{2+}/CaM from nNOS. This is reminiscent of the mechanism discovered for several other CaM-dependent enzymes that contain an insert with sequence similarity to CaM, which binds to the CaM-recognition site in competition with CaM. The inhibitory insert in NOS has no sequence similarity to CaM; it was suggested that instead of binding to the CaM recognition site, it may occupy a different position on the surface of the NOS protein, such that it inhibits electron transfer until it is displaced by CaM.

7.3. iNOS Needs Less Ca^{2+}

It may be presumed that iNOS without bound CaM would also be inactive: to date, it has not been possible to isolate a correctly folded iNOS without bound CaM. Expression of functional iNOS in *E. coli* was only achieved with co-expression of CaM *(32)*. If CaM was not co-expressed, the iNOS polypeptide failed to incorporate heme and could catalyze neither citrulline formation nor cytochrome-c reduction. This suggests that cotranslational binding of CaM to the nascent iNOS polypeptide may inhibit wrong interactions with the other domains that can otherwise interfere with their folding. CaM cannot be removed from the enzyme by exchange onto other target proteins, and does not dissociate even under strongly denaturing conditions.

What is the difference in the iNOS that underlies the Ca^{2+}-independent CaM binding? The most obvious difference in the sequence is the lack of the inhibitory insert identified by Salerno et al. *(85)*. However, the inhibitory-insert peptides tested by Salerno et al. did not inhibit iNOS as strongly as nNOS or eNOS, and did not appear to cause dissociation of CaM from the enzyme. An elegant study by Venema et al. *(83)*, using chimeric NOS enzymes in which the CaM recognition sequences were exchanged between the isoenzymes, sheds more light on this question. Replacing the CaM recognition sequence in eNOS (493–512) with residues 501–523 from iNOS resulted in an enzyme which was sensitive to both added CaM and Ca^{2+}. Substitution of a slightly longer iNOS sequence (501–532) at the same position in eNOS conferred irreversible CaM binding, but the enzyme activity was still fully dependent on added Ca^{2+}. An iNOS enzyme, in which residues 501–523 were replaced by residues 493–512 of eNOS, bound CaM reversibly and Ca^{2+}-dependently, like native eNOS.

In the same study, the dissociation constant for the complex between a synthetic peptide corresponding to the full CaM-recognition sequence from iNOS was found to be about 1.5 n*M*, rather similar to the other peptide studies: this represents a much lower affinity of binding than is observed between CaM and the whole enzyme. The difference suggests that the irreversible, Ca^{2+}-independent binding of CaM to iNOS is partly dependent on other determinants beyond the CaM recognition sequence: perhaps through extra, direct contacts between the enzyme and CaM, or through an effect of the surrounding structure of the enzyme on the conformation of the recognition sequence. The achievement of Ca^{2+}-independent CaM binding in eNOS by the insertion of the iNOS-recognition sequence suggests that these determinants are conserved in eNOS, and further, underlines that the inhibitory insert present in eNOS does not have a decisive effect on this high-affinity CaM binding.

7.4. A Two-Stage Mechanism

One more intriguing aspect of CaM activation of NOS was brought to light by these experiments: namely, that CaM binding alone is not always sufficient to fully activate the enzyme. In the native eNOS and nNOS, this is masked by the fact that CaM binding and enzyme activation both occur in the same range of Ca^{2+} concentrations. Native iNOS was found to be about twice as active in the presence of 2.5 m*M* Ca^{2+} as in the presence of 10 m*M* ethylene glycol-bis-(2-aminoethyl)-tetraacetic acid (EGTA) ***(83,86)***. These two extreme conditions are obviously beyond the range of Ca^{2+} concentrations experienced by the enzyme in vivo: nevertheless, they show that even iNOS has the potential for some activation by Ca^{2+}, even though binding and dissociation of CaM are not involved. The eNOS chimera containing the full length (501–532) CaM recognition sequence was even more revealing: although the CaM binding was irreversible and completely Ca^{2+}-independent, Ca^{2+} was absolutely required for enzyme activity.

It is therefore very tempting to propose a two-stage model for Ca^{2+}/CaM activation of NOS (**Fig. 5**). CaM binding is largely determined by the recognition sequence and some other as-yet-unidentified interactions with the enzyme. However, activation of the enzyme is fully dependent on a movement of the inhibitory insert, which can only be effected by the Ca^{2+}-saturated conformation of CaM. The absence of the inhibitory insert from iNOS is the reason for its physiological insensitivity to Ca^{2+}. In eNOS and nNOS , because of differences in the CaM-recognition sequence, only the Ca^{2+}-saturated form of CaM can bind, so that the effects of Ca^{2+} and CaM binding mask each other.

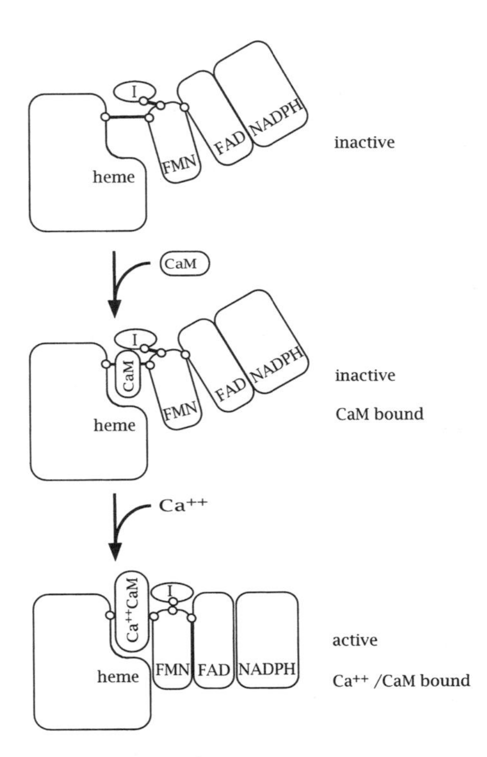

Fig. 5. Model for activation by Ca^{2+}/CaM. The figure shows a two-stage model of Ca^{2+}/CaM activation of NOS. (**Top**) NOS in the resting state without CaM; eNOS and nNOS are in this state in the absence of CaM or Ca^{2+}. The enzyme is inactive because the separation between the redox centers is too great for electron transfer. (**Center**) The chimeric eNOS enzyme containing the CaM recognition sequence from iNOS *(83)* is able to bind CaM in the absence of Ca^{2+}; this does not result in activation of the enzyme because this conformation of CaM cannot displace the inhibitory insert (I). (**Bottom**) When Ca^{2+}/CaM is bound to the enzyme, the inhibitory insert (I) is displaced and the enzyme shifts into an active structure, in which the redox centers are closer together.

How might Ca^{2+} activate electron transfer in NOS? It is possible that the conformational change induced by Ca^{2+} binding to CaM is passed on to the inhibitory insert, which, as proposed by Salerno et al. *(85)*, lies nearby. Electron transfer between the redox centers of FAD, FMN, and heme is probably very sensitive to their separation, so that the ultimate effect of this movement might be to bring two of these centers—perhaps FAD and FMN—closer together. The retention of the irreversibly-bound CaM by iNOS may also reflect the sensitivity of electron-transfer between the FMN and the heme to the relative orientation of the oxygenase and reductase domains.

Acknowledgments

Work in the authors laboratory was supported by Grants 10,655, 10,859, and 11,478 of the “Fonds zur Förderung der Wissenschaftlichen Forschung in Österreich.”

References

1. Griffith, O. W. and Stuehr, D. J. (1995) Nitric oxides synthases: Properties and catalytic mechanism. *Annu. Rev. Physiol.* **57,** 707–736.
2. Mayer, B. (1995) Biochemistry and molecular pharmacology of nitric oxide synthases, in *Nitric Oxide in the Nervous System* (S. R. Vincent, ed.), Academic, New York, 21–42.
3. Masters, B. S. S., McMillan, K., Sheta, E. A., Nishimura, J. S., Roman, L. J., and Martasek, P. (1996) Cytochromes P450. 3. Neuronal nitric oxide synthase, a modular enzyme formed by convergent evolution: Structure studies of a cysteine thiolate-liganded heme protein that hydroxylates L-arginine to produce NO center dot as a cellular signal. *FASEB J.* **10,** 552–558.
4. Marletta, M. A., Yoon, P. S., Iyengar, R., Leaf, C. D., and Wishnok, J. S. (1988) Macrophage oxidation of L-arginine to nitrite and nitrate: nitric oxide is an intermediate. *Biochemistry* **27,** 8706–8711.
5. Kwon, N. S., Nathan, C. F., Gilker, C., Griffith, O. W., Matthews, D. E., and Stuehr, D. J. (1990) L-citrulline production from L-arginine by macrophage nitric oxide synthase. The ureido oxygen derives from dioxygen. *J. Biol. Chem.* **265,** 13,442–13,445.
6. Mayer, B., John, M., Heinzel, B., Werner, E. R., Wachter, H., Schultz, G., and Böhme, E. (1991) Brain nitric oxide synthase is a biopterin- and flavin-containing multi-functional oxido-reductase. *FEBS Lett.* **288,** 187–191.
7. Stuehr, D. J., Kwon, N. S., Nathan, C. F., Griffith, O. W., Feldman, P. L., and Wiseman, J. (1991) N omega-hydroxy-L-arginine is an intermediate in the biosynthesis of nitric oxide from L-arginine. *J. Biol. Chem.* **266,** 6259–6263.
8. Klatt, P., Schmidt, K., Uray, G., and Mayer, B. (1993) Multiple catalytic functions of brain nitric oxide synthase. Biochemical characterization, cofactor-requirement and role of NG-hydroxy-L-arginine as an intermediate. *J. Biol. Chem.* **268,** 14,781–14,787.
9. Feldman, P. L., Griffith, O. W., and Stuehr, D. J. (1993) The surprising life of nitric oxide. *Chem. Eng. News* **71,** 26–38.
10. Marletta, M. A. (1994) Nitric oxide synthase: Aspects concerning structure and catalysis. *Cell* **78,** 927–930.
11. Pufahl, R. A. and Marletta, M. A. (1993) Oxidation of NG-Hydroxy-L-arginine by nitric oxide synthase—evidence for the involvement of the heme in catalysis. *Biochem. Biophys. Res. Commun.* **193,** 963–970.
12. Clement, B., Schultze-Mosgau, M. H., and Wohlers, H. (1993) CytochromeP450 dependent N-hydroxylation of a guanidine (debrisoquine), microsomal catalysed reduction and further oxidation of the N-hydroxy-guanidine metabolite to the urea

derivative—similarity with the oxidation of arginine to citrulline and nitric oxide. *Biochem. Pharmacol.* **46,** 2249–2267.
13. Feelisch, M. and Noack, E. A. (1987) Correlation between nitric oxide formation during degradation of organic nitrates and activation of guanylate cyclase. *Eur. J. Pharmacol.* **139,** 19–13.
14. Pryor, W. A. and Squadrito, G. L. (1995) The chemistry of peroxynitrite: A product from the reaction of nitric oxide with superoxide. *Am. J. Physiol. Lung Cell. Mol. Physiol.* **12,** L699–L722.
15. Obrien, A. J., Young, H. M., Povey, J. M., and Furness, J. B. (1995) Nitric oxide synthase is localized predominantly in the Golgi apparatus and cytoplasmic vesicles of vascular endothelial cells. *Histochem. Cell Biol.* **103,** 221–225.
16. Sessa, W. C., Garcia-Cardena, G., Liu, J. W., Keh, A., Pollock, J. S., Bradley, J., Thiru, S., Braverman, I. M., and Desai, K. M. (1995) The Golgi association of endothelial nitric oxide synthase is necessary for the efficient synthesis of nitric oxide. *J. Biol. Chem.* **270,** 17,641–17,644.
17. Garcia-Cardena, G., Oh, P., Liu, J., Schnitzer, J. S., and Sessa, W. C. (1996) Targeting of nitric oxide synthase to endothelial cell caveolae via palmitoylation: Implications for nitric oxide signaling. *Proc. Natl. Acad. Sci. USA* **93,** 6448–6453.
18. Shaul, P. W., Smart, E. J., Robinson, L. J., German, Z., Yuhanna, I. S., Ying, Y. S., Anderson, R. G. W., and Michel, T. (1996) Acylation targets endothelial nitric-oxide synthase to plasmalemmal caveolae. *J. Biol. Chem.* **271,** 6518–6522.
19. Michel, T., Li, G. K., and Busconi, L. (1993) Phosphorylation and subcellular translocation of endothelial nitric oxide synthase. *Proc. Natl. Acad. Sci. USA* **90,** 6252–6256.
20. Robinson, L. J., Busconi, L., and Michel, T. (1995) Agonist-modulated palmitoylation of endothelial nitric oxide synthase. *J. Biol. Chem.* **270,** 995–998.
21. Garthwaite, J. and Boulton, C. L. (1995) Nitric oxide signaling in the central nervous system. *Annu. Rev. Physiol.* **57,** 683–706.
22. Rand, M. J. and Li, C. G. (1995) Nitric oxide as a neurotransmitter in peripheral nerves: Nature of transmitter and mechanism of transmission. *Annu. Rev. Physiol.* **57,** 659–682.
23. Bredt, D. S. and Snyder, S. H. (1990) Isolation of nitric oxide synthetase, a calmodulin-requiring enzyme. *Proc. Natl. Acad. Sci. USA* **87,** 682–685.
24. Mayer, B., John, M., and Böhme, E. (1990) Purification of a Ca^{2+}/calmodulin-dependent nitric oxide synthase from porcine cerebellum. Cofactor-role of tetrahydrobiopterin. *FEBS Lett.* **277,** 215–219.
25. Schmidt, H. H. H. W., Pollock, J. S., Nakane, M., Gorsky, L. D., Förstermann, U., and Murad, F. (1991) Purification of a soluble isoform of guanylyl cyclase-activating-factor synthase. *Proc. Natl. Acad. Sci. USA* **88,** 365–369.
26. Kröncke, K. D., Fehsel, K., and Kolb-Bachofen, V. (1995) Inducible nitric oxide synthase and its product nitric oxide, a small molecule with complex biological activities. *Biol. Chem. Hoppe - Seyler* **376,** 327–343.
27. Kleinert, H., Euchenhofer, C., Ihrig-Biedert, I., and Förstermann, U. (1996) In murine 3T3 fibroblasts, different second messenger pathways resulting in the

induction of NO synthase II (iNOS) converge in the activation of transcription factor NF-kappa B. *J. Biol. Chem.* **271,** 6039–6044.
28. Cho, H. J., Xie, Q. W., Calaycay, J., Mumford, R. A., Swiderek, K. M., Lee, T. D., and Nathan, C. (1992) Calmodulin is a subunit of nitric oxide synthase from macrophages. *J. Exp. Med.* **176,** 599–604.
29. Regulski, M. and Tully, T. (1995) Molecular and biochemical characterization of dNOS: A Drosophila Ca^{2+}/calmodulin-dependent nitric oxide synthase. *Proc. Natl. Acad. Sci. USA* **92,** 9072–9076.
30. Lin, A. W., Chang, C. C., and McCormick, C. C. (1996) Molecular cloning and expression of an avian macrophage nitric-oxide synthase cDNA and the analysis of the genomic 5'-flanking region. *J. Biol. Chem.* **271,** 11,911–11,919.
31. Bredt, D. S., Hwang, P. M., Glatt, C. E., Lowenstein, C., Reed, R. R., and Snyder, S. H. (1991) Cloned and expressed nitric oxide synthase structurally resembles cytochrome P-450 reductase. *Nature* **351,** 714–718.
32. Wu, C. Q., Zhang, J. G., Abu-Soud, H., Ghosh, D. K., and Stuehr, D. J. (1996) High-level expression of mouse inducible nitric oxide synthase in *Escherichia coli* requires coexpression with calmodulin. *Biochem. Biophys. Res. Commun.* **222,** 439–444.
33. Sheta, E. A., McMillan, K., and Masters, B. S. S. (1994) Evidence for a bidomain structure of constitutive cerebellar nitric oxide synthase. *J. Biol. Chem.* **269,** 15,147–15,153.
34. Lowe, P. N., Smith, D., Stammers, D. K., Riveros-Moreno, V., Moncada, S., Charles, I., and Boyhan, A. (1996) Identification of the domains of neuronal nitric oxide synthase by limited proteolysis. *Biochem. J.* **314,** 55–62.
35. Renaud, J. P., Boucher, J. L., Vadon, S., Delaforge, M., and Mansuy, D. (1993) Particular ability of liver P450s3A to catalyze the oxidation of N(omega)-hydroxyarginine to citrulline and nitrogen oxides and occurrence in NO synthases of a sequence very similar to the heme-binding sequence in P450s. *Biochem. Biophys. Res. Commun.* **192,** 53–60.
36. Degtyarenko, K. N. and Archakov, A. I. (1993) Molecular evolution of P450 superfamily and P450-containing monooxygenase systems. *FEBS Lett.* **332,** 1–8.
37. Richards, M. K. and Marletta, M. A. (1994) Characterization of neuronal nitric oxide synthase and a C415H mutant, purified from a baculovirus overexpression system. *Biochemistry* **33,** 14,723–14,732.
38. Chen, P. F., Tsai, A. L., and Wu, K. K. (1994) Cysteine 184 of endothelial nitric oxide synthase is involved in heme coordination and catalytic activity. *J. Biol. Chem.* **269,** 25,062–25,066.
39. Sari, M. A., Booker, S., Jaouen, M., Vadon, S., Boucher, J. L., Pompon, D., and Mansuy, D. (1996) Expression in yeast and purification of functional macrophage nitric oxide synthase. Evidence for cysteine-194 as iron proximal ligand. *Biochemistry* **35,** 7204–7213.
40. Chen, P. F., Tsai, A. L., and Wu, K. K. (1995) Cysteine-99 of endothelial nitric oxide synthase (NOS-III) is critical for tetrahydrobiopterin-dependent NOS-III stability and activity. *Biochem. Biophys. Res. Commun.* **215,** 1119–1129.

41. Cho, H. J., Martin, E., Xie, Q. W., Sassa, S., and Nathan, C. (1995) Inducible nitric oxide synthase: Identification of amino acid residues essential for dimerization and binding of tetrahydrobiopterin. *Proc. Natl. Acad. Sci. USA* **92,** 11,514–11,518.
42. Nishimura, J. S., Martasek, P., McMillan, K., Salerno, J. C., Liu, Q., Gross, S. S., and Masters, B. S. S. (1995) Modular structure of neuronal nitric oxide synthase: Localization of the arginine binding site and modulation by pterin. *Biochem. Biophys. Res. Commun.* **210,** 288–294.
43. Fulco, A. J. (1991) P450BM-3 and other inducible bacterial P450 cytochromes: Biochemistry and regulation. *Annu. Rev. Pharmacol. Toxicol.* **31,** 177–203.
44. Sevrioukova, I., Shaffer, C., Ballou, D. P., and Peterson, J. A. (1996) Equilibrium and transient state spectrophotometric studies of the mechanism of reduction of the flavoprotein domain of P450BM-3. *Biochemistry* **35,** 7058–7068.
45. Hendriks, W. (1995) Nitric oxide synthase contains a discs-large homologous region (DHR) sequence motif. *Biochem. J.* **305,** 687–688.
46. Brenman, J. E., Chao, D. S., Xia, H. H., Aldape, K., and Bredt, D. S. (1995) Nitric oxide synthase complexed with dystrophin and absent from skeletal muscle sarcolemma in Duchenne muscular dystrophy. *Cell* **82,** 743–752.
47. Brenman, J. E., Chao, D. S., Gee, S. H., Mcgee, A. W., Craven, S. E., Santillano, D. R., Wu, Z. Q., Huang, F., Xia, H. H., Peters, M. F., Froehner, S. C., and Bredt, D. S. (1996) Interaction of nitric oxide synthase with the postsynaptic density protein PSD-95 and alpha 1-syntrophin mediated by PDZ domains. *Cell* **84,** 757–767.
48. Sessa, W. C., Barber, C. M., and Lynch, K. R. (1993) Mutation of N-myristoylation site converts endothelial cell nitric oxide synthase from a membrane to a cytosolic protein. *Circ. Res.* **72,** 921–924.
49. Busconi, L. and Michel, T. (1993) Endothelial nitric oxide synthase—N-terminal myristoylation determines subcellular localization. *J. Biol. Chem.* **268,** 8410–8413.
50. Liu, J. W., Garcia-Cardena, G., and Sessa, W. C. (1995) Biosynthesis and palmitoylation of endothelial nitric oxide synthase: Mutagenesis of palmitoylation sites, cysteines-15 and/or -26, argues against depalmitoylation-induced translocation of the enzyme. *Biochemistry* **34,** 12,333–12,340.
51. Arnet, U. A., McMillan, A., Dinerman, J. L., Ballermann, B., and Lowenstein, C. J. (1996) Regulation of endothelial nitric-oxide synthase during hypoxia. *J. Biol. Chem.* **271,** 15,069–15,073.
52. Sessa, W. C., Pritchard, K., Seyedi, N., Wang, J., and Hintze, T. H. (1994) Chronic exercise in dogs increases coronary vascular nitric oxide production and endothelial cell nitric oxide synthase gene expression. *Circ. Res.* **74,** 349–353.
53. Ayajiki, K., Kindermann, M., Hecker, M., Fleming, I., and Busse, R. (1996) Intracellular pH and tyrosine phosphorylation but not calcium determine shear stress-induced nitric oxide production in native endothelial cells. *Circ. Res.* **78,** 750–758.
54. Inoue, N., Ramasamy, S., Fukai, T., Nerem, R. M., and Harrison, D. G. (1996) Shear stress modulates expression of Cu/Zn superoxide dismutase in human aortic endothelial gels. *Circ. Res.* **79,** 32–37.

55. Baek, K. J., Thiel, B. A., Lucas, S., and Stuehr, D. J. (1993) Macrophage nitric oxide synthase subunits—purification, characterization, and role of prosthetic groups and substrate in regulating their association into a dimeric enzyme. *J. Biol. Chem.* **268,** 21,120–21,129.
56. List, B. M., Klösch, B., Völker, C., Gorren, A. C. F., Sessa, W. C., Werner, E. R., Kukovetz, W. R., Schmidt, K., and Mayer, B. (1996) Characterization of bovine endothelial nitric oxide synthase as a homodimer with down-regulated uncoupled NADPH oxidase activity: Tetrahydrobiopterin binding kinetics and role of haem in dimerization. *Biochem. J.* **323,** 159–165.
57. List, B. M., Klatt, P., Werner, E. R., Schmidt, K., and Mayer, B. (1996) Overexpression of neuronal nitric oxide synthase in insect cells reveals requirement of heme for tetrahydrobiopterin binding. *Biochem. J.* **315,** 57–63.
58. Klatt, P., Pfeiffer, S., List, B. M., Lehner, D., Glatter, O., Werner, E. R., Schmidt, K., and Mayer, B. (1996) Characterization of heme-deficient neuronal nitric oxide synthase reveals role for heme in subunit dimerization and binding of the amino acid substrate and tetrahydrobiopterin. *J. Biol. Chem.* **271,** 7336–7342.
59. Xie, Q. W., Leung, M., Fuortes, M., Sassa, S., and Nathan, C. (1996) Complementation analysis of mutants of nitric oxide synthase reveals that the active site requires two hemes. *Proc. Natl. Acad. Sci. USA* **93,** 4891–4896.
60. Klatt, P., Schmidt, K., Lehner, D., Glatter, O., Bächinger, H. P., and Mayer, B. (1995) Structural analysis of porcine brain nitric oxide synthase reveals novel role of tetrahydrobiopterin and L-arginine in the formation of an SDS-resistant dimer. *EMBO J.* **14,** 3687–3695.
61. Rodriguez-Crespo, I., Gerber, N. C., and Ortiz DeMontellano, P. R. (1996) Endothelial nitric-oxide synthase Expression in Escherichia coli, spectroscopic characterization, and role of tetrahydrobiopterin in dimer formation. *J. Biol. Chem.* **271,** 11,462–11,467.
62. Wang, J. L., Stuehr, D. J., and Rousseau, D. L. (1995) Tetrahydrobiopterin-deficient nitric oxide synthase has a modified heme environment and forms a cytochrome P-420 analogue. *Biochemistry* **34,** 7080–7087.
63. Gorren, A. C. F., List, B. M., Schrammel, A., Pitters, E., Hemmens, B., Werner, E. R., Schmidt, K., and Mayer, B. (1996) Pteridine-free neuronal nitric oxide synthase: evidence for two identical highly anticooperative tetrahydrobiopterin binding sites. *Biochem.* **35,** 16,735–16,745.
64. McMillan, K. and Masters, B. S. S. (1995) Prokaryotic expression of the heme- and flavin-binding domains of rat neuronal nitric oxide synthase as distinct polypeptides: Identification of the heme-binding proximal thiolate ligand as cysteine-415. *Biochemistry* **34,** 3686–3693.
65. Heinzel, B., John, M., Klatt, P., Böhme, E., and Mayer, B. (1992) Ca2+/calmodulin-dependent formation of hydrogen peroxide by brain nitric oxide synthase. *Biochem. J.* **281,** 627–630.
66. Pou, S., Pou, W. S., Bredt, D. S., Snyder, S. H., and Rosen, G. M. (1992) Generation of superoxide by purified brain nitric oxide synthase. *J. Biol. Chem.* **267,** 24,173–24,176.

67. Culcasi, M., Lafon-Cazal, M., Pietri, S., and Bockaert, J. (1994) Glutamate receptors induce a burst of superoxide via activation of nitric oxide synthase in arginine-depleted neurons. *J. Biol. Chem.* **269,** 12,589–12,593.
68. Abu-Soud, H. M. and Stuehr, D. J. (1993) Nitric oxide synthases reveal a novel role for calmodulin in controlling electron transfer. *Proc. Natl. Acad. Sci. USA* **90,** 10,769–10,772.
69. Mayer, B., Klatt, P., Werner, E. R., and Schmidt, K. (1995) Kinetics and mechanism of tetrahydrobiopterin-induced oxidation of nitric oxide. *J. Biol. Chem.* **270,** 655–659.
70. Xia, Y., Dawson, V. L., Dawson, T. M., Snyder, S. H., and Zweier, J. L. (1996) Nitric oxide synthase generates superoxide and nitric oxide in arginine-depleted cells leading to peroxynitrite-mediated cellular injury. *Proc. Natl. Acad. Sci. USA* **93,** 6770–6774.
71. Huang, Z. H., Huang, P. L., Panahian, N., Dalkara, T., Fishman, M. C., and Moskowitz, M. A. (1994) Effects of cerebral ischemia in mice deficient in neuronal nitric oxide synthase. *Science* **265,** 1883–1885.
72. Duch, D. S. and Smith, G. K. (1991) Biosynthesis and function of tetrahydrobiopterin. *J. Nutr. Biochem.* **2,** 411–423.
73. Mayer, B. and Werner, E. R. (1995) In search of a function for tetrahydrobiopterin in the biosynthesis of nitric oxide. *Naunyn-Schmiedeberg's Arch. Pharmacol.* **351,** 453–463.
74. Klatt, P., Schmid, M., Leopold, E., Schmidt, K., Werner, E. R., and Mayer, B. (1994) The pteridine binding site of brain nitric oxide synthase—tetrahydrobiopterin binding kinetics, specificity, and allosteric interaction with the substrate domain. *J. Biol. Chem.* **269,** 13,861–13,866.
75. Werner, E. R., Pitters, E., Schmidt, K., Wachter, H., Werner-Felmayer, G., and Mayer, B. (1996) Identification of the 4-amino analogue of tetrahydrobiopterin as dihydropteridine reductase inhibitor and potent pteridine antagonist of rat neuronal nitric oxide synthase. *Biochem. J.* **320,** 193–196.
76. Abu-Soud, H. M., Wang, J. L., Rousseau, D. L., Fukoto, J. M., Ignarro, L. J., and Stuehr, D. J. (1995) Neuronal nitric oxide synthase self-inactivates by forming a ferrous-nitrosyl complex during aerobic catalysis. *J. Biol. Chem.* **270,** 22,997–23,006.
77. Abu-Soud, H. M., Mayer, B., and Stuehr, D. J. (1996) Kinetic analysis of NO and CO binding to ferric and ferrous neuronal NO synthase, in *2nd International Conference Biochemistry* and *Molecular Biology of Nitric Oxide*, Abstract A1, Los Angeles, CA, July 13–17.
78. Klatt, P., Heinzel, B., John, M., Kastner, M., Böhme, E., and Mayer, B. (1992) Ca2+/calmodulin-dependent cytochrome c reductase activity of brain nitric oxide synthase. *J. Biol. Chem.* **267,** 11,374–11,378.
79. Abu-Soud, H. M., Yoho, L. L., and Stuehr, D. J. (1994) Calmodulin controls neuronal nitric-oxide synthase by a dual mechanism—Activation of intra- and interdomain electron transfer. *J. Biol. Chem.* **269,** 32,047–32,050.

80. Vorherr, T., Knopfel, L., Hofmann, F., Mollner, S., Pfeuffer, T., and Carafoli, E. (1993) The calmodulin binding domain of nitric oxide synthase and adenylyl cyclase. *Biochemistry* **32,** 6081–6088.
81. Zhang, M. J. and Vogel, H. J. (1994) Characterization of the calmodulin-binding domain of rat cerebellar nitric oxide synthase. *J. Biol. Chem.* **269,** 981–985.
82. Anagli, J., Hofmann, F., Quadroni, M., Vorherr, T., and Carafoli, E. (1995) The calmodulin-binding domain of the inducible (macrophage) nitric oxide synthase. *Eur. J. Biochem.* **233,** 701–708.
83. Venema, R. C., Sayegh, H. S., Kent, J. D., and Harrison, D. G. (1996) Identification, characterization, and comparison of the calmodulin-binding domains of the endothelial and inducible nitric oxide synthases. *J. Biol. Chem.* **271,** 6435–6440.
84. Zhang, M. J., Yuan, T., Aramini, J. M., and Vogel, H. J. (1995) Interaction of calmodulin with its binding domain of rat cerebellar nitric oxide synthase—A multinuclear NMR study. *J. Biol. Chem.* **270,** 20,901–20,907.
85. Salerno, J. C., Harris, D. E., Irizzary, K., Smith, S. M. E., McMillan, K., Martasek, P., Roman, L. J., Masters, B. S. S., Jones, C., Weissman, B. A., Liu, Q., and Gross, S. S. (1996) The inhibitory polypeptide of constitutive NOS (cNOS) is the missing control site element of the inducible isoform (iNOS), in *2nd International Conference Biochemistry and Molecular Biology of Nitric Oxide*, Abstract A20, Los Angeles, CA, July 13–17.
86. Stevens-Truss, R. and Marletta, M. A. (1995) Interaction of calmodulin with the inducible murine macrophage nitric oxide synthase. *Biochemistry* **34,** 15,638–15,645.
87. Charles, I. G., Chubb, A., Gill, R., Clare, J., Lowe, P. N., Holmes, L. S., Page, M., Keeling, J. G., Moncada, S., and Riveros-Moreno, V. (1993) Cloning and expression of a rat neuronal nitric oxide synthase coding sequence in a baculovirus/insect cell system. *Biochem. Biophys. Res. Commun.* **196,** 1481–1489.
88. Black, S. M. and Ortiz de Montellano, P. R. (1995) Characterization of rat neuronal nitric oxide synthase expressed in Saccharomyces cerevisiae. *DNA Cell Biol.* **14,** 789–794.
89. Ogura, T., Yokoyama, T., Fujisawa, H., Kurashima, Y., and Esumi, H. (1993) Structural diversity of neuronal nitric oxide synthase messenger RNA in the nervous system. *Biochem. Biophys. Res. Commun.* **193,** 1014–1022.
90. Nakane, M., Schmidt, H. H. H. W., Pollock, J. S., Forstermann, U., and Murad, F. (1993) Cloned human brain nitric oxide synthase is highly expressed in skeletal muscle. *FEBS Lett.* **316,** 175–180.
91. Sessa, W. C., Harrison, J. K., Barber, C. M., Zeng, D., Durieux, M. E., Dangelo, D. D., Lynch, K. R., and Peach, M. J. (1992) Molecular cloning and expression of a cDNA encoding endothelial cell nitric oxide synthase. *J. Biol. Chem.* **267,** 15,274–15,276.
92. Nishida, K., Harrison, D. G., Navas, J. P., Fisher, A. A., Dockery, S. P., Uematsu, M., Nerem, R. M., Alexander, R. W., and Murphy, T. J. (1992) Molecular cloning

and characterization of the constitutive bovine aortic endothelial cell nitric oxide synthase. *J. Clin. Invest.* **90,** 2092–2096.

93. Lamas, S., Marsden, P. A., Li, G. K., Tempst, P., and Michel, T. (1992) Endothelial nitric oxide synthase—molecular cloning and characterization of a distinct constitutive enzyme isoform. *Proc. Natl. Acad. Sci. USA* **89,** 6348–6352.
94. Janssens, S. P., Shimouchi, A., Quertermous, T., Bloch, D. B., and Bloch, K. D. (1992) Cloning and Expression of a cDNA-Encoding Human Endothelium-Derived Relaxing Factor Nitric-Oxide Synthase. *J. Biol. Chem.* **267,** 4519–4522.
95. Marsden, P. A., Schappert, K. T., Chen, H. S., Flowers, M., Sundell, C. L., Wilcox, J. N., Lamas, S., and Michel, T. (1992) Molecular cloning and characterization of human endothelial nitric oxide synthase. *FEBS Lett.* **307,** 287–293.
96. Lowenstein, C. J., Glatt, C. S., Bredt, D. S., and Snyder, S. H. (1992) Cloned and expressed macrophage nitric oxide synthase contrasts with the brain enzyme. *Proc. Natl. Acad. Sci. USA* **89,** 6711–6715.
97. Lyons, C. R., Orloff, G. J., and Cunningham, J. M. (1992) Molecular cloning and functional expression of an inducible nitric oxide synthase from a murine macrophage cell line. *J. Biol. Chem.* **267,** 6370–6374.
98. Xie, Q. W., Cho, H. J., Calaycay, J., Mumford, R. A., Swiderek, K. M., Lee, T. D., Ding, A. H., Troso, T., and Nathan, C. (1992) Cloning and characterization of inducible nitric oxide synthase from mouse macrophages. *Science* **256,** 225–228.
99. Adachi, H., Iida, S., Oguchi, S., Ohshima, H., Suzuki, H., Nagasaki, K., Kawasaki, H., Sugimura, T., and Esumi, H. (1993) Molecular cloning of a cDNA encoding an inducible calmodulin-dependent nitric-oxide synthase from rat liver and its expression in COS-1 cells. *Eur. J. Biochem.* **217,** 37–43.
100. Galea, E., Reis, D. J., and Feinstein, D. L. (1994) Cloning and expression of inducible nitric oxide synthase from rat astrocytes. *J. Neurosci. Res.* **37,** 406–414.
101. Geng, Y. J., Almqvist, M., and Hansson, G. K. (1994) cDNA cloning and expression of inducible nitric oxide synthase from rat vascular smooth muscle cells. *Biochim. Biophys. Acta* **1218,** 421–424.
102. Karlsen, A. E., Andersen, H. U., Vissing, H., Larsen, P. M., Fey, S. J., Cuartero, B. G., Madsen, O. D., Petersen, J. S., Mortensen, S. B., and Mandrup-Poulsen, T. (1995) Cloning and expression of cytokine-inducible nitric oxide synthase cDNA from rat islets of Langerhans. *Diabetes.* **44,** 753–758.
103. Geller, D. A., Lowenstein, C. J., Shapiro, R. A., Nussler, A. K., Disilvio, M., Wang, S. C., Nakayama, D. K., Simmons, R. L., Snyder, S. H., and Billiar, T. R. (1993) Molecular cloning and expression of inducible nitric oxide synthase from human hepatocytes. *Proc. Natl. Acad. Sci. USA* **90,** 3491–3495.
104. Sherman, P. A., Laubach, V. E., Reep, B. R., and Wood, E. R. (1993) Purification and cDNA sequence of an inducible nitric oxide synthase from a human tumor cell line. *Biochemistry* **32,** 11,600–11,605.
105. Charles, I. G., Palmer, R. M. J., Hickery, M. S., Bayliss, M. T., Chubb, A. P., Hall, V. S., Moss, D. W., and Moncada, S. (1993) Cloning, characterization, and expression of a cDNA encoding an inducible nitric oxide synthase from the human chondrocyte. *Proc. Natl. Acad. Sci. USA* **90,** 11,419–11,423.

106. Maier, R., Bilbe, G., Rediske, J., and Lotz, M. (1994) Inducible nitric oxide synthase from human articular chondrocytes: cDNA cloning and analysis of mRNA expression. *Biochimica et Biophysica Acta* **1208,** 145–150.
107. Chartrain, N. A., Geller, D. A., Koty, P. P., Sitrin, N. F., Nussler, A. K., Hoffman, E. P., Billiar, T. R., Hutchinson, N. I., and Mudgett, J. S. (1994) Molecular cloning, structure, and chromosomal localization of the human inducible nitric oxide synthase gene. *J. Biol. Chem.* **269,** 6765–6772.
108. Adler, H., Frech, B., Thony, M., Pfister, H., Peterhans, E., and Jungi, T. W. (1995) Inducible nitric oxide synthase in cattle. Differential cytokine regulation of nitric oxide synthase in bovine and murine macrophages. *J. Immunol.* **154,** 4710–4718.
109. McMillan, K., Bredt, D. S., Hirsch, D. J., Snyder, S. H., Clark, J. E., and Masters, B. S. S. (1992) Cloned, expressed rat cerebellar nitric oxide synthase contains stoichiometric amounts of heme, which binds carbon monoxide. *Proc. Natl. Acad. Sci. USA* **89,** 11,141–11,145.
110. Nakane, M., Pollock, J. S., Klinghofer, V., Basha, F., Marsden, P. A., Hokari, A., Ogura, T., Esumi, H., and Carter, G. W. (1995) Functional expression of three isoforms of human nitric oxide synthase in baculovirus-infected insect cells. *Biochem. Biophys. Res. Commun.* **206,** 511–517.
111. Harteneck, C., Klatt, P., Schmidt, K., and Mayer, B. (1994) Expression of rat brain nitric oxide synthase in baculovirus-infected insect cells and characterization of the purified enzyme. *Biochem. J.* **304,** 683–686.
112. Riveros-Moreno, V., Heffernan, B., Torres, B., Chubb, A., Charles, I., and Moncada, S. (1995) Purification to homogeneity and characterisation of rat brain recombinant nitric oxide synthase. *Eur. J. Biochem.* **230,** 52–57.
113. Roman, L. J., Sheta, E. A., Martasek, P., Gross, S. S., Liu, Q., and Masters, B. S. S. (1995) High-level expression of functional rat neuronal nitric oxide synthase in Escherichia coli. *Proc. Natl. Acad. Sci. USA* **92,** 8428–8432.
114. Gerber, N. C. and Ortiz DeMontellano, P. R. (1995) Neuronal nitric oxide synthase—Expression in Escherichia coli, irreversible inhibition by phenyldiazene, and active site topology. *J. Biol. Chem.* **270,** 17,791–17,796.
115. Busconi, L. and Michel, T. (1995) Recombinant endothelial nitric oxide synthase: Post-translational modifications in a baculovirus expression system. *Mol. Pharmacol.* **47,** 655–659.
116. Seo, H. G., Fujii, J., Soejima, H., Niikawa, N., and Taniguchi, N. (1995) Heme requirement for production of active endothelial nitric oxide synthase in baculovirus-infected insect cells. *Biochem. Biophys. Res. Commun.* **208,** 10–18.
117. Chen, P. F., Tsai, A. L., Berka, V., and Wu, K. K. (1996) Endothelial nitric-oxide synthase - Evidence for bidomain structure and successful reconstitution of catalytic activity from two separate domains generated by a baculovirus expression system. *J. Biol. Chem.* **271,** 14,631–14,635.
118. Martasek, P., Liu, Q., Liu, J., Roman, L. J., Gross, S. S., Sessa, W. C., and Masters, B. S. S. (1996) Characterization of bovine endothelial nitric oxide synthase expressed in E. coli. *Biochem. Biophys. Res. Commun.* **219,** 359–365.

119. Laubach, V. E., Garvey, E. P., and Sherman, P. A. (1996) High-level expression of human inducible nitric oxide synthase in Chinese hamster ovary cells and characterization of the purified enzyme. *Biochem. Biophys. Res. Commun.* **218,** 802–807.
120. Xie, Q. W., Cho, H., Kashiwabara, Y., Baum, M., Weidner, J. R., Elliston, K., Mumford, R., and Nathan, C. (1994) Carboxyl terminus of inducible nitric oxide synthase—Contribution to NADPH binding and enzymatic activity. *J. Biol. Chem.* **269,** 28,500–28,505.
121. Moss, D. W., Wei, X. Q., Liew, F. Y., Moncada, S., and Charles, I. G. (1995) Enzymatic characterisation of recombinant murine inducible nitric oxide synthase. *Eur. J. Pharmacol. Mol. Pharmacol. Sec.* **289,** 41–48.
122. Fossetta, J. D., Niu, X. D., Lunn, C. A., Zavodny, P. J., Narula, S. K., and Lundell, D. (1996) Expression of human inducible nitric oxide synthase in *Escherichia coli. FEBS Lett.* **379,** 135–138.

2

Purification of the Inducible Nitric Oxide Synthase

Edward P. Garvey

1. Introduction

Regulation of the inducible nitiric oxide synthase (iNOS) isozyme is distinct from the constitutively expressed endothelial (e) and neuronal (n) NOS isozymes in that it occurs at the level of gene expression and not by Ca^{+2}/calmodulin (CaM) (reviewed in **ref.** ***1***). These distinctions have direct bearing on the purification of iNOS. First, the source of iNOS is typically a cell line that can be induced to synthesize iNOS in response to one or more of a number of factors. Second, because iNOS binds CaM essentially irreversibly in the absence of Ca^{+2}, the purification of iNOS does not include CaM-affinity chromatography. As is true for the constitutive NOS isozymes, the purification of iNOS depends on 2',5'-adenosine diphosphate (ADP)-affinity chromatography. And because iNOS cannot bind CaM resins, purification of iNOS to homogeneity requires either an anionic-exchange step or gel-filtration step, or both *(2,3)*.

1.1. Induction of iNOS in Tissue-Culture Cells

The 5' untranslated region of the gene for iNOS contains many response elements that regulate gene expression (e.g., **ref.** ***4*** for the human gene). The first reported study of a factor stimulating production of NO in tissue cells was the effect of *Escherichia coli* lipopolysaccharide (LPS) on macrophages *(5)*. Subsequently, a multitude of studies have reported a wide range of cytokines, pathogens, or components of pathogens that stimulate, often in a synergistic fashion, the expression of NOS (listed in **ref.** ***4***). Two systems of note are murine macrophage-cell lines *(5)* and the human colorectal-adenocarcinoma cell line DLD-1 *(6)*. Murine-macrophage cells generate, after stimulation by LPS

From: *Methods in Molecular Biology, Vol. 100. Nitric Oxide Protocols*
Edited by: Michael A. Titheradge © Humana Press Inc., Totowa, NJ

and interferon-γ (IFN-γ), the highest reported levels of iNOS from a tissue-culture cell line; 100 μg of homogenous iNOS can be purified from 6L of cells. The human cancer-cell line is attractive because it is a source of authentic human iNOS. The synthesis of human iNOS is induced by IFN-γ, interleukin-6 (IL-6), tumor necrosis factor α (TNF α), and interleukin-β (IL-1β). In contrast to the murine macrophage-cell lines, DLD-1 cells generate approximately 10-fold less iNOS.

1.2. Stability of NOS

NOS shows lability especially in the latter stages of purification and at dilute protein concentrations (less than 0.1 mg/mL). Thus, many purification schemes contain additives (glycerol, sucrose, detergents) and relatively high-protein concentrations [rapid concentration, addition of bovine serum albumin (BSA)], and are done as rapidly as possible (with some rapid freezing steps allowed). Under optimal conditions, highly pure NOS is stable indefinitively at –70°C.

2. Materials (all buffers stored and used at 4°C)

1. Tissue-culture cells can be obtained from American Type Culture Collection (ATCC, Rockville, MD). Two typically used mouse monocyte-macrophage cells are J774A.1 and RAW264.7. The human colon-adenocarcinoma cells line DLD-1 is a source for human iNOS. This chapter will describe the induction and purification of iNOS from DLD-1 cells *(7)*.
2. RPMI 1640 medium (Gibco BRL).
3. L-glutamine, penicillin, streptomycin, and 10% (v/v) heat-inactivated fetal bovine serum (FBS) (Hyclone).
4. Human cytokines: IFN-γ, IL-6, and TNF α (Boehringer Mannheim), and IL-1β (Genzyme).
5. Phosphate-buffered saline (PBS): 50 m*M* sodium phosphate, pH 7.4, 150 m*M* NaCl.
6. TDGB buffer: 20 m*M* Tris-HCl, pH 7.5 (pH at room temperature), 2 m*M* dithiothreitol (DTT) 10% (v/v) glycerol, 2 μ*M* tetrahydrobiopterin (BH_4).
7. CHAPS {3-[(3-cholamidopropyl)dimethylammonio]-1-propane sulfonate}.
8. 2',5'-ADP agarose (Sigma) or 2',5'-ADP sepharose (Pharmacia).
9. 0.5 *M* NaCl in TDGB buffer.
10. 3 m*M* malic acid, 0.15 m*M* nicotinamide adenine dinucleotide phosphate ($NADP^+$) in TDGB buffer.
11. 1 m*M* 2'-AMP in TDGB.
12. 1 m*M* nicotinamide adenine dinucleotide, reduced form (NADH) in TDGB.
13. 2 m*M* nicotinamide adenine dinucleotide phosphate, reduced form (NADPH), 0.5 mg/mL BSA in TDGB.
14. DEAE Bio-Gel A-agarose (BioRad).
15. 50 m*M* NaCl in TDGB.
16. Centriprep 30 (Amicon).
17. PD-10 desalting columns (Pharmacia).

3. Methods

1. DLD-1 cells are grown in ten 150 cm^2 flasks (20 mL/flask) at 37°C, 5% CO_2 in RPMI 1640 medium, supplemented with L-glutamine (4 m*M*), penicillin (50 units/mL), streptomycin (50 μg/mL), and 10% heat-inactivated FBS. Cells are grown to confluence, and treated with 100 units/mL IFN-γ, 200 units/mL IL-6, 10 ng/mL TNF α, and 0.5 ng/mL IL-1β (*see* **Note 1**). At 18–24 h post-induction, cells are harvested by scraping and washed with PBS. Pelleted cells are quick frozen and stored at –70°C.
2. Purification of iNOS is performed at 4°C on a single day. Cells are rapidly thawed in 5 mL of TDGB. Cells are disrupted by three cycles of freeze/thawing. Solid CHAPS is added to make solution 5 m*M* in CHAPS (*see* **Note 2**). Solution is left on ice for 30 min, and then centrifuged at 100,000*g* for 45 min.
3. 0.2 mL of 2',5'-ADP agarose (or sepharose) that has been equilibrated with TDGB is added to supernatant. This slurry is rotated at 4°C for 30 min.
4. The slurry is packed into a column (5 × 7 mm). The resin is washed by gravity flow sequentially with 3 mL each of TDGB, 500 m*M* NaCl in TDGB, 3 m*M* malic acid plus 0.15 m*M* $NADP^+$ in TDGB, 1 m*M* 2'-AMP in TDGB, 1 m*M* NADH in TDGB, and TDGB. NOS is eluted with 2 m*M* NADPH plus 0.5 mg/mL BSA in TDGB.
5. Active fractions are pooled and applied at a flow of 0.2 mL/min to a 1-mL column of DEAE Bio-Gel A agarose that had been equilibrated in TDGB. The resin is washed with 5 mL each of TDGB and 50 m*M* NaCl in TDGB. NOS is eluted with a linear gradient (10 mL, 50 to 500 m*M*) of NaCl in TDGB. NOS elutes at about 200 m*M* NaCl (*see* **Note 3**). BSA is added immediately to give a final concentration of 0.5 mg/mL. Aliquots are quick frozen and stored at –70°C (*see* **Notes 4–6**).

4. Notes

1. The specific activity of iNOS in crude extracts increases when cytokines are added to the growth medium after confluence has been reached. Typically, cells reach confluence on d 4 after seeding. Maximum specific activity of iNOS is observed if cells are induced on d 6 after seeding (fivefold increase in specific activity at d 6 vs d 4).
2. CHAPS is typically added to preparations of eNOS to release eNOS that is membrane-bound. There is no reported documentation that iNOS is membrane-bound; however, we have reproducibly observed that in the absence of detergent a significant proportion (20–30%) of NOS activity pellets during an 100,000*g* centrifugation. Therefore, CHAPS is added to maximize the yield of iNOS.
3. NADPH that is present from the 2',5'-ADP-chromatography step partially coelutes during the salt gradient of the ion-exchange step. If NADPH needs to be completely removed, iNOS can be desalted using PD-10 Sephadex G-25 columns.
4. If NOS needs to be concentrated, centriprep-30 concentrators can be used. Prepare concentrator for use by thoroughly washing filter with TDGB buffer (i.e., pass buffer through filter).

5. Protein concentration can be determined by a Bradford assay (Pierce). NOS activity is determined by following the conversion of radiolabeled L-arginine to L-citrulline *(8)*, *see* Chapter 7, this volume.
6. Approximately 20 µg of homogenous-human iNOS can be purified from approx 2 g of wet DLD-1 cells. This preparation of iNOS from DLD-1 cells can be used to purify iNOS from induced mouse-macrophage cells or recombinant iNOS from mammalian cells *(9)*.

References

1. Morris, S. M., Jr. and Billiar, T. R. (1994) New insights into the regulation of inducible nitric oxide synthesis. *Am. J. Physiol.* **266,** E829–E839.
2. Stuehr, D. J., Cho, H. J., Kwon, N. S., Weise, M. F., and Nathan, C. F. (1991) Purification and characterization of the cytokine-induced macrophage nitric oxide synthase: an FAD- and FMN-utilizing flavoprotein. *Proc. Natl. Acad. Sci. USA* **88,** 7773–7777.
3. Hevel, J. M., White, K. A., and Marletta, M. A. (1991) Purification of the inducible murine macrophage nitric oxide synthase. *J. Biol. Chem.* **266,** 22,789–22,791.
4. Chartrain, N. A., Geller, D. A., Koty, P. P., Sitrin, N. F., Nussler, A. K., Hoffman, E. P., Billiar, T. R., Hutchinson, N. I., and Mudgett, J. S. (1994) Molecular cloning, structure, and chromosomal localization of the human inducible nitric oxide synthase gene. *J. Biol. Chem.* **269,** 6765–6772.
5. Marletta, M. A., Yoon, P. S., Iyengar, R., Leaf, C. D., and Wishnok, J. S. (1988) Macrophage oxidation of L-arginine to nitrite and nitrate: nitric oxide is an intermediate. *Biochem.* **27,** 8706–8711.
6. Nathan, C. and Xie, Q-w. (1994) Nitric oxide synthases: roles, tolls, and controls. *Cell* **78,** 8706–8711.
7. Sherman, P. A., Laubach, V. E., Reep, B. R., and Wood, E. R. (1993) Purification and cDNA sequence of an inducible nitric oxide synthase from a human tumor cell line. *Biochem.* **32,** 11,600–11,605.
8. Schmidt, H. H. H. W., Pollock, J. S., Nakane, M., Gorsky, L. D., Forstermann, U., and Murad, R. (1991) Purification of a soluble isoform of guanyl cylcase-activating-factor synthase. *Proc. Natl. Acad. Sci. USA* **88,** 365–369.
9. Laubach, V. E., Garvey, E. P., and Sherman, P. A. (1995) High-level expression of human inducible nitric oxide synthase in chinese hamster ovary cells and characterization of the purified enzyme. *Biochem. Biophys. Res. Commun.* **218,** 802–807.

3

Purification of the Constitutive Nitric Oxide Synthase

Edward P. Garvey

1. Introduction

Constitutive nitric oxide synthase (NOS) exists as two isoforms: endothelial (e) and neuronal (n) (reviewed in **ref. *1***). The first reports on the purification of eNOS and nNOS were by Schmidt et al. *(2)* and Bredt and Snyder *(3)*, respectively; all subsequent reports have been modifications of these procedures. The purification of both of the constitutive isoforms and the inducible (i) NOS depends on the use of 2',5'-adenosine diphosphate (ADP) affinity chromatography. The 2',5'-ADP ligand is a structural mimic of the cofactor nicotinamide adenine dinucleotide phosphate (NADPH) that is required for NOS activity. Therefore, the biospecific elution is accomplished by the presence of NADPH. The purification of both constitutive isoforms, unlike iNOS, also depends on a second affinity-chromatography step, calmodulin (CaM)-affinity chromatography. Exogenously-added CaM is essential for both purified eNOS and nNOS activity. In the presence of calcium, CaM binds both constitutive isoforms in a tight (K_d ~ 10 n*M*) but reversible complex, thereby activating NOS. In contrast, CaM binds to iNOS in the absence of added calcium and dissociates only in the presence of a protein denaturant. Thus, during the 2',5'-ADP-affinity step, CaM is removed from eNOS and nNOS, but not iNOS. The binding of eNOS and nNOS to the CaM resin is performed in the presence of calcium, and elution of NOS is achieved by the absence of calcium and presence of ethylenediaminetetraacetic acid (EDTA) and high salt. Depending on the source of enzyme, near homogenous NOS is often obtained by utilizing these two affinity steps. A common addition to reach the desired purity is anionic-exchange chromatography.

From: *Methods in Molecular Biology, Vol. 100. Nitric Oxide Protocols*
Edited by: Michael A. Titheradge © Humana Press Inc., Totowa, NJ

1.1. Subcellular Location of eNOS and nNOS

A significant fraction of eNOS is often isolated in the membrane fraction of a given crude homogenate. The amino acid sequence of eNOS has a consensus sequence for N-terminal myristoylation as well as for phosphorylation at several serine sites, and biochemical studies *(4)* have shown that post-translational modifications at these sites play important roles in directing subcellular location of eNOS. Thus, purification of eNOS often involves initial isolation of eNOS in the particulate fraction of a crude homogenate. To date, no primary sequence of nNOS contains such sites and the location of nNOS has always been described as cytosolic.

1.2. Stability of NOS

On the whole, eNOS and nNOS are reasonably stable enzymes. However, they do show lability especially in the latter stages of purification and at dilute protein concentrations (less than 0.1 mg/mL). Thus, many purification schemes have evolved to contain additives (glycerol, sucrose, detergents) and relatively high-protein concentrations [rapid concentration, addition of bovine serum albumin (BSA)], and to be done as rapidly as possible (with some rapid-freezing steps allowed). Under optimal conditions, highly pure NOS is stable indefinitively at –70°C.

2. Materials (all buffers stored and used at 4°C)

1. eNOS has been purified both from cultured-endothelial cells and tissue, and nNOS has been purified from brain tissue from many different species. In addition, both isoforms have been purified as recombinant proteins expressed in a number of systems. This chapter will describe the purification of eNOS from human placenta *(5)*. Human placenta can often be obtained from the birthing center at a local hospital.
2. 0.9% (w/v) NaCl.
3. HEDS buffer: 20 m*M* HEPES, pH 7.8 (pH at room temperature), 0.1 m*M* EDTA, 5 m*M* DTT, and 0.2 *M* sucrose.
4. 0.1 *M* PMSF (in isoproponal).
5. CHAPS {3-[(3-cholamidopropyl)dimethylammonio]-1-propane sulfonate}.
6. 500 m*M* NaCl in HEDS buffer.
7. 2 m*M* nicotinamide adenine dinucleotide phosphate (NADPH) in HEDS buffer.
8. 2',5'-ADP agarose (Sigma) or 2',5'-ADP sepharose (Pharmacia).
9. 1 m*M* $CaCl_2$ in HEDS buffer.
10. 1 m*M* $CaCl_2$, 1% (v/v) NP-40 in HEDS buffer.
11. 5 m*M* EDTA, 1 *M* NaCl in HEDS buffer.
12. CaM agarose (Sigma).
13. Tween-20.

14. Centriprep 30 and Centricon 30 (Amicon).
15. PD-10 desalting columns (Pharmacia).
16. 0.1% (v/v) Tween-20 in HEDS buffer.
17. 0.1% (v/v) Tween-20, 0.5 *M* NaCl in HEDS buffer.
18. Mono Q FPLC HR 5/5 column.

3. Methods (all steps done at 4°C)

3.1. Day One

1. Amnion and chorion are removed (≤ 1 h after delivery) from fresh placenta (typically, ~600 gm), which is then rinsed with 0.9% NaCl until excess blood is removed, and homogenized in a Waring blender in three volumes of HEDS buffer plus 0.1 m*M* PMSF (*see* **Notes 1–3**).
2. Solid CHAPS is added to a concentration of 5 m*M*, and the homogenate is slowly stirred for 30 min; then centrifuged at 27,500*g* for 30 min (*see* **Note 4**).
3. The 2',5'-ADP agarose (10 mL equilibrated with HEDS buffer) is added to the supernatant. The slurry is mixed slowly overnight.

3.2. Day Two

4. The slurry is packed into a column (*see* **Note 5**). The resin is sequentially washed at a flow of 1 mL/min with 50 mL each of HEDS, 500 m*M* NaCl in HEDS, and HEDS. The NOS is then eluted with 100 mL of 2 m*M* NADPH in HEDS (*see* **Notes 6** and **7**).
5. $CaCl_2$ is added to the pooled fractions to a final concentration of 1 m*M*, and this is loaded at an appropriately slow-flow rate overnight onto a 0.5 mL column of CaM agarose (equilibrated with 1 m*M* $CaCl_2$ in HEDS buffer).

3.3. Day Three

6. At a flow of 1 mL/min, the resin was washed with 10 mL each of 1 m*M* $CaCl_2$ plus 1% NP-40 in HEDS buffer, then with 1 m*M* $CaCl_2$ in HEDS buffer (*see* **Note 8**). The NOS is eluted with 5 m*M* EDTA, 1 *M* NaCl in HEDS buffer.
7. Pooled fractions are made 0.1% with Tween-20. The pool is concentrated using a Centriprep 30 concentrator to 2 mL (*see* **Note 9**) and then desalted using PD-10 prepacked columns (as described by Pharmacia) and 0.1% Tween-20 in HEDS buffer as an eluent.
8. The enzyme is then applied at a flow rate of 1 mL/min to a Mono Q HR 5/5 FPLC column (equilibrated with 0.1% Tween-20 in HEDS buffer). After it is washed with 5 mL equilibration buffer, the column is developed with a linear gradient (14 mL total volume, 0–500 m*M*) of NaCl in equilibration buffer. NOS elutes at about 300 m*M* NaCl.
9. Pooled fractions are desalted in 0.1% Tween-20 in HEDS buffer and then concentrated using a Centricon 30 microconcentrator to a protein concentration ≥ 0.5 mg/mL. Aliquots are quick frozen and stored at –70°C (*see* **Notes 10–12**).

4. Notes

1. Although any potential medical hazard should be flagged by the collaborating hospital, placental tissue should be treated as hazardous. For example, tissue homogenization should be carried out inside a hooded area, and all waste material should be either autoclaved or treated with bleach.
2. PMSF is extremely harmful and gloves and mask should be worn when preparing stock solutions. As an isoproponal solution, it is stable for about 2 mo.
3. Tissue should be placed on ice as soon as possible after delivery.
4. The addition of CHAPS releases any membrane-associated NOS. Thus, in this procedure, a mixture of soluble and particulate eNOS is purified. If one wished to purify only the soluble eNOS, one could centrifuge (27,000*g* for 30 min) the homogenate without adding CHAPS and simply use the supernatant. If one wished to purify only the particulate form, one could isolate the pellet after centrifugation (27,000*g* for 30 min) and redissolve with 5 m*M* CHAPS in HEDS buffer, do a high-speed centrifugation (100,000*g* for 60 min), and then use the supernatant. When such separation procedures are done, no differences are observed in the kinetic parameters of substrate and a wide variety of inhibitors for the two forms of placental eNOS.
5. Pouring homogenate/2',5'-ADP-resin slurry into a column is very slow because the material is extremely thick. If material is poured into a number of columns (e.g., four) and then each settled resin combined using HEDS buffer, time can be reduced significantly.
6. NOS eluted with a gradient of 0–10 m*M* NADPH in HEDS buffer does not increase the purification of NOS in this step.
7. 2'5'-ADP resin can be reused. To clean resin, wash with at least 10 times column volume of 6 *M* urea to fully remove nonspecific protein from resin.
8. The 1% NP-40 wash of the CaM agarose step will remove residual cytochrome P450 reductase that is nonspecifically bound to NOS.
9. When using Amicon concentrators, prepare concentrator for use by thoroughly washing filter with 0.1% Tween-20 in HEDS buffer (i.e., pass buffer through filter).
10. Protein concentration is determined by a Bradford assay (Pierce). NOS activity is determined by following the conversion of radiolabeled L-arginine to L-citrulline *(2)*, as described in Chapter 7 of this volume.
11. Approximately 100–200 μg of homogenous eNOS could be obtained from a human placenta weighing about 600 gm.
12. The preparation of nNOS from human-brain tissue is very similiar. The following differences are noted. A potential tissue source is the Cooperative Human Tissue Network (Birmingham, AL). Purification buffer (HED) does not contain sucrose. There is no addition of CHAPS. An ammonium sulfate fractionation is added prior to the 2',5'-ADP-agarose step. Solid ammonium sulfate is added to 30% of saturation, and the mixture is slowly stirred for 30 min. The precipitate is collected by centrifugation at 13,000*g* for 30 min and is then redissolved in 1/5 the original volume with HED, which includes 4 μ*M* tetrahydrobiopterin (BH_4),

1 μM flavin adenine dinucleotide (FAD), and 1 μM flavin mononucleotide (FMN). Solution is centrifuged at 41,000*g* for 60 min and the supernatant is removed, quick frozen, and stored at –70°C. The 2',5'-ADP agarose step is performed by loading the thawed material directly onto the resin. Chromatography is as described for eNOS except for the use of HED. The pooled eluted enzyme is made 15% glycerol, 1 m*M* $CaCl_2$, 10 μM BH_4, 0.1% Tween-20, 1 μM FAD, and 1 μM FMN. This material is applied to a CaM-agarose column as described for eNOS except for: the use of HED, the 1% NP-40 wash is omitted, and 15% glycerol is added to the elution buffer (i.e., elution buffer is 5 m*M* EDTA, 1 *M* NaCl, and 15% glycerol in HED). The eluted enzyme is made 10 μM BH_4, 1 μM FAD, 1 μM FMN, and 0.1 % Tween-20. As with eNOS, the purified nNOS is concentrated to reach a protein concentration of ≥ 0.5 mg/mL prior to quick freezing and storage at –70°C.

References

1. Stuehr, D. J. and Griffith, O. W. (1992) Mammalian nitric oxide synthases. *Adv. Enzymol. Relat. Areas Mol. Biol.* **65,** 287–346.
2. Schmidt, H. H. H. W., Pollock, J. S., Nakane, M., Gorsky, L. D., Forstermann, U., and Murad, R. (1991) Purification of a soluble isoform of guanyl cylcase-activating-factor synthase. *Proc. Natl. Acad. Sci. USA* **88,** 365–369.
3. Bredt, D. S. and Synder, S. H. (1990) Isolation of nitric oxide synthetase, a calmodulin-requiring enzyme. *Proc. Natl. Acad. Sci. USA* **87,** 682–685.
4. Michel, T., Li, G. K., and Busconi, L. (1993) Phosphorylation and subcellular translocation of endothelial nitric oxide synthase. *Proc. Natl. Acad. Sci. USA* **90,** 6252–6256.
5. Garvey, E. P., Tuttle, J. V., Covington, K., Merrill, B. M., Wood, E. R., Baylis, S. A., and Charles, I. G. (1994) Purification and characterization of the constitutive nitric oxide synthase from human placenta. *Arch. Biochem. Biophys.* **311,** 235–241.

4

Cloning and Expression of Human Inducible Nitric Oxide Synthase

Mark L. Johnson, Richard A. Shapiro, and Timothy R. Billiar

1. Introduction

Much of the excitement over nitric oxide (NO) is owing to its diverse physiologic and pathophysiologic functions. Induced NO production has been shown to have both beneficial and detrimental consequences. The inducible or high-output NO synthase (NOS) pathway was first characterized and cloned in murine macrophages *(1–3)*. Activation with lipopolysaccharide (LPS) and interferon-γ (IFN-γ) was uniformly used in these studies to increase inducible (i) NOS expression and improve cloning efficiency. Lyons et al. *(1)* injected mRNA from stimulated RAW 264.7 cells into *Xenopus* oocytes. RNA fractions from oocytes that expressed measurable nitrite by the Griess reaction were extracted and used to create a cDNA library that was ligated into a phage vector, lambda ZAP II. DNA from phage pools was used as a template in a polymerase chain reaction (PCR) using primers from suspected cofactor-binding sites. The radiolabeled PCR product was then used to screen the cDNA library by plaque hybridization. Recognized phages were isolated and plasmids containing the inserts were rescued by helper-phage superinfection. Two overlapping clones were combined into the expression vector pGEM. In vitro transcribed cRNA from the expression vector was then inserted back into oocytes to confirm functional expression. Nitrite production did not require the presence of Ca^{2+} and was not inhibited by trifluoperazine, a calmodulin (CaM), and brain-NOS inhibitor.

Xie et al. *(2)* stimulated RAW 264.7 cells and created a cDNA-expression library which was used to clone the murine iNOS cDNA by immunoscreening with an iNOS specific polyclonal antibody. Liquid chromatography-mass spec-

From: *Methods in Molecular Biology, Vol. 100. Nitric Oxide Protocols*
Edited by: Michael A. Titheradge © Humana Press Inc., Totowa, NJ

troscopy was used to confirm amino-acid sequence and showed that macrophage NOS (mac-NOs) differs extensively from cerebellar NOS.

Lowenstein et al. *(3)* also stimulated RAW 264.7 cells, isolated RNA, and created cDNA using a degenerate primer for the consensus nicotinamide adenine dinucleotide phosphate (NADPH)-binding region. A PCR product was then used to probe a cDNA library constructed from stimulated RAW 264.7 cells. Isolated clones were sequenced and found to display 50% sequence identity to the neuronal (n) NOS. The mac-NOS cDNA was ligated into the expression-vector pCIS and transfected into 293 embryonic human-kidney cells to confirm NOS-catalytic activity. mac-NOS activity was both NADPH and Ca^{2+} independent.

Rat hepatocytes express an iNOS following exposure to the combination of LPS and tumor-necrosis factor (TNF), interleukin-1 (IL-1), and IFN-γ, or to high concentrations of IL-1 alone *(4,5)*. Our observation that cultured human hepatocytes expressed iNOS-like activity in response to a similar combination of cytokines *(6)* provided the first conclusive evidence that a human cell type could express an induced NOS. Confirmation that the activity was owing to an enzyme in humans homologous to the cloned rodent iNOS required that we clone, sequence, and express the human iNOS cDNA. This protocol describes the cloning of the human-hepatocyte iNOS cDNA using a cross-species hybridization method employing the mouse-macrophage NOS cDNA *(7)*. So far, only a single complete iNOS gene has been identified in humans. Therefore, this approach can be used to isolate cDNA from any cell which expresses iNOS. The human-hepatocyte iNOS cDNA predicts an enzyme of 1153 amino acids which displays 80% sequence homology to the murine-macrophage NOS cDNA. Like other NOS isoforms, recognition sites for flavin mononucleotide (FMN), flavin adenine dinucleotide (FAD), and NADPH are present as well as a consensus-CaM binding site.

Recently, we have shown that human iNOS functional activity is also dependent on tetrahydrobiopterin (BH_4) availability *(8)*. The human-iNOS cDNA obtained from the following protocol was constitutively expressed in National Institutes of Health (NIH) 3T3 cells which are deficient in *de novo* BH_4 synthesis. BH_4 appears to catalyze dimerization of iNOS protein into its active form. This is an important consideration in any cloning and expression strategy because an exogenous source of BH_4 may be required to demonstrate enzyme activity (*see* **Note 1**).

2. Materials

2.1. Isolation and Culture of Human Hepatocytes

1. Human-liver samples obtained at laparotomy (following appropriate institutional approval) from normal appearing liver-wedge resections and stored in cold University of Wisconsin preservation fluid.

2. Percoll (Pharmacia, Piscataway, NJ) stored at 4°C.
3. Collagenase Type P (Boehringer-Ingelheim, Germany).
4. Perfusion solution I (PER I): Dilute 50 mL of a 10X stock solution containing 1.42 *M* NaCl, 6.7 m*M* KCl, 100 m*M* HEPES (pH 7.4) with 450 mL of deionized water. Dissolve EGTA (final concentration 2.5 m*M*) in the solution and adjust the pH to 7.4 with 10 *M* NaOH. This solution is then filter sterilized through a 0.2 µm filter (Gelman, Ann Arbor, MI) and warmed to 37°C.
5. Perfusion solution II: The stock solution consists of 67 m*M* NaCl, 0.67 m*M* KCl, 4.8 m*M* $CaCl_2$, 100 m*M* HEPES, and 0.5% bovine serum albumin (BSA) (Hyclone Laboratories). Add to this collagenase type P at a final concentration of 20 mg/100 mL. Filter the solution through a 0.2 µm filter and warm to 37°C.
6. 16 g Jelco iv catheter (Critikon, Tampa, FL).
7. 2-0 silk suture (Ethicon, Johnson & Johnson, Somerville, NJ).
8. Heated water bath (Shel-Lab, Cornelius, OR).
9. Single-channel variable-speed peristaltic pump (Cole-Palmer, Niles, IL).
10. Williams' media E (GIBCO/BRL, Grand Island, NY) containing 0.50 m*M* L-arginine, 1 µ*M* insulin, 15 m*M* HEPES buffer solution, 2 m*M* L-glutamine, 100,000 U/L penicillin, 100 mg/L streptomycin, and 10% (vol/vol) low-endotoxin calf serum.
11. Cytokine/LPS-containing media. To the above media, add 5 units/mL recombinant human IL-1β (Cistron, Pine Brook, NJ), 500 units/mL recombinant human TNF-α (Genzyme), 100 units/mL recombinant human IFN-γ (Amgen), and 10 µg/mL LPS (*Escherichia coli* 0111:B4; Sigma, St. Louis, MO).

2.2. RNA Extraction and Poly A^+ Isolation

1. RNAZol (Biotecx Laboratories, Houston, TX) stored at 4°C.
2. Chloroform (ACS grade).
3. Isopropanol (ACS grade).
4. 75% Ethanol (ACS grade).
5. Diethylpyrocarbonate (DEPC)-treated deionized water.
6. Oligo-dT cellulose (Collaborative Research).
7. Buffer I: 0.5 *M* NaCl, 10 m*M* Tris-HCl, 1 m*M* ethylenediaminetetra-acetic acid (EDTA), 0.5% sodium dodecyl sulfate (SDS), pH 7.4.

2.3. cDNA Library Construction and Screening for iNOS

1. ZAP II phage (Stratagene, La Jolla, CA).
2. *E. coli* Sure cells (Stratagene).
3. NZY media: To 950 mL of deionized water, add 10 g casein hydrolysate enzymatic amine (NZ amine, ICN Biochemicals, Cleveland, OH), 5 g NaCl, and 5 g Bacto-yeast extract.
4. SM buffer.
5. Nylon hybridization transfer membranes (Genescreen, NEN Research Products, Boston, MA).
6. Membrane-denaturing solution: 1.5 *M* NaCl, 0.5 *M* NaOH.

7. Membrane-neutralizing solution: 1.5 *M* NaCl, 0.5 *M* Tris-HCl, pH 8.0.
8. 20X SSPE: 3 *M* NaCl, 0.2 *M* NaH_2PO_4, 0.02 *M* EDTA, pH 7.4.
9. 2X SSPE.
10. Membrane hybridization solution: 0.25 *M* $NaHPO_4$, pH 7.2, 0.25 *M* NaCl, 1 m*M* EDTA, 100 μg/mL denatured salmon-sperm DNA, 7% SDS.
11. 20X SSC: 0.3*M* trisodium citrate, 3 *M* NaCl, pH 7.2.
12. Rinse solution A: 2X SSC, 0.1% SDS.
13. Rinse solution B: 25 m*M* $NaHPO_4$, pH 7.2, 1 m*M* EDTA, 0.1% SDS.
14. Rinse solution C: 25 m*M* $NaHPO_4$, pH 7.2, 1 m*M* EDTA, 1% SDS.

2.4. Expression of Recombinant Human-Hepatocyte iNOS Protein

1. 2X HEPES-buffered saline (HBS): 280 m*M* NaCl, 10 m*M* KCl, 1.5 m*M* Na_2HPO_4, 12 m*M* dextrose, 50 m*M* HEPES. Adjust to pH 7.05 with 0.5 *N* NaOH, and filter sterilize by passage through a 0.2 μm filter. Store at –20°C.
2. 2 *M* $CaCl_2$.
3. TE buffer: 10 m*M* Tris-HCl, 1 m*M* EDTA, pH 8.0.
3. 0.1X TE, pH 8.0.
4. 2 m*M* NADPH in H_2O.
5. 10 m*M* Tris-HCl, pH 7.4.
6. 2 m*M* EDTA in H_2O.
7. 2 m*M* EGTA in H_2O.
8. Trifluoperazine 40–60 μ*M*.

3. Methods

3.1. Isolation and Culture of Human Hepatocytes

1. Cannulate visible vessels within the wedge resection with 16 g IV catheters. Place a 2-0 silk suture circumferentially to secure the catheter.
2. Isolate hepatocytes from nonparenchymal cells by first perfusing them with 0.8–1.5 L of PER I solution at 45 mL/min at 37°C. Then perfuse samples with 1.0–2.0 L of PER II solution at 45 mL/min at 37°C. Per II solution can be recycled by placement of liver samples inside a sterile beaker.
3. After perfusion, dissociate hepatocytes by scraping the sample with a disposable cell scraper (Costar, Cambridge, MA) in Williams' media E. Pour the cell suspension through gauze-lined funnels into 50 mL centrifuge tubes and keep on ice throughout the subsequent washing process.
4. Separate the nonparenchymal cells from hepatocytes by differential centrifugation at 50*g* for 4 min at 4°C.
5. Resuspend hepatocytes in a 30% Percoll gradient and recentrifuge at 50*g* for 4 min at 4°C. Lastly, wash the cells by centrifuging three additional times at 50*g* for 4 min.
6. Confirm hepatocytes purity with light microscopy, and assess viability by trypan blue exclusion.

7. Plate the hepatocytes (5×10^6 cells) onto gelatin-coated tissue culture dishes in 6 mL of culture media. After 24 h, change the media to the cytokine/LPS-containing mixture for 8 h (*see* **Note 2**).

3.2. RNA Extraction and Poly A^+ Isolation

1. Discard media and add 1.2 mL of RNAzol for every 5×10^6 hepatocyte. Shear the cultured hepatocytes by passage through a 23-g needle three times.
2. Extract the RNA using chloroform, precipitate with isopropanol, and wash the precipitate twice in 75% EtOH in DEPC-treated water.
3. Dry the pellet briefly under vacuum and resolubilize in DEPC-treated water.
4. Separate poly A^+ RNA from transfer and ribosomal RNA using an Oligo dT column.
5. Add 0.5 mL of H_2O to 0.5 g of Oligo-dT cellulose. Pour the slurry into a 10-mL syringe plugged with sterile-glass wool.
6. Equilibrate the column with Buffer 1. Heat denature approx 1 mg of RNA at 70°C for 1 min and then dilute into 40 mL of Buffer 1.
7. Pass the diluted RNA through the Oligo dT column twice, then wash the column with 40 mL of SDS-free Buffer 1. Discard the effluent. Elute the poly A^+ RNA from the resin by adding 10 mL of 10 m*M* Tris-HCl, pH 7.4. Lyophylize the RNA to dryness, wash three times in 75% EtOH, and resolubilize the RNA in DEPC-treated H_2O.
8. Quantify the yield of poly A^+ RNA using a spectrophotometer by measuring the absorbance at 260 nm. The ratio of absorbance at 260 and 280 nm should exceed 1.8.
9. To confirm the presence of human-iNOS mRNA in poly A^+ RNA sample, perform a Northern-blot analysis with 0.5 µg of poly A^+ RNA, and probe with radiolabeled mac-NOS cDNA.

3.3. cDNA Library Construction and Screening for iNOS

1. The induced human-cell cDNA library can be custom constructed by Stratagene in the phage vector, Zap II, by reverse transcription using Oligo-dT and random priming from 20 µg of poly A^+ mRNA or constructed using a commercially available kit.
2. To produce a confluent lysis, plate 1×10^6 plaque forming units (pfu) with 1.6×10^8 *E. coli* Sure cells in NZY top agarose media at a density of 2×10^5 pfu / 24 × 24 cm plates at 37°C.
3. Once confluence is achieved, place a positively-charged nylon membrane over the top agar so that it comes in contact with the plaques. Be careful not to trap air bubbles. Handle the filter with gloved hands. Mark the filter for orientation purposes.
4. After 30–60 s, peel off the filter and immerse it in denaturing solution for 30–60 s. Transfer the filter into neutralizing solution for 5 min. Rinse the filter in 2X SSPE and place it on Whatman paper.

5. Fix the DNA to the filter by baking for 30 min at 80°C in a vacuum oven.
6. Radiolabel the murine-macNOS cDNA using the random prime-labeling method (Boehringer Mannheim).
7. Soak the filter in just enough hybridization solution to allow free movement for at least 30 min before adding radiolabeled probe.
8. Add greater than 2×10^6 cpm/mL of the radiolabeled macNOS-cDNA probe and hybridize overnight at 43°C.
9. Briefly rinse the filter in rinse solution A.
10. Wash the filter twice in 200 mL of rinse solution A at 53°C for 15 min.
11. Wash the filter twice in 200 mL of rinse solution B at 53°C for 15 min.
12. Wash the filter twice in 200 mL of rinse solution C at 53°C for 15 min.
13. Replate putative-positive plaques and rescreen by plaque-lift hybridization in 100 mm Petrie dishes until clonal plaques are isolated. At least two additional rounds of screening are necessary to produce clonal plaques. Identify hybridization positive plaques, core from the dish using a pipet end, transfer to 1 mL SM buffer, and allow to incubate at room temperature for 1 h to elute the phage from the agar.
14. Rescue plasmids containing the inserts by coinfection with helper phage as per manufacturerís instructions.

3.4. DNA Sequencing and Analysis

1. Determine the size of each insert by restriction-digest analysis and Southern blot and sequence the longest clones on both the sense and antisense strands using overlapping internal primers.
2. Plasmid inserts are sequenced by an automated dideoxy chain-termination. The human-iNOS cDNA sequence is compared to the murine-iNOS cDNA and endothelial-cNOS cDNA.

3.5. Expression of Recombinant-Human-Hepatocyte iNOS Protein

1. Ligate the full length 4.1 Kb human-hepatocyte iNOS cDNA downstream from the CMV enhancer-promoter in the mammalian-expression vector pCIS (Genentech) as described in the manufacturer's instructions.
2. Dissolve the expression vector in $0.1 \times$ TE, pH 8.0 at a concentration of 40 µg/mL.
3. Place 220 µL of DNA and 250 µL of 2X HBS in a disposable , sterile 5-mL plastic tube. Slowly add 31 µL of 2 *M* $CaCl_2$ while gently mixing. Incubate for 20–30 min at room temperature. Resuspend the resultant precipitate.
4. Transfer 0.5 mL of the calcium phosphate-DNA suspension into 5 mL of fresh media overlying cell cultures containing human embryonic-kidney 293 cells (American Type Culture Collection).
5. Incubate the cells for 24 h at 37°C in a humidified incubator in an atmosphere of 5–7% CO_2.
6. Assay NOS activity by conversion of [^{3}H] arginine to [^{3}H] citrulline as described in Chapter 7, this volume.

7. Determine the calcium/calmodulin (Ca/CAM) requirements by adding EDTA, ethylene glycol-bis (β-aminoethyl ether) N,N,N′,N′-tetraacetic acid (EGTA), and trifluoperazine to culture media and rechecking NOS activity.

4. Notes

1. BH_4 is a necessary cofactor for the dimerization, and thus activation, of monomeric iNOS protein *(8)*. Cell lines lacking de novo BH_4 biosynthesis produce minimal amounts of active iNOS protein.
2. Inducible NOS, unlike the cNOS isoforms, may be difficult to demonstrate in cell culture in the absence of cytokine and LPS stimulation, although constitutive expression of iNOS has been shown in several cell lines.

References

1. Lyons, C. R., Orloff, G. J., and Cunningham, J. M. (1992) Molecular cloning and functional expression of an inducible nitric oxide synthase from a murine macrophage cell line. *J. Biol. Chemistry* **267,** 6370–6374.
2. Xie, Q., Cho H. J., Calaycay, J., and Mumford, R. A., et al. (1992) Cloning and characterization of inducible nitric oxide synthase from mouse macrophages. *Science* **256,** 225–228.
3. Lowenstein, C. J., Glatt, C. S., Bredt, D. S., and Snyder, S.H. (1992) Cloned and expressed nitric oxide synthase contrasts with the brain enzyme. *Proc. Natl. Acad. Sci. USA* **89,** 6711–6715.
4. Geller, D. A., deVera, M. E., and Russell, D. A., et al. (1995) A central role for IL-1 beta in the in vitro and in vivo regulation of hepatic inducible nitric oxide synthase. IL-1 beta induces hepatic nitric oxide synthesis. *J. Immunol.* **155,** 4890–4898.
5. Geller, D. A., Nussler, A. K., Di Silvio, M., et al. (1993) Cytokines, endotoxin, and glucocorticoids regulate the expression of inducible nitric oxide synthase in hepatocytes. *Proc. Natl. Acad. Sci. USA* **90,** 522–526.
6. Nussler, A. K., Di Silvio, M., and Billiar, T. R., et al. (1992) Stimulation of the nitric oxide synthase pathway in human hepatocytes by cytokines and endotoxin. *J. Exp. Med.* **176,** 261–264.
7. Geller, D. A., Lowenstein, C. J., Shapiro, R. A., et al. (1993) Molecular cloning and expression of inducible nitric oxide synthase from human hepatocytes. *Proc. Natl. Acad. Sci. USA* **90,** 3491–3495.
8. Tzeng, E., Billiar, T. R., Robbins, P. D., Loftus, M., and Stuehr, D. J. (1995). Expression of human inducible nitric oxide synthase in a tetrahydrobiopterin (H_4B)-deficient cell line: H_4B promotes assembly of enzyme subunits into an active dimer. *Proc. Natl. Acad. Sci. USA* **93,** 11,771–11,775.

Barry W. Allen

5

Cloning and Expression of Human eNOS and nNOS Using the Baculovirus-Insect Cell System

Ian G Charles, Neale Foxwell, and Ann Chubb

1. Introduction

To date, the cDNAs and genes encoding three distinct forms of the enzyme nitric oxide synthase (NOS) have been described (reviewed in **ref.** *1*). The family of NOS enzymes generate nitric oxide (NO) from arginine, producing citrulline as a co-product and requires the presence of heme, tetrahydrobiopterin (BH_4), nicotinamide adenine dinucleotide phosphate (NADPH), calmodulin (CaM), flavin mononucleotide (FMN), and flavine adenine nucleotide (FAD) *(2)*. The NO generated by the family of NO synthases has been found to play an important role in many biological processes involved in health and disease including neurotransmission, the mediation of vasodilation, and host defense *(2)*.

In order to aid in the characterization of this important family of enzymes, high levels of protein are required; in order to achieve this, we have expressed all three forms of NOS in a baculovirus-insect cell system *(3–6)*. This article describes in detail the procedures required for the cloning and expression of both the human endothelial (e) and neuronal (n) NOS isoforms.

The insect baculovirus used for expression studies is the 128 kDa *Autographa Californica* Nuclear Polyhedrosis Virus (AcNPV). The virus infects lepidopteran species with a biphasic-life cycle. Initial infection (10–24 h) involves early gene-expression events associated with formation of mature nucleocapsids, whereas the next phase involves high-level expression of the polyhedrin protein, essential for lateral-viral transmission. The polyhedrin protein encapsulates mature-virus particles within the nucleus of infected cells, giving rise to discrete viral "occlusion" bodies. Polyhedrin itself, however, is not required for viral transmission in tissue-culture cells and,

From: *Methods in Molecular Biology, Vol. 100. Nitric Oxide Protocols*
Edited by: Michael A. Titheradge © Humana Press Inc., Totowa, NJ

consequently, can be deleted and replaced with heterologous sequences. One of the advantages of this system is that the cDNAs of foreign proteins can be engineered adjacent to the very strong polyhedrin promoter, which can direct expression to levels approaching 50% of the total-protein content of the insect cell. A second advantage of the use of the polyhedrin system is that recombinant plaques, generated by cloning into the polyhedrin site, are "occlusion body"-negative and can be identified easily. In the initial series of baculovirus-insect cell vectors, this "occlusion body"-negative phenotype was used to identify recombinant AcNPV from a background of wild-type plaques. Current cloning strategies rely on more rapid procedures to generate recombinants and this chapter is aimed at introducing one such approach, using a positive-selection vector, for the expression of recombinant-human eNOS and nNOS in insect cells.

2. Materials

1. The insect cell line *Spodoptera frugiperda* clone 21 (*Sf* 21) can be obtained from Invitrogen (The Netherlands).
2. Cells are maintained in spinner culture at 18°C in TC-100 medium supplemented with 10% heat inactivated fetal calf serum (FCS), 2 m*M* glutamine, antibiotic-antimycotic solution, and 20 μg/mL gentamicin (all from GIBCO, BRL).
3. For NOS expression, the following co-factors are added to the TC-100 medium described above: hemin (0.5 μg/mL), nicotinic acid (5 μ*M*), riboflavin (1 μ*M*), and BH_4 (10 μ*M*).
4. NOS extraction buffer: 320 m*M* sucrose, 50 m*M* Tris-HCl, pH 7.0, 1 m*M* EDTA, 1 m*M* DL-dithiothreiol, 0.1 m*M* phenylmethylsulphonyl fluoride (PMSF), 10 μg/mL leupeptin, 10 μg/mL soybean trypsin inhibitor, 30 μg/mL pepstatin, and 2 μg/mL aprotinin.
5. Adenosine diphosphate (ADP)-agarose affinity chromatography, buffer A: 50 m*M* HEPES, pH 7.5, 10% (v/v) glycerol, 0.5 μ*M* leupeptin, 10 μ*M* chymostatin, and 1 μ*M* PMSF.
6. In order to analyze the recombinant nNOS and eNOS protein expressed in the insect cell-baculovirus expression system, we routinely carry out gel electrophoresis followed by Western blotting. All reagents are from Sigma unless otherwise stated.
   a. Acrylamide (Biorad): A stock solution of 30% acrylamide/bis acrylamide 37.5:1 is prepared (CARE; acrylamide is a neurotoxin) in H_20. The gel solutions require the following stock buffers: 1 *M* Tris-HCl, pH 8.8 (30.25 g Tris base and approx 4.7 mL conc. HCl, in 250 mL H_2O); 1 *M* Tris-HCl, pH 6.8 (30.25 g Tris base and approx 20.8 mL conc. HCl in 250 mL H_2O).
   b. Running buffer pH 8.3: 0.025 *M* Tris, 0.2 M glycine, 0.1% SDS. To make 2 L of 10x buffer, mix: 60.6 g Tris base, 300 g glycine, and 20 g sodium dodecylsulfate (SDS). Check that the pH is 8.3.

c. Make the following stock sample buffer (4x stock). Buffer A: 3% SDS, 15% glycerol, 0.12 *M* Tris–HCl, ph 6.8, 0.15 *M* dithiothreitol (DTT) made from the following mix: 12.0 mL of 10% SDS solution, 4.8 mL of 1 *M* Tris-HCl, pH 6.8, 6.0 mL of glycerol, and 1.2 mL of distilled water (to give a total of 24 mL). Buffer B: 1 *M* DTT (1.545 g/10 mL). Distribute into 0.3 mL aliquots and freeze. To make the buffer (on the day of use) add 1.2 mL solution A + 0.3 mL solution B.
d. Staining solution: methanol: acetic acid: water, 50:10:40. Add 10 mL of a 1% (w/v) Coomassie blue (Sigma) solution in water with 190 mL of the aforementioned solution.
e. Destaining solution: Methanol: Acetic acid: Water, 5:10:85.

7. For Western blotting, proteins are transferred onto membranes using transfer buffer: 0.05 *M* Tris base, 0.04 *M* glycine, 3.7% SDS, 20% methanol.

3. Methods

3.1. Cloning of Human eNOS and nNOS cDNA Sequences into a Baculovirus-Transfer Vector

To obtain high levels of expression of any recombinant protein in the baculovirus-insect cell system, the heterologous DNA encoding the sequence is placed under the control of a viral promoter. In most cases, the polyhedrin promoter is selected, and we have chosen to express both human eNOS and nNOS in the baculovirus transfer-vector pVL1393 (Pharmingen) that utilizes this promoter.

The full-length genome of the AcNPV baculovirus used in the experiments is approx 128 kDa. This large size makes direct cloning strategies unfeasible. Instead, a series of smaller 'transfer vectors' have been developed that have multiple-cloning sites and can direct in vivo homologous recombination into the AcNPV genome. The system we have chosen uses Baculo Gold linearized DNA (Pharmingen). The baculovirus DNA contains a lethal deletion that can be complemented by polyhedrin-based transfer vectors (such as pVL1393). Using this baculovirus polyhedrin transfer-vector system, the majority of plaques recovered from a transfection experiment express the target cDNA.

A number of groups have now published the cDNA sequences for various mammalian eNOS and nNOS cDNAs (reviewed in **ref. *1***). For our cDNA clones, human eNOS may be excised from the parent plasmid pBluescript II SK (+) (Stratagene) as a 4-kb *Eco*RI fragment, whereas human nNOS may be excised as a 4.5-kb *Xba*I fragment ***(6)***.

1. Digest the vector pVL1393 with *Eco*RI (for eNOS) and *Xba*I (for nNOS) and phosphatase the protruding 5'-overhangs by including calf intestinal phosphatase in the digest buffer. This procedure reduces (or eliminates) the background of self-ligated vector.

2. The pBluescript plasmids containing the cDNA sequences for human eNOS and nNOS are separately digested with *Eco*RI and *Xba*I respectively and ligated into the phosphatased pVL1393 vector preparations.
3. Transform competent *Escherichia coli* with the ligation mix and select on ampicillin plates (100 µg/mL).
4. Mini-prep DNA is prepared and digested with appropriate restriction endonucleases in order to confirm the orientation of the cDNA insert in the expression plasmid.
5. The same stock mini-prep DNA can be used to co-transfect with linearized Baculo Gold DNA to generate recombinant eNOS and nNOS baculovirus.

3.2. Co-Transfection of Linearized Baculovirus DNA (Baculo Gold) with Recombinant eNOS and nNOS in pVL1393

1. Set up one T25 flask per transfection experiment, seeded to 70/80% confluence (approx 5×10^6 cells). Leave the cells to settle and adhere for 1 h.
2. Mix 3 µg of eNOS or nNOS pVL1393 construct DNA with 0.2 µg linearized Baculo Gold DNA (Pharmingen). Make up to 50 µL with H_2O.
3. In a separate tube, mix 40 µL Lipofectin (Gibco BRL) with 10 µL H_2O. Mix the contents of the two tubes and leave at room temperature for 10 min.
4. Wash the cells attached to the bottom of the T25 flasks (one for eNOS and one for nNOS transfections) gently with two changes of serum-free TC-100 medium, then gently cover the cells with 2 mL of serum-free TC-100.
5. Add the Lipofectin/DNA complex to the cells drop-wise, rock the flask gently to distribute the DNA evenly over the cells. Incubate for 5 h at 28°C.
6. Remove the media from the cells and add 2.5 mL complete media (i.e., TC-100 with 10% serum).
7. Leave for 3 d at 28°C, at which stage the medium is removed and virus assayed in the plaque assay.

3.3. Plaque Assay

1. Prepare ten 30 mm tissue culture grade dishes (Falcon) seeded with approx 10^6 cells/dish. (In practice, this is approx 1.5 mL of the stock culture maintained in the spinner flask.) Leave the cells to settle and adhere for one h.
2. Make a series of 10-fold dilutions of the virus in complete TC-100 media from neat to 10^{-10}.
3. Add 200 µL of each dilution to cells attached to the bottom of a 30-mm plate from which the media has been removed, rock gently to evenly distribute the virus, then leave for two h at 28°C.
4. Remove the medium from the dish and overlay with 2 mL of a solution containing 1% Sea Plaque ultra pure agarose (Gibco BRL), (make a 3% stock solution in H_2O, autoclave, cool to 37°C, then add complete TC-100 medium to dilute to 1%). Leave to set for 15 min.
5. Overlay with 1 mL of complete TC-100 medium and leave in a humid box for 4 d at 28°C. (We use a small snap-shut Tupper Ware box, and place the tissue-culture dishes on tissue paper soaked in water.)

3.4. Plaque Identification by Staining

1. Add 1 mL of a neutral red solution (Sigma, 3.3 g/L stock) diluted 1 in 10 in phosphate-buffered solution (PBS) to each tissue-culture dish harboring virus.
2. Leave for 1–2 h at 28°C.
3. Drain off the neutral red stain and leave for a further 1–2 h; the plates can then be examined for viral plaques.

3.5. Plaque Purification and the Generation of High-Titer Stocks

1. Plaques identified in the aforementioned isolation and staining protocols can be re-purified by repeating the procedure through several rounds, resulting in a virus population that is homogenous.
2. When eNOS and nNOS expressing virus have been identified, high titer stocks are required for the infection of insect cells and the generation of recombinant protein. Pick a plaque corresponding to an eNOS or nNOS baculovirus clone into 500 μL of TC-100 medium plus 10% fetal calf serum (FCS), and incubate at 37°C for 30 min to release the virus.
3. Plate between 0.3–1 × 10^6 cells in a 30 mm well of a tissue-culture dish and leave to attach for a minimum one h at 28°C
4. Infect the cells with between 50–200μL of the virus stocks from the eNOS and nNOS preparations. Add 1 mL of complete TC-100 medium and leave for 4 d.
5. Following infection, transfer between 0.3–1 mL of the virus to a T25 flask containing approx 5 × 10^6 cells (that have been left to attach as described above). Add 5 mL of complete TC-100 medium and incubate the cells at 28°C for 4 d.
6. Add between 1–5 mL of stock virus from the above infection to a T75 flask (containing approx 10^7 attached cells). Add 10 mL of complete TC-100 medium and incubate the cells at 28°C for 4 d.
7. Transfer between 2.5–10 mL of virus stock to a T175 flask (containing 5 × 10^7 cells). Add 40 mL of complete TC-100 medium and incubate for 4 d at 28°C. The viral stocks prepared in this way can be stored at 4°C in the dark, and should maintain viability for many years.

3.6. Human eNOS Expression

We have empirically determined optimal conditions for the growth of insect cells infected with recombinant baculovirus expressing both human eNOS and nNOS (*see* **Notes 1–3**).

1. Human eNOS is expressed in *Sf*21 cells grown at 28°C monolayers in T500 plates (500 cm^2 tissue-culture plates, NUNC, GIBCO).
2. Grow the cells to confluence to give approx 10^8 cells/plate.
3. Add the recombinant virus (prepared from stocks as described above) to give 5 pfu/cell, i.e., 5 × 10^8 total phage.
4. Add co-factors to the TC-100 medium as described in the Materials section.
5. Carry out the infection for 48 h.

3.7. Human nNOS Expression

1. We have determined that nNOS can be grown under the same conditions described above for eNOS. Interestingly, human nNOS expression appears to be less dependent on the presence of additional co-factors and as a consequence, human nNOS can be expressed in *Sf*21 cells grown in suspension at 28°C in complete TC-100 medium with no additional co-factors.
2. To prepare recombinant nNOS in a spinner culture, grow 100 mL of cells in a 250 mL spinner at approx 10^6 cells/mL.
3. Recover the cells by centrifugation in 2 × 50 mL sterile screw-cap centrifuge tubes (NUNC, GIBCO) at 200*g* for 5 min.
4. Remove the medium and add recombinant nNOS baculovirus at a multiplicity of infection of 5 at 28°C for 1 h.
5. Following incubation, dilute the cells with fresh complete TC-100 medium to give approx 10^6 cells/mL and incubate in a spinner flask for a further 48 h.

3.8. Extraction of Recombinant NOS

Insect cells containing recombinant human eNOS and nNOS grown in monolayer or suspension culture can be recovered, lyzed, and the NOS purified by affinity chromatography on an ADP-agarose column.

1. Scrape cells containing eNOS or nNOS grown in monolayers into NOS extraction buffer on ice to give a concentration of approx 10^8 cells/mL.
2. Recover cells containing nNOS grown in suspension culture by centrifugation at 200*g* for 5 min, and wash twice with PBS. Resuspend the cell pellet in the NOS-extraction buffer to give approx 10^8 cells/mL.
3. Lyse the cells by sonication using an MSE Soniprep 150 Ultrasonic Disintegrator (three 10-s bursts of 14 mm amplitude at 4°C with 30-s cooling time between each burst).
4. Remove insoluble material by centrifugation at 105,000*g* for 30 min at 4°C.
5. Add the supernatant to 2',5'-ADP-agarose (Sigma) equilibrated in buffer A at a ratio of 50 mL original culture:0.1 mL ADP-agarose. Mix the sample gently at 4°C for 45 min and pour into a column.
6. Wash the ADP agarose extensively with buffer A to remove any nonspecific binding. Elute NOS-specific bands in buffer A containing 10 m*M* NADPH. Collect fractions and analyze by sodium dodecylsulfate-polyacrylamide electrophoresis (SDS-PAGE) and Western blotting as described below.

3.9. SDS-PAGE and Western Blotting

In order to analyze the recombinant nNOS and eNOS protein expressed in the insect cell baculovirus expression system, we routinely carry out gel electrophoresis followed by Western blotting as described below.

1. Typically, for the analysis of NOS protein samples, a 10% gel is run. The following parameters are described for a Hoefer SE400 gel kit (California). For the

running-gel solution, mix 10 mL of 30% acrylamide stock with 11.2 mL of Tris-HCl, pH 8.8, and 8.8 ml H_2O (total 30 mL). Add 100 µL of a fresh solution of 10% ammonium persulphate and 25 µL TEMED to polymerize the gel.

2. Prepare a 4.5% stacking gel (1.5 mL stock 30% acrylamide, 1.25 mL Tris-HCl, pH 6.8, and 7.25 mL H_2O). Polymerize the gel with the addition of 50 µL 10% ammonium persulphate and 10 µL TEMED.
3. Pour the running gel in between the glass plates of the gel apparatus, leaving 2 cm clear at the top. Carefully layer water on top and leave the gel to polymerize. (At this stage, gels can be left for several days if wrapped in clingfilm at 4°C.) To use, pour the water off, add the stacking gel, and insert the gel-comb of choice.
4. Boil the protein preparations with sample buffer (generally 30 µL of sample plus 10 µL of 4x sample buffer) for 5 min and load onto the gel.
5. Electrophorese the samples at 30–50 mA until the blue dye has reached the bottom of the gel.
6. Remove the gel from the plates, and place in staining solution (see above) for 1 h. Protein bands can be resolved following destaining. This destaining step can usually be carried out overnight.

3.10. Western Blotting

For Western blotting, proteins separated by SDS-PAGE are transferred to Immobilon membrane (Whatman), and NOS-specific band(s) identified with specific antibodies.

1. Wet the Immobilon membrane in methanol, then soak in transfer buffer.
2. At the same time, soak six pieces of Whatman 3 MM paper in the transfer buffer, (these paper sheets have been cut to the same size, or slightly larger than the gel).
3. Assemble the transfer in the following order: three pieces of 3 MM paper, the Immobilon membrane, the gel, and finally a further three pieces of 3 MM paper. These are placed into the semi-dry electro-blotter (ATTO) with the membrane on the anode side.
4. Transfer for one h at 200 mA.
5. Remove the membrane and rinse for 30 min (with gentle shaking) in PBS, pH 7.2 + 0.1% Tween 20 (PBS/Tween).
6. Pour off the solution, and add 30 mL of primary antibody [diluted 1/500 in 3% BSA (Sigma) in PBS/Tween)]. For eNOS and nNOS, use the relevant monoclonal antibody (MAb) (we have had good results with those produced by Affiniti Research laboratories). Shake slowly for 2 h at room temperature or 4°C overnight.
7. Wash three times for 5 min with PBS/Tween.
8. Add anti-mouse alkaline phosphatase conjugate (Sigma) at a 1/3000 dilution in PBS/Tween for one h at room temperature (with gentle shaking).
9. Wash three times for 5 min with PBS/Tween.
10. Add the detection substrate [BCIP-fast; (Sigma)] in 10 mL water and shake until the bands show. Rinse with water, then air-dry the blot. It is best to monitor this step carefully as bands can appear rapidly.

An example of the analysis of recombinant NOS is shown in **Fig. 1**. Both eNOS and nNOS purification profiles are shown alongside the profile of iNOS for comparison. In all cases, extracts of recombinant human-NOS samples were prepared as described above, run on SDS-polyacrylamide gels, and electroblotted for Western analysis with NOS-specific antibodies. In **Fig. 1A**, track 1 shows inducible (i) NOS-baculovirus infected *Sf*21-cell supernatant loaded onto the ADP-sepharose column. Tracks 2–8 correspond to fractions eluted from this column with 10 m*M* NADPH. Track 9 shows a Western blot of infected cell supernatant with an anti-iNOS specific antibody, revealing a 131-kDa band (shown with an arrow). **Fig. 1B** shows the same procedure carried out for protein prepared from eNOS baculovirus-infected cells. Track 1 corresponds to protein from a total-cell lysis, track 2 shows the supernatant applied to the ADP-sepharose column, track 3 shows the flow-through, and tracks 4–18 show the fractions eluted from the column with 10 m*M* NADPH. A Western blot (track 19) is shown of the total lysate probed with an anti-eNOS antibody, the 133 kDa eNOS band is marked with an arrow. **Fig. 1C** shows a similar experiment carried out with lysates prepared from nNOS-baculovirus infected insect cells. In this case, samples were applied to an 8% SDS-polyacrylamide gel. Track 1 shows the total cell lysate and track 2 is the column flow-through. Tracks 3–15 are fractions eluted from the ADP-sepharose column with 10 m*M* NADPH. Track 16 shows a Western blot of the cell lysate with an anti-nNOS specific antibody, showing the presence of a 161-kDa nNOS band (marked with an arrow). In all the panels, gel molecular weight markers are indicated (M).

4. Notes

1. We have found that the production of different NOS proteins can lead to cross-contamination unless care is taken in the infection procedures. It is normal practice to passage uninfected cells in a separate tissue-culture hood than that used for infected cells. We have extended this practice to use a separate hood for each of the infections with different NOS-baculovirus samples.
2. Although the baculovirus-insect cell expression system appears to be able to generate recombinant NOS, there are quite large differences in the expression levels for different family members. For example, in a typical experiment infecting 2×10^8 cells, the yield from the ADP-agarose column of eNOS is 2 mg, for nNOS it is 5 mg, whereas for iNOS it is much lower at 50 μg. This difference in expression may either reflect intrinsic differences between the three proteins, or may be a consequence of the different 5' and 3'-noncoding regions (that are retained in each of the different cDNA sequences) contributing to aspects of mRNA stability/turnover. It is possible that removing noncoding regions of the cDNA inserts may generate recombinant NOS-baculovirus preparations capable of higher expression levels.

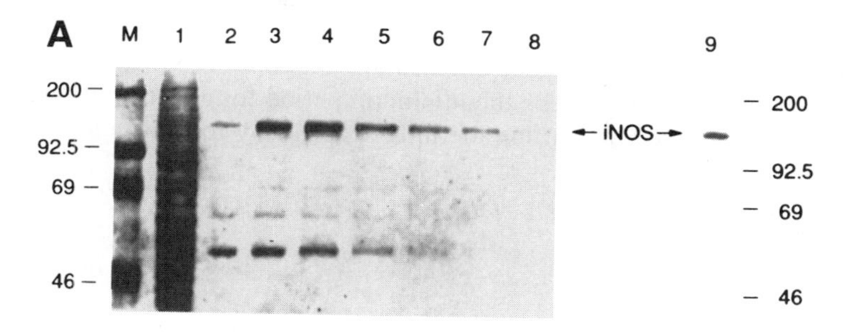

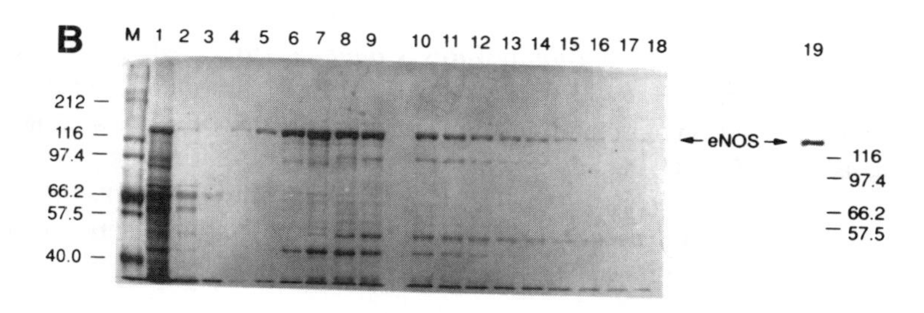

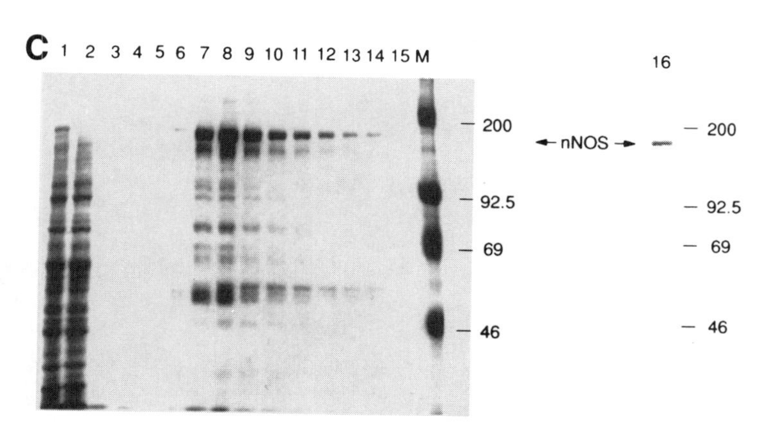

Fig. 1. Analysis of the expression of recombinant NOS by SDS-PAGE and Western blotting. Expression and purification profiles for human iNOS (**A**), eNOS (**B**), and nNOS (**C**) following lysis of insect cells infected with NOS-baculovirus constructs and ADP-sepharose chromatography.

3. We have found that the most important feature for the expression of NOS is the primary step of allowing the cells to attach overnight prior to the addition of the virus. Although 1 h seems to be the minimum period for attachment, yields are significantly improved by starting the infection the day following an overnight attachment.

References

1. Knowles, R. G. and Moncada, S. (1994) Nitric oxide synthases in mammals. *Biochem J.* **298,** 249–258.
2. Moncada, S., Palmer, R. M. J., and Higgs, A. (1991) Nitric oxide: physiology, pathophysiology and pharmacology. *Pharmacol. Rev.* **43,** 109–141.
3. Charles, I. G., Chubb, A., Gill, R., Clare, J., Lowe, P., Holmes, L. S., Page, M., Keeling, J. G., Moncada, S., and Riveros-Moreno, V. (1993) Cloning and expression of a rat neuronal nitric oxide synthase coding sequence in a baculovirus/insect cell system. *Biochem. Biophys. Res. Commun.* **196,** 1481–1489.
4. Moss, D. W., Wei, X., Liew, F. Y., Moncada, S., and Charles, I. G. (1995) Enzymatic characterisation of recombinant murine inducible nitric oxide synthase. *Eur. J. Pharmacol.* **289,** 41–48.
5. Riveros-Moreno, V., Heffernan, B., Torres, B., Chubb, A., Charles, I., and Moncada, S. (1995) Purification to homogenicity and characterisation of recombinant nitric oxide synthase. *Eur. J. Biochem.* **230,** 52–57.
6. Charles, I. G., Scorer, C. A., Moro, M. A., Fernandez, C., Chubb, A., Dawson, J., Foxwell, N., Knowles, R. G., and Baylis, S. A. (1996) Expression of human nitric oxide synthase isozymes. *Methods Enzymol.* **268,** 449–460.

Barry W. Allen

6

Assay of NOS Activity by the Measurement of Conversion of Oxyhemoglobin to Methemoglobin by NO

Mark Salter and Richard G. Knowles

1. Introduction

Nitric oxide synthase (NOS) catalyzes the conversion of L-arginine, molecular oxygen, and nicotinamide adenine dinucleotide phosphate (NADPH) to NO, citrulline, and $NADP^+$ (reviewed in **ref. *1***). This chapter describes the measurement of NOS activity by utilizing the conversion of oxyhemoglobin to methemoglobin by NO while monitoring the absorption difference between the wavelengths 401 and 421 nm or at just 401 nm. This assay was first described by Feelisch and Noack *(2)* and has since been modified to allow not only the measurement of NOS activity in vitro but also the degree of in vivo inhibition of neuronal (n) NOS (and also possibly endothelial (e) NOS) by ex vivo analysis after the administration of slowly dissociating inhibitors such as N^{ω}-nitro-L-arginine (L-NNA) and its methyl ester, L-NAME *(3,4)*. Using the protocol described, this technique is sensitive (limit of detection approx 20 pmol/min per g tissue), specific for NOS in conjunction with the use of NOS inhibitors, applicable to most preparations (with the exception of samples containing large amounts of hemoglobin) and suitable for continual monitoring.

2. Materials

1. 0.144 m*M* oxyhemoglobin: Commercially available hemoglobin consists largely of the met form and must be converted to its reduced oxy-form. Oxyhemoglobin can be easily prepared from commercially available crystallized hemoglobin by slowly and gently dissolving 20 mg of hemoglobin in 1 mL of distilled water, followed by the addition of sodium dithionite in a 2–3-fold molar excess changing the color from brown (met) to dark red (deoxy). Note that hemoglobin con-

From: *Methods in Molecular Biology, Vol. 100. Nitric Oxide Protocols*
Edited by: Michael A. Titheradge © Humana Press Inc., Totowa, NJ

tains 4 heme groups/mol. Blowing oxygen over the surface with gentle swirling changes the color to bright red (oxy); leaving the solution in air will also convert to the oxy-form, albeit more slowly. Purification and desalting are carried out by running the solution through a Sephadex G-25 column. The oxyhemoglobin is eluted as a single bright-red band. The front and back edges of the band should be discarded. The concentration of the hemoglobin can be determined by dividing the absorbance at 415 nm by the extinction coefficient for oxyhemoglobin at 415 nm, 131,000 M^{-1} cm^{-1}. This method of purification has been described in detail elsewhere ***(5)***. Samples should be aliquoted out and kept at –20°C.
2. 50 m*M* potassium phosphate buffer, pH 7.2. This can be stored for several weeks at 4°C.
3. 25 m*M* $CaCl_2$. Store at 4°C.
4. 100 m*M* $MgCl_2$. Store at 4°C.
5. 2 m*M* L-arginine: prepare in water and keep as a frozen stock solution at –20°C.
6. 10 m*M* NADPH: prepare in water and keep as a frozen stock solution at –20°C.
7. 0.1 m*M* flavin mononucleotide (FMN): prepare in water and keep as a frozen stock solution at –20°C.
8. 0.1 m*M* flavin adenine dinucleotide (FAD): prepare in water and keep as a frozen stock solution at –20°C.
9. 10 μ*M* calmodulin (CaM): prepare in water and keep as a frozen stock solution at –20°C.
10. 1 m*M* tetrahydrobiopterin (BH_4): must be made up freshly on the day of use and kept on ice until used.
11. Assay buffer: Prior to the assay of NOS, the oxyhemoglobin and $CaCl_2$ (e-NOS and nNOS) or $MgCl_2$ (iNOS) are combined in the potassium phosphate buffer to give final concentrations of 1.44 μ*M*, 250 μ*M*, and 1 m*M* respectively.

3. Methods

3.1. Extraction Buffer and Preparation of Tissue Sample

Preparation of the extraction buffer and tissue sample is as described by Knowles and Salter (*see* Chapter 7). However, an important distinction from the radiochemical-NOS assay is that the homogenate or sonicate must be first centrifuged (at least 12,000*g* for 5 min) to minimize turbidity.

3.1.1. Assay Procedure

1. Prewarm the oxyhemoglobin, $CaCl_2$, and buffer mixture to the required assay temperature and add to the cuvet within the spectrophotometer to achieve at least 80% of the final reaction-mixture volume.
2. Add L-arginine and NADPH to final concentrations of 20 μ*M* (this can be changed if required) and 100 μ*M*, respectively.
3. In the case of partially or fully purified NOS, add the other cofactors as a "cocktail." This can be made and kept on ice prior to the assay to simplify additions during the assay. The final concentrations of the effectors are 1 μ*M* FMN, 1 μ*M* FAD, 100 n*M* CaM, and 10 μ*M* BH_4.

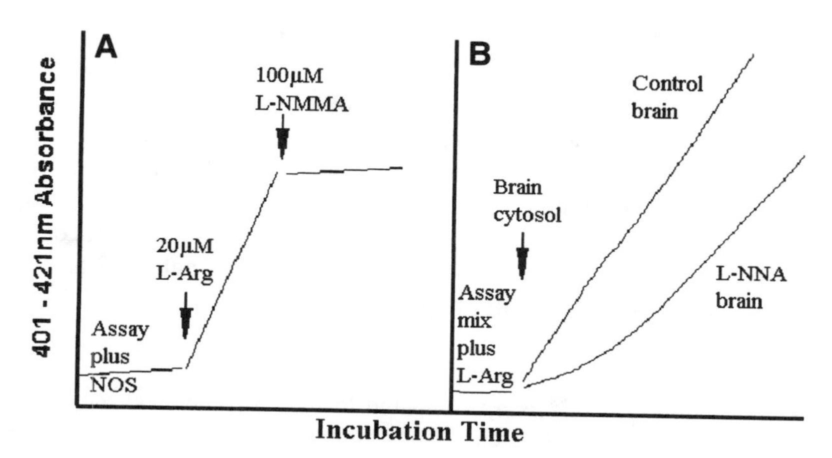

Fig. 1. **(A)** Schematic representation of a typical activity trace showing the NOS reaction being initiated by the addition of 20 µ*M* L-arginine and stopped by the addition of 100 µ*M* of the nonselective NOS inhibitor L-NMMA. **(B)** Schematic representation of a typical NOS-activity trace seen with rat-brain cytosol after the pretreatment of rats with the slowly-dissociating inhibitor of nNOS L-NNA or vehicle (control).

4. Start the reaction by the addition of the tissue extract (*see* **Notes 1** and **2**) or the purified NOS preparation.
5. Follow the reaction by reading the difference in absorbance between 421 nm and 401 nm with a slit-width of 5 nm using a dual-wavelength or diode-array spectrophotometer, or at just 401 nm using a single-wavelength spectrophotometer if the former are not available. The change in absorbance with time should be read at the initial linear phase of the reaction (however, *see* **Notes 3** and **4**). This is particularly important for ex vivo assays of nNOS, after the in vivo administration of the slowly dissociating inhibitors such as L-NNA and its methyl ester L-NAME, where there is a rapid dissociation of the inhibitor and a concomitant increase in NOS-activity rate within 2 min at 37°C (*see* **Notes 1** and **5**).
6. Ensure that the conversion of oxyhemoglobin to methemoglobin is owing to NOS activity by adding an excess of an NOS inhibitor such as L-NNA (final concentration > 10 µ*M* for nNOS and eNOS) or N^{ω}-monomethyl-L-arginine (L-NMMA) (final concentration > 100 µM for nNOS, iNOS, and eNOS) to inhibit NOS (**Fig. 1A**). A greater concentration of these inhibitors should be used if a higher concentration of L-arginine is used in the assay owing to the competitive mechanism of action of most NOS inhibitors and the low µ*M* Km of the NOS isozymes for L-arginine. The calcium dependence of the NOS under examination can be determined by the addition of ethylene glycol bis (β-aminoethyl ether) N,N,N′,N′-tetraacetic acid (EGTA) (final concentration 1 m*M*). If the NOS isozyme is unknown, the property of calcium dependence can help to determine whether the NOS activity is owing to iNOS (calcium independent) or eNOS and nNOS (calcium dependent) or both.

3.1.2. Calculation of Results

The change in absorbance/unit time is divided by the extinction coefficient (77,200 M^{-1} cm^{-1} for 421–401 nm readings or 38,600 M^{-1} cm^{-1} for 401 nm only) using a 1 cm path-length within the reaction mixture in the cuvet to yield a NOS activity in units of molar change/unit time. This value is now multiplied by the volume of the reaction mixture within the cuvet and divided by the weight of tissue or protein within that volume to give a final NOS activity of moles/unit time per weight of sample/protein.

4. Notes

1. The disadvantage of this assay compared to the radiochemical method is the potential for negative interference from samples with substantial quantities of hemoglobin present such as the spleen. A high concentration of hemoglobin in samples diminishes the signal-to-noise ratio (SNR) of the assay and may completely obscure enzyme activity. These samples should be assayed using the radiochemical-NOS assay described in this volume. For example, administration of lipopolysaccharide produces substantial induction of iNOS in the spleen as determined by the radiochemical-NOS assay, but no activity is observed when the oxyhemoglobin method is used *(3)*. In addition, with some tissues there can be a nonlinear relationship between tissue extract and measured activity *(3)*. In these cases, the radiochemical-NOS assay should be used. A further disadvantage may be the limitation in the numbers of assays that can be simultaneously carried out using this method, compared to the radiochemical-NOS assay. However, this is negated somewhat by the speed of the oxyhemoglobin assay, and the possibility of starting a number of oxyhemoglobin NOS assays and transferring the reaction mixture to a cuvet for activity determination or indeed carrying out the assay in a microtiter-plate format (*see* Chapter 22, this volume).
2. An important part of assaying an enzyme for the first time is to have a source of known activity. The brain has one of the highest NOS activities seen in animals and, therefore, a brain-cytosol preparation represents a simple and robust positive control when setting up the assay.
3. A major advantage of this assay over the radiochemical-NOS assay is that there is an instantaneous readout of NOS activity that the observer can react to and examine further effects of other effectors on that reaction. An additional advantage is the speed of readout. With reasonable levels of NOS activity, measurements can be made in the first 30 s of assay.
4. The continuous nature of the assay makes it particularly suitable for the analysis of progressive inhibition, which is shown by a significant number of NOS inhibitors.
5. As mentioned above, this is particularly useful for ex vivo studies with L-NNA (and L-NAME) and nNOS in which the dissociation of inhibitor-enzyme complex is slow enough to reflect the degree of binding in vivo in the first min of reaction at 37°C (**Fig. 1B**). The rate of dissociation is somewhat slower if nNOS is assayed at nonphysiologically lower temperatures. The degree of ex vivo

inhibition will only reflect the in vivo situation if the enzyme is kept at 4°C during preparation prior to assay. Many studies have severely underestimated the effect of these inhibitors in vivo owing to the length of time of their ex vivo assay. The other commonly used inhibitors, such as L-NMMA, cannot be used for ex vivo studies with NOS as they dissociate rapidly from NOS even at 4°C, and will, therefore, reach a new dissociated equilibrium immediately upon homogenization/dilution.

References

1. Knowles, R. G. and Moncada, S. (1994) Nitric oxide synthases in mammals. *Biochem. J.* **298,** 249–258.
2. Feelisch, M. and Noack, E. A. (1987) Correlation between nitric oxide formation during degradation of organic nitrates and activation of guanylate cyclase. *Eur. J. Pharmacol.* **139,** 19–30.
3. Salter, M., Knowles, R. G., and Moncada, S. (1991) Widespread tissue distribution, species distribution and changes in activity of calcium-dependent and calcium-independent nitric oxide synthases. *FEBS Lett.* **291,** 145–149.
4. Salter, M., Duffy, C., and Hazelwood, R. (1995) Determination of brain nitric oxide synthase inhibition in vivo: ex vivo assays of nitric oxide synthase can give incorrect results. *Neuropharmacology* **34,** 327–334.
5. Feelisch, M., Kubitzek, D., and Werringloer, J. (1996). The oxyhaemoglobin assay in *Methods in Nitric Oxide Research,* (Feelisch, M. and Stammler, J. S., eds.), Wiley, London, pp. 455–478.

7

Measurement of NOS Activity by Conversion of Radiolabeled Arginine to Citrulline Using Ion-Exchange Separation

Richard G. Knowles and Mark Salter

1. Introduction

Nitric oxide synthase (NOS) catalyzes a complex reaction utilizing L-arginine, nicotinamide adenine dinucleotide phosphate (NADPH) and oxygen to synthesize NO, and with citrulline and $NADP^+$ being produced as co-products (reviewed in **ref.** ***1***). This chapter describes the measurement of NOS-functional activity by determining the NOS inhibitor-sensitive conversion of radiolabeled (^{14}C or ^{3}H) arginine to citrulline, with separation of the labeled product from substrate by ion-exchange techniques. The earliest forms of this assay used rather slow and labor-intensive separations, e.g., on an amino-acid analysis ion-exchange column *(2)*. However, a simple method of separating arginine from citrulline using the Na^+ form of the strongly acidic-cation exchanger Dowex 50 (which has sulphonic-acid functional groups) was described by Gopalakrishna and Nagarajan *(3)*, and this technique was applied by Bredt and Snyder *(4)* to assays of purified NOS. We have subsequently modified this assay *(5–7)* to simplify it further and to make it suitable for the study of NOS activity in a wide variety of settings.

Using the protocol described, this is a highly sensitive [limit of detection ~50 pmol/min per g tissue or ~0.7 pmol/min per mg protein *(6,7)*], specific (given that NOS-inhibitor controls are carried out), and broadly applicable (from purified enzymes to crude-tissue homogenates, unaffected by sample color or turbidity) end-point assay.

From: *Methods in Molecular Biology, Vol. 100. Nitric Oxide Protocols*
Edited by: Michael A. Titheradge © Humana Press Inc., Totowa, NJ

2. Materials

1. Extraction buffer (EB): The extraction buffer should be prepared in advance. The basic buffer is first prepared by dissolving sucrose (320 m*M*), HEPES (20 m*M*), and ethylenediaminetetra-acetic acid (EDTA) (1 m*M*) in double-distilled or milliQ-grade water and bringing its pH to 7.2 at room temperature by the addition of HCl. The following constituents are then added to the final concentrations indicated: 1 m*M* D/L-dithiothreitol (DTT), 10 μg/mL leupeptin, 10 μg/mL soybean-trypsin inhibitor, and 2 μg/mL aprotinin; the buffer is then made up to its final volume with water. This EB is then distributed into aliquots (typically 50 mL per aliquot) and stored at –20°C until required. The composition of this EB is designed to permit extraction of NOS from tissues without breaking intracellular organelles and minimizing proteolysis by chelating-divalent cations with EDTA and the inclusion of protease inhibitors.
2. 10 mg/mL phenylmethylsulphonyl fluoride (PMSF): Because it is unstable in aqueous solution, PMSF is not included in the buffer at this stage, but prepared as a solution in absolute ethanol, stored at –20°C and added to the EB during the extraction procedure (*see* below).
3. ^{14}C-arginine-assay buffer: The assay buffer is based on 50 m*M* potassium phosphate buffer, pH to 7.2. This can be made in bulk and stored at 4°C. In this are dissolved or diluted the other assay components, to obtain an assay buffer containing 1.2 m*M* L-citrulline, 1.2 m*M* $MgCl_2$, 0.24 m*M* $CaCl_2$, 120 μ*M* NADPH, and L-[U-^{14}C]arginine (Amersham, 0.7 μCi/ml, i.e., 25 kBq/mL, *see* **Note 1**) plus L-arginine (to give a total concentration in this buffer of 24 μ*M*). The final concentrations in the assay after addition of enzyme, etc., will be 20% lower than in this buffer, e.g., 20 μ*M* L-arginine. This assay buffer can be stored frozen at –20°C until used.
4. ^{3}H-arginine-assay buffer: When ^{3}H-arginine is used (*see* **Note 1**), the buffer is made up as above, but is stored without the labeled arginine. This is added on the day of the assay by placing the required amount in a vial, drying it under a stream of dry nitrogen (or air), and then adding the rest of the assay-buffer components. This removes any ^{3}H exchanged with water.
5. NOS inhibitors: 100 m*M* N^{ω}-monomethyl-L-arginine (L-NMMA), N^{ω}-nitro-L-arginine (L-NNA), or S-ethylisothiourea (SEITU) (*see* **Note 2**). Store frozen.
6. 100 m*M* EGTA: Prepare by neutralizing ethylene glycol bis (β-aminoethyl ether) N,N,N′,N′-tetraacetic acid (EGTA) with KOH and store at 4°C.
7. Ion-exchange resin: The strongly acidic-cation exchange resin used to separate the arginine substrate from citrulline product is Dowex 50W, 200–400 mesh, 8% cross-linked (Sigma 50X8-400). This must be converted from the H^+ form provided to the Na^+ form by washing the resin with 1 *M* NaOH. Four washes with at least 10 volumes of NaOH are carried out by adding the NaOH to resin, thoroughly mixing, allowing the resin to settle under gravity (~5 min), and pouring off the supernatant. The Na^+ form resin is then washed with water until the pH is <8 and finally brought to 50% resin in water (approximately by the height of the settled resin). This resin can be prepared in bulk and stored at 4°C.

3. Methods

3.1. Sample Preparation

3.1.1. Preparation of Tissue Extracts

Tissues should either be assayed immediately on collection or else rapidly frozen and stored at –70°C. This is ideally carried out by freeze-clamping the tissue between blocks of aluminium cooled in liquid nitrogen, although if this is not possible then simple immersion in liquid nitrogen will usually suffice.

Extraction and storage of tissue extracts prior to assay has to be carried out at 0–4°C to avoid loss of enzyme activity.

1. In order to prepare the extract, weigh a portion of each tissue in a 10 mL clear-plastic tube that has been precooled on solid carbon dioxide and place on ice.
2. Add a measured volume of EB (typically 5 mL/g tissue) and then snip the tissue into small fragments with (cooled) scissors.
3. Homogenize the sample in the same tube using a mechanical homogenizer of the Ystral or Polytron type. The homogenizer probe is precooled in ice/water, rinsed in ice-cold distilled water, wiped dry with a tissue, and then placed in the tube containing tissue plus EB. After 10 s of homogenization, PMSF (10 μL per mL of EB) is added, followed by a further 10 s homogenization. This should result in an extract with a homogenous appearance: the presence of "lumps" or inhomogeneous color would indicate that the homogenization is incomplete.

3.1.2. Preparation of Cell Extracts

As is the case with tissues, cell pellets should either be assayed immediately on collection or else rapidly frozen and stored at –70°C.

1. Because cell-culture media contain high concentrations of L-arginine (typically >0.5 m*M*), wash the cells with an arginine-free medium (e.g., phosphate-buffered saline).
2. Suspend adherent cells directly in a small volume of ice-cold EB using a rubber policeman or a purpose-made cell scraper with the plate or flask standing on ice. These are best extracted and assayed immediately.
3. Centrifuge cells in suspension to collect them as a cell pellet and resuspend the cell pellet in a small volume, typically 2 mL per mL of pellet, of EB.
4. Disrupt the cell extracts by sonication. As with the procedures for tissue extraction, it is important to keep the temperature at 0–4°C. Most probe sonicators should be suitable for this purpose; we have routinely used a Soniprep 150 with a precooled 5 mm tip probe, sonicating twice for 10 s at 10 μm amplitude with 30 s cooling in ice/water in between.

3.1.3. Subcellular Fractionation

The radiochemical-NOS assay can be carried out on whole homogenates/sonicates, and if the purpose of the assay is to measure the total activity present then this may be the most appropriate. However, the composition of the EB is compatible with standard subcellular fractionation procedures, so that if it is preferred the assay can be carried out on a postmitochondrial supernatant (cytosol + microsomes, containing most of the NOS in cells, e.g. 10,000*g* for 30 min) or a cytosol (NOS in the soluble fraction of cells only, e.g., 100,000*g* for 60 min).

3.1.4. Removal of Endogenous Arginine

The arginine concentration in the assay (20 μ*M*) has been chosen as a compromise between having a high concentration, in order to minimize the effect of endogenous contributions and having a low concentration to maximize the sensitivity of the assay.

However, arginine can sometimes be present in cells, tissues, and media in sufficiently high concentrations that it significantly reduces the specific activity of the radiolabeled arginine in an uncontrolled way. In such instances, the endogenous arginine can be depleted by a pre-treatment with the ion-exchange resin, prepared as described above and then washed in EB.

1. Add 2 volumes of enzyme extract to 1 vol of ice-cold packed resin in microfuge tubes and mix.
2. Briefly centrifuge the mixture (e.g., 10,000*g* for 1 min at 0–4°C) and collect the supernatant for assay. It is crucial that no resin is carried forward into the assay: it will adsorb the labeled substrate!

3.2. The Assay Procedure

1. Prepare 10 mL clear plastic test-tubes by addition of 1.2 μL of water (controls), 100 m*M* EGTA (calcium-free), or L-NMMA (blanks), plus any other additions required for the specific experiment (*see* **Notes 2** and **3**).
2. Add 100 μl of assay buffer and prewarm the tubes to 37°C in a water bath at this temperature.
3. Initiate the assay by the addition of 18 μL of enzyme extract and mixing.
4. Continue the incubation for (typically) 10 min at 37°C (*see* **Note 4**).
5. Terminate the incubation by adding 1.5 mL of the ion-exchange resin, which stops the reaction by removing the arginine substrate, cooling, and dilution (*see* **Note 5**). The stopped assays can be left at room temperature until the batch of incubations has been completed.
6. Add 5 mL of water to the stopped incubations and mix.
7. Allow the resin to settle for approx 10 min before aspirating 4 mL of supernatant for scintillation counting. Typically, the 4 mL will need to be mixed with 10 mL of scintillation fluid to achieve a homogenous solution.

8. In order to calculate the specific activity of the ^{14}C- or ^{3}H-arginine, count the radioactivity in 100 μL of the assay buffer to determine the total dpm added.

3.3. Calculation of Results

The arginine conversion to citrulline will routinely be expressed as pmol/min per mg protein or as pmol/min per g tissue weight. The radiolabel present in the L-NMMA blanks is subtracted from that present in the other incubations to leave the radiolabel corresponding to NOS-dependent citrulline formation. This is divided by the specific activity of the labeled arginine and multiplied by 6.62 mL/4.0 mL to obtain the pmol citrulline formed in the assay. This is then divided by the incubation time and the weight of tissue (or sample protein) in the assay to give a value in units of pmol/min per g tissue (or per mg protein). For example, for a typical assay of rat-forebrain cytosol NOS in which 9000 dpm are measured in uninhibited-control incubations:

(9000 dpm—1000 dpm *blank*) ÷ 77 dpm/pmol *specific activity* × 6.62 mL/4.0 mL ÷ 10 min *incubation time* ÷ 0.00316 g *tissue in the 18 μL assayed* = NOS activity of 5440 pmol/min per g. The total and calcium-independent NOS activities are determined from the incubations in the absence and presence of EGTA to chelate free-calcium ions, respectively.

4. Notes

1. Either ^{14}C-arginine or ^{3}H-arginine can be used. In general, we have preferred ^{14}C-arginine because the blank values obtained are smaller and the signal-to-noise ratio (SNR) correspondingly larger. The efficiency of the separation of arginine from citrulline can be checked by carrying out the procedure with no enzyme and with radiolabeled arginine and citrulline (both available from Amersham). With either ^{14}C-arginine or ^{3}H-arginine, it may be necessary to purify the radiolabel if it gives high "no enzyme" blanks (above about 1.5%). This can be achieved by adding the radiolabel in 1 mL of 20 m*M* HEPES buffer, pH 5.5 to a 1 mL column of the Dowex resin, washing with 8 mL of water, and eluting with 4 mL of 0.5 *M* ammonia. The eluant is freeze-dried, redissolved in 2% ethanol, and stored at 4°C.
2. Every set of assays will require some control incubations in the presence of a NOS inhibitor such as L-NMMA, L-NNA, or SEITU. SEITU is listed as "2-ethyl-2-thiopseudourea" in the Aldrich catalog (Cat. # 30,131-0); it is the cheapest potent, entirely nonselective-NOS inhibitor available. L-NNA is also cheap, but is not very potent against inducible (i) NOS. In order to assess the calcium-dependence or independence of the measured NOS, 1 m*M* EGTA is added to some incubations to chelate free-calcium ions. With some sources of enzyme (especially purified NOS), other additions are required for maximal activity: (6R)-5,6,7,8-tetrahydro-L-biopterin (BH_4), (Dr. B. Schirks, Jona, Switzerland; 10 μ*M*, made fresh in dilute HCl), flavin adenine dinucleotide (FAD), and flavin

mononucleotide (FMN) (both 1 μ*M*, 50-100-fold stocks stored frozen), and calmodulin (CaM) (100 nM, 100-fold stock stored frozen). In instances where arginase may be present (especially liver tissue or cells, but also some other cells and tissues), an inhibitor of this enzyme should be added: L-valine (60 m*M*, dissolved directly in the assay buffer on the day of assay).

3. The additions of 1.2 μL are naturally not very quantitatively accurate, but this does not matter because the L-NMMA or EGTA are in any case being added in considerable excess. When accuracy is needed (e.g., in concentration-response studies with inhibitors), the addition will need to be made in a larger volume which can be added with appropriate accuracy.
4. It is crucial to ensure that NOS assays under the particular conditions of the assay buffer, temperature, and the enzyme source being used, are linear with both time and concentration of enzyme extract added. These can vary quite widely: at 37°C under the conditions described here, for example, neuronal (n) NOS is linear for only approx 10 min, whereas iNOS is linear for approx 30 min; if the assay is carried out at 25°C, however, nNOS is linear for approx 60 min. It is also sensible to check whether "NOS inhibitor blanks" are significantly higher than "zero time blanks." When L-[U-^{14}C]-arginine is used as the substrate, this is a useful indication of whether arginase is present and could be competing for the arginine substrate; if so, valine should be added to inhibit the arginase.
5. The ion-exchange separation can also be carried out in small columns rather than as batch additions, but this is much more labor-intensive and in our hands does not result in significantly lower blanks; with the batch procedure described above, hundreds of assays can be carried out in 1 d. If the separation is to be carried out in columns, then the assay incubations will need to be terminated by addition of a "stop reagent": HEPES (20 m*M*), EDTA (1 m*M*), SEITU (1 m*M*), pH to 7.2 with HCl. The stopped incubations are applied to columns (0.5 mL bed volume) of the Dowex resin, typically in plastic, disposable chromatography columns (Biorad), and the labeled citrulline is eluted with 2 × 2 mL of water. The combined eluate is then mixed with scintillation fluid for scintillation counting. In either case, it is possible to re-use the ion-exchange resin following extensive washing in 1 *M* NaOH. The pipeting of the Dowex resin can be made easier by cutting the end off the pipet tip used with a sharp blade, leaving a larger aperture.
6. It is valuable to have a source of reference NOS to set up the assay and to periodically check for any problems; because the brain has the highest activity of NOS *(6)* we have used rat-brain cytosol, prepared as described above and stored in aliquots at –70°C, for this purpose.
7. When not specified, the reagents used can be obtained from Sigma, Aldrich, or Calbiochem.

References

1. Knowles, R. G. and Moncada, S. (1994) Nitric oxide synthases in mammals. *Biochem. J.* **298,** 249–258.

2. Knowles, R. G., Palacios, M., Palmer, R. M. J., and Moncada, S. (1989) Formation of nitric oxide from L-arginine in the central nervous system: a transduction mechanism for stimulation of the soluble guanylate cyclase. *Proc. Natl. Acad. Sci. USA* **86,** 5159–5162.
3. Gopalakrishna, R. and Nagarajan, B. (1980) *Anal. Biochem.* **107,** 318–323.
4. Bredt, D. S. and Snyder, S. H. (1990) Isolation of nitric oxide synthase, a calmodulin-requiring enzyme. *Proc. Natl. Acad. Sci. USA* **87,** 682–685.
5. Knowles, R. G., Merrett, M., Salter, M., and Moncada, S. (1990) Differential induction of brain, lung, and liver nitric oxide synthase by endotoxin in the rat. *Biochem. J.* **270,** 833–836.
6. Salter, M., Knowles, R. G., and Moncada, S. (1991) Widespread tissue distribution, species distribution and changes in activity of calcium-dependent and calcium-independent nitric oxide synthases. *FEBS Lett.* **291,** 145–149.
7. Thomsen, L. L., Miles, D. W., Happerfield, L., Bobrow, L. G., Knowles, R. G., and Moncada, S. (1995) Nitric oxide synthase activity in human breast cancer. *Br. J. Cancer* **72,** 41–44.

8

Radiochemical Measurement of NOS Activity by Conversion of [^{14}C]L-Arginine to Citrulline Using HPLC Separation

James M. Cunningham and Richard C. Rayne

1. Introduction

Monitoring conversion of radiolabeled arginine to citrulline, the co-product of the nitric oxide synthase (NOS)-catalyzed reaction, is a widely used method to assay for NOS activity. In cases where NOS is the sole or overwhelmingly predominant arginine-metabolizing enzyme present in a cell or tissue preparation, it is convenient to separate arginine from citrulline using a Dowex cation-exchange chromatography system (*see* Chapter 7, this volume). In some cells or tissues, however, other pathways of arginine metabolism are present and the products formed may fail to be separated from arginine, from each other, or from citrulline (if it is even present) using the simple Dowex method. For example, the widely distributed enzyme arginase *(1)* can hydrolyze arginine to ornithine and urea. Using the Dowex method, it is not possible to distinguish ornithine from arginine or urea from citrulline.

A more rigorous separation regime using high-pressure liquid chromatography (HPLC) can be employed to identify directly the products of arginine metabolism in such cases. The method described here is based on a protocol developed by Vallance et al. for separation of arginine analogs *(2)* and, using an ion-pairing reagent (hexanesulphonic acid) to aid retention of amino acids, requires no sample derivitization prior to HPLC. This separation procedure allows quantification of the relative activities of NOS and arginase in cell and tissue extracts by permitting identification of radiolabeled arginine-derived citrulline, ornithine, and urea. This method is useful, therefore, as a means to identify arginine-derived citrulline, to assess the contribution of pathways other

From: *Methods in Molecular Biology, Vol. 100. Nitric Oxide Protocols*
Edited by: Michael A. Titheradge © Humana Press Inc., Totowa, NJ

than that catalyzed by NOS to the generation of arginine-derived products, and to simultaneously monitor NOS and arginase activities in a particular cell or tissue type.

Here, we provide a detailed account for the measurement of inducible (i) NOS activity present in cultured, cytokine-stimulated RINm5F cells, a line of insulin-secreting cells which contain arginase *(3)*. We have included protocols for NOS preparation and enzyme assay (*see* **Subheading 3.1.**), and for separation by HPLC of the arginine-derived products of these enzyme-catalyzed reactions (*see* **Subheading 3.2.**).

2. Materials

2.1. NOS Preparation and Enzyme Assay

Items 1–8 are stock solutions that can be combined on the d of use to produce the working "sonication buffer" (*see* **item 9**).

1. Leupeptin stock: 10 m*M* in dH_2O. The solution is stable for 1 mo at –20°C.
2. Pepstatin-A stock: 1 m*M* in methanol. The solution is stable for several months at –20°C.
3. Bestatin stock: 4 mg/mL in methanol. The solution is stable for approx 1 mo at –20°C.
4. Dithiothreitol (DTT) stock: 500 m*M* in dH_2O. The solution is stable for several mo at 4°C.
5. *trans*-Epoxysuccinyl-L-leucylamido-(4-guanidino)butane (E-64) stock: 1 m*M* in 50% v/v EtOH (in dH_2O). The solution is stable for several months at –20°C.
6. Ethylene diaminetetraacetic acid (EDTA) stock: 500 m*M* in dH_2O, pH approx 8.0. To prepare 10 mL of solution, add 1 mL of 5 *M* NaOH to 8 mL of dH_2O, then add 1.86 gm of disodium, dihydrate EDTA. Dissolve the EDTA and adjust the volume to 10 mL with dH_2O. Store at 4°C.
7. Egtazic acid (EGTA) stock: 100 m*M* in dH_2O. To prepare 10 mL of solution, follow the procedure for EDTA (*see* **item 6**), except add 0.38 gm of EGTA. Store at 4°C.
8. Tris-HCl stock: 0.5 *M*, pH 7.4, in dH_2O. Store at 4°C.
9. Sonication buffer: 50 m*M* Tris-HCl, pH 7.4, 0.5 m*M* EDTA, 0.5 m*M* EGTA, 0.5 m*M* DTT, 20 μ*M* leupeptin, 2 μ*M* pepstatin A, 10 μ*M* E-64, and 40 μg/mL bestatin. Prepare fresh just before use by combining aliquots from the respective stock solutions (*see* **items 1–8**) in an appropriate volume of dH_2O and adjusting the pH as required.
10. Assay buffer: 50 m*M* Tris-HCl, pH 7.4, 20 μ*M* L-arginine, 10^7 dpm/mL L-[U-^{14}C]-arginine monohydrochloride (supplied at 11.2 GBq/mmol; Amersham International, Amersham, Berks, UK). Prepare fresh just before use (*see* **Notes 1** and **2**).
11. Trifluoroacetic acid (TFA) solution: 50% (v/v) HPLC-grade TFA in dH_2O.
12. Sonicator: e.g., MSE 150-W ultrasonic disintegrator Mk. 2. (Fisher Scientific UK, Loughborough, Leics., UK).

13. Ultracentrifuge and fixed-angle rotor capable of 100,000*g*.
14. Ultracentrifuge tubes to match the rotor.
15. Microfuge capable of 14,000*g*.
16. Microfuge tubes, 1.5-mL capacity.
17. Waterbaths: 37°C and boiling.

2.2. HPLC Separation and Sample Analysis

1. HPLC-buffer stock: 25 m*M* orthophosphoric acid, 10 m*M* hexanesulphonic acid, pH 2.5. For 250 mL, dissolve 0.47 g of hexanesulphonic acid in approx 220 mL of HPLC-grade water and add 0.424 mL of HPLC-grade orthophosphoric acid. Adjust the pH to 2.5 using 5 *M* KOH. Make the volume up to 250 mL with HPLC-grade water and pass the solution through a 0.22 μm filter. Use this solution immediately to make HPLC buffers A and B (*see* **Subheading 2.2., items 2** and **3**) (*see* **Note 3**).
2. HPLC buffer A: 25 m*M* orthophosphoric acid, 10 m*M* hexanesulphonic acid, 0.5% (v/v) acetonitrile, pH 2.5. Add 0.85 mL of HPLC-grade acetonitrile to 170 mL of HPLC buffer stock (*see* **Subheading 2.2., item 1**). Purge the buffer for approx 15 min with inert gas (e.g., argon or helium) immediately prior to use. Use within 2–3 d (*see* **Note 3**).
3. HPLC buffer B: 25 m*M* orthophosphoric acid, 10 m*M* hexanesulphonic acid, 20% (v/v) acetonitrile, pH 2.5. Add 20 mL of HPLC-grade acetonitrile to 80 mL of HPLC-buffer stock (*see* **Subheading 2.2., item 1**). Purge the buffer for approx 15 min with inert gas (e.g., argon or helium) immediately prior to use. Use within 2–3 d (*see* **Note 3**).
4. HPLC system comprising an injector, a binary-pumping system, programmable pump controller, and UV detector (*see* **Note 4**).
5. Fraction collector.
6. HPLC column: C18, 2.1 mm × 220 mm; 5-μm Brownlee column (Applied Biosystems Ltd., Warrington, UK) (*see* **Note 5**).
7. Liquid-scintillation counter.
8. Mini-vials and cocktail (e.g., Wallac Optiphase "Safe," Fisher Scientific UK, Loughborough, UK) for liquid-scintillation counting.

3. Methods

3.1. NOS Preparation and Enzyme Assay

1. Add ice-cold sonication buffer to cytokine-stimulated RINm5F cells in a 2.5 mL plastic tube, on ice (approx 2×10^7 cells/0.5 mL buffer).
2. Keeping the tube on ice, sonicate the mixture to homogenize the cells (*see* **Note 6**).
3. Centrifuge the homogenate at 100,000*g* for 1 h at 4°C. Save the cytosolic (supernatant) fraction for assay (*see* **Note 7**).
4. Add 50 μL of the supernatant to 50 μL of the assay buffer in a 1.5 mL microfuge tube. Prepare several "blank" (no-enzyme) samples, replacing tissue/cell supernatant with sonication buffer (*see* **Note 8**). Mix the tubes by vortexing.

5. Incubate the mixtures for 45 min in a 37°C water bath (*see* **Note 9**).
6. Stop the reactions by adding to each 10 μL of the TFA solution (i.e., to 5%, v/v, final concentration). Vortex mix, then heat samples in a boiling-water bath for 3–5 min to precipitate protein (*see* **Note 10**).
7. Centrifuge the samples in a microfuge at full speed (approx 14,000*g*) for 5 min.
8. Remove the supernatant and store it at –20°C until analysis by HPLC, as described in **Subheading 3.2.**

3.2. HPLC Separation and Sample Analysis

1. Enter a solvent-delivery program as follows: deliver only HPLC buffer A (0% B) for 8 min; increase HPLC buffer B to 100% over the following 2 min; hold at 100% HPLC buffer B for 3 min; return to 0% HPLC buffer B over 1 min and hold for 4 min. Use a flow rate of 200 μL/min throughout (*see* **Notes 11** and **12**).
2. Equilibrate the HPLC column with HPLC buffer A for at least 15 min prior to the first sample run of the day (*see* **Note 13**).
3. Meanwhile, set up the fraction collector to obtain 20-s fractions and set the UV absorbance detector to 200 nm (*see* **Notes 14** and **15**).
4. Dilute the sample 1:1 with HPLC buffer A, mix, and inject the sample onto the equilibrated column. Initiate the solvent-delivery program and start the fraction-collector coincident with sample injection (*see* **Note 16**).
5. At the end of the solvent program, deliver only HPLC buffer A (0% B) for at least 6 min prior to injection of another sample (*see* **Note 17**).
6. Add 3 mL of scintillation cocktail to each fraction and analyze the fractions by liquid-scintillation counting (*see* **Note 18**).

4. Notes

1. Some cells and tissues will contain sufficient NOS cofactors to allow maximal enzyme activity in the assay procedure as described. This is the case for the cultured RINm5F cells on which this protocol is based. In some cases, supplementation of the assay buffer with NOS cofactors will be essential. In particular, when the constitutive calcium-requiring type I and III NOS (brain- and endothelial-cell type) are being assayed, addition of calcium (e.g., calcium chloride, final concentration 1 m*M*) is essential when EDTA and EGTA have been used in the sonication buffer. Recommended final concentrations of other components that may be required are: nicotinamide adenine dinucleotide phosphate (NADPH) (1 m*M*), flavin adenine dinucleotide (FAD) (5 μ*M*), flavin mononucleotide (FMN) (5 μ*M*), tetrahydrobiopterin (BH_4) (10 μ*M*), and calmodulin (CaM) (50 U/mL).
2. L-[2,3,4,5-^{3}H]-arginine monohydrochloride (Amersham International) is available at higher-specific activity than the [^{14}C]-labeled arginine and, therefore, may be preferred as a substrate for use with samples containing low levels of NOS.
3. It is essential to produce HPLC buffers using HPLC-grade reagents, including water (e.g., "Milli-Q" or equivalent). Filtration of the completed buffers prior to use is also essential to prevent particulate matter from entering the HPLC apparatus and/or column. De-gassing the solvents is important to prevent bubble forma-

tion in the detector cell. We have not systematically investigated the shelf-life of these buffers, but have noticed that they sometimes deteriorate (especially buffer A) after approx 3–5 d. For this reason, and because hexanesulphonic acid is not inexpensive, we routinely make up volumes of the HPLC buffers that we anticipate using completely within 2 d. Also, it is worth noting that the acetonitrile percentages quoted in **Subheading 2.2.**, **items 2** and **3** are not strictly correct as v/v percentages, as acetonitrile undergoes a solvent contraction upon addition to water.

4. We have used an HPLC system from Applied Biosystems Ltd. (Warrington, UK) comprising a Model 112A oven/injector, Model 140B Solvent Delivery System, and Model 785A Programmable Absorbance Detector. This is a syringe-pump system optimized for narrow-bore and microbore LC applications. Although we have not tested the method on other HPLC apparatus, we have no reason to believe that it could not be replicated on any quality HPLC system.
5. We have experience of only a few (C18) HPLC columns for this application and our approach to choosing them was hardly systematic! Initially, two C18 columns (of the same dimensions, particle size, and pore diameter reported for **item 6** in **Subheading 2.2.**) which were "cast-offs" from our peptide sequencer were used. There was little difference in performance between them. Other brands of column may work better than the one we report here—it is probably worth testing a few "old" C18 columns, if they are available in the lab, before committing to buying a new one. One column, bought new and currently in use, has been used for 200–300 separations without any noticable loss in quality of performance.
6. **Steps 1** and **2** in **Subheading 3.1.** must be carried out at 4°C because of the low stability of NOS at room temperature and to limit proteolysis during cell disruption. For sonication of cells, we have used a MSE 150-W ultrasonic disintegrator Mk. 2. set at 15 μm. Perform sonication in a plastic rather than glass tube and place the tube in ice during sonication to avoid any rise in temperature. Sonication may be inappropriate in some cases; for other tissues or cells, a teflon/glass homogenizer with motor-driven pestle or a Polytron-type homogenizer may be more suitable.
7. The 100,000*g* supernatant should be used in a NOS assay as soon as possible, as NOS is notoriously labile. Keep the supernatant on ice until just before use. Retain a sample (stored at –20°C) for determination of total-protein concentration.
8. The cytosolic extract will contain some arginine derived from the cell/tissue sample. To obtain quantitative values for the conversion of arginine to citrulline, the endogenous arginine must be accounted for or removed from the extract prior to incubation with the radiolabeled substrate. Endogenous arginine can be removed by passage of 100,000*g* supernatant from cell or tissue preparations over a column of Dowex AG50WX-8 (Na^+ form) as described in ***(4)***. "Blank" (no enzyme) samples are necessary to quantify enzyme-independent metabolism of arginine in otherwise complete reaction mixtures (*see* also **Note 18**).
9. The time period over which conversion of arginine to citrulline is linear should be determined for each particular preparation.

10. Addition of TFA and heating the samples will cause most of the protein in the reactions to precipitate. Apart from stopping the enzyme reaction(s), removal of this material reduces the protein load onto the HPLC column. This is an important consideration, as the conditions used to elute arginine and its metabolites (*see* **Subheading 3.2.**) are unlikely to effectively elute hydrophobic components (e.g., proteins) from a C18 column. (A build-up of bound material on the column can degrade its performance. *See* **Note 17** for a column-washing procedure.) High-speed centrifugation of the samples and avoidance of the pelleted material when collecting the supernatant (*see* **Subheading 3.1., item 7**) also protects the HPLC apparatus by removing particulate material from the samples.
11. The solvent program will separate isocratically, in order of elution, urea, citrulline and ornithine. Typically, the elution times are 2–3 min, 4–5 min, and 6–9 min. Arginine is strongly retained under these conditions, and the linear increase to 100% HPLC-buffer B (a stronger eluent, containing 20% acetonitrile) speeds its elution (approx 12–14 min).

    The times of elution and quality of separation of the individual components are most strongly influenced by the pH of buffer A. Increasing the pH to 3–3.5 may give narrower bands for each analyte (sharper peaks), but reduce the band separation (peak resolution). At pH values below 2.5, the citrulline and ornithine may appear as double rather than single peaks. The pH that we recommend (2.5) should be considered a starting point for optimization; optimum conditions should be established using as markers authentic radiolabeled arginine (as **Subheading 2.1., item 10**), L-[U-^{14}C]-ornithine hydrochloride, [^{14}C]-urea (Amersham International), and L-[ureido-^{14}C]-citrulline (NEN-Du Pont, Dreiech, Germany). Dilute the standards in sonication buffer containing 5% (v/v) TFA and mix an aliquot 1:1 with HPLC buffer A.
12. We normally use a flow rate of 200 μL/min for 2.1 mm columns on our HPLC system. This flow rate is not obligatory and may require alteration depending on the characteristics of other HPLC systems and/or columns.
13. We routinely store our HPLC columns in 60% MeOH (v/v with water). When preparing a column for use, it is wise to first pump out the MeOH using water (pump for about 15 min at the recommended flow rate for the column) then pump on the HPLC-buffer A. (At the end of the day, flush the HPLC system thoroughly with water before pumping in the 60% MeOH.) If your work will carry over onto the next day, it is convenient to leave the system running on 0% B at a very low-flow rate (e.g., 10–20 μL/min).

    It is advisable also to perform a "blank" HPLC run before the first sample run. Inject a sample of HPLC-buffer A, or better yet, a sample containing a 1:1 mixture of HPLC-buffer A and sonication buffer containing 5% (v/v) TFA. We also advise performing runs of radiolabeled standards (*see* **Note 11**) prior to and after each series of "unknown" samples.
14. To minimize the number of fractions to be analyzed, we collect 20 s fractions only over the period from 1–10 min postinjection. After this period, the fraction time can be increased to 1 min until arginine is eluted.

15. We routinely monitor the eluate at 200 nm to provide a chromatographic record of each run. Although the complexity of the samples leads to many spurious peaks (which often obscure peaks attributable to arginine and/or its metabolites), a "typical" chromatogram is produced for each set of samples. This record—essentially useless for quantifying the material eluted—is nonetheless sometimes useful for troubleshooting and method optimization.
16. Typically, we load only a fraction of the total available sample, normally 50 μL. It is very important that the samples are acidified prior to loading as the elution times of urea, citrulline, and ornithine are highly sensitive to pH. If the reaction-stopping protocol recommended here is used (*see* **Subheading 3.1., item 6**), sample pH will not be a problem. If samples for HPLC analysis have been otherwise treated, add TFA to 5% v/v (centrifuge to pellet protein if necessary) and proceed.
17. To prevent accumulation of strongly retained material on the column, we periodically perform a washing procedure. First, remove the buffered solvent (i.e., HPLC-buffer A) from the system by flushing with water for at least 15 min. Change the A solvent to 0.1% TFA, v/v, in water, and the B solvent to 0.08% TFA, v/v, in acetonitrile. Pump in solvent A alone (0% B) for about 15 min, then run a linear gradient to 100% B over 30–60 min and hold at 100% B for 3–5 min. Ramp the system back to 0% B over 2–3 min and hold for about 5 min. At this point, the column may either be stored in 60% MeOH (v/v with water) or re-equilibrated in the HPLC buffer A for arginine/arginine metabolite separations. We have no hard and fast rule as to how frequently this procedure should be performed, but it is probably wise to employ a wash after approximately every 50 runs as a preventative measure, or sooner if degradation of the separation is noticed.
18. Batches of radiolabeled arginine typically contain a variable proportion of radiolabeled impurities/arginine degradation products which elute before the arginine peak. The [^{3}H]-labeled arginine is, in our experience, less stable than the [^{14}C]-labeled variety and the quantity of these nonarginine impurities increases with storage time. It is, therefore, important to subtract the activity recovered in fractions from "blank" (no-enzyme) samples from that found in corresponding fractions from the enzyme-containing samples.

References

1. Jenkinson, C. P., Grody, W. W., and Cederbaum, S. D. (1996) Comparative properties of arginases. *Comp. Biochem. Physiol.* **114,** 107–132.
2. Vallance, P., Leone, A., Calver, A., Collier, J., and Moncada, S. (1992) Accumulation of an endogenous inhibitor of nitric-oxide synthesis in chronic renal failure. *Lancet* **339,** 572–575.
3. Cunningham, J. M., Mabley, J. G., and Green, I. C. (1997) Interleukin 1β-mediated inhibition of arginase in RINm5F cells. *Cytokine*, 9, in press.
4. Bredt, D. S. and Snyder, S. H. (1989) Nitric oxide mediates glutamate-linked enhancement of cGMP levels in the cerebellum. *Proc. Natl. Acad. Sci. USA* **86,** 9030–9033.

Barry W. Allen

9

The Enzymatic Measurement of Nitrate and Nitrite

Michael A. Titheradge

1. Introduction

The involvement of nitric oxide (NO) as an effector molecule in many physiological and pathological situations is now firmly established. Although NO can be measured directly using NO-specific electrodes and chemiluminescence, the measurement of the stable end-products of NO metabolism, nitrite (NO_2^-) and nitrate (NO_3^-), is commonly used as a rapid and simple way to assess NO production when large numbers of samples are to be processed. NO can undergo a series of reactions within biological tissues, including:

(a)

$$2NO + O_2 \rightarrow 2NO_2$$

$$2NO_2 \leftrightarrow N_2O_4$$

$$N_2O_4 + 2OH^- \rightarrow NO_2^- + NO_3^- + H_2O$$

(b)

$$NO_2 + NO \rightarrow N_2O_3$$

$$N_2O_3 + 2OH^- \rightarrow 2NO_2^- + H_2O$$

Though the measurement of nitrite alone is often used as a monitor of NO metabolism, nitrate is also produced in these reactions and much of the nitrite in biological systems may be further oxidized to nitrate in the presence of heme iron-containing proteins (reviewed in **ref. *1***). Therefore, the relative proportions of these two end-products are variable and dependent upon the origin of the sample, and the best indication of the extent of NO metabolism can only be obtained when both nitrite and nitrate are measured.

From: *Methods in Molecular Biology, Vol. 100. Nitric Oxide Protocols*
Edited by: Michael A. Titheradge © Humana Press Inc., Totowa, NJ

1.1. The Griess Reaction

The concentration of nitrate plus nitrite in samples is conveniently measured by a two-step procedure. The nitrate is first converted to nitrite using either enzymatic conversion *(2–4)*, or reduction with metallic cadmium *(5,6)*, followed by measurement of the total nitrite in the sample by a stoichiometric diazotization reaction using the Griess reagent *(6)* to form a purple-azo product. The original nitrite concentration in the samples can be determined in samples in which the nitrate has not been converted, thus allowing the measurement of nitrate by difference (*see* **Scheme 1**).

Two enzymatic methods are described for the conversion of nitrate to nitrite, one using nicotinamide adenine dinucleotide phosphate (NADPH)-dependent nitrate reductase from *Aspergillus* species, which is readily available commercially, and the second using formate-nitrate reductase prepared from *Escherichia coli*. These methods have the advantage over the conversion of nitrate to nitrite with metallic cadmium, in that they have been adapted to perform both the nitrate conversion and nitrite determination in 96-well plates without the need for any sample transfer *(3)*. In addition, they remove the problem of exposure of the laboratory worker to highly toxic cadmium and eliminate the difficulties in disposing of cadmium waste. Both methods produce comparable results in our laboratory; the disadvantage of the former being the high cost of the commercial enzyme if large numbers of samples are to be processed, whereas the formate-nitrate reductase is easily and inexpensively prepared.

1.2. NADPH-Dependent Nitrate Reductase

The simultaneous measurement of nitrate and nitrite using the NADPH-nitrate reductase (EC 1.6.6.2) was first described by Wu and Brosnan *(2)* and is based upon the reduction of nitrate to nitrite using NADPH and measurement of the total nitrite formed with the Griess reagent. This method was subsequently modified by Verdon et al. *(3)* by the inclusion of a glucose-6-phosphate/glucose-6-phosphate dehydrogenase NADPH-regenerating system to prevent the interference with color development in the Griess reaction caused by the high concentrations of $NADP^+$.

$$NO_3^- + NADPH + H^+ \xrightarrow{\textit{Nitrate reductase}} NO_2^- + NADP^+ + H_2O$$

$$G\text{-}6\text{-}P + NADP^+ + H_2O \xrightarrow{\textit{G-6-PDH}} 6\text{-}PG + NADPH + H^+$$

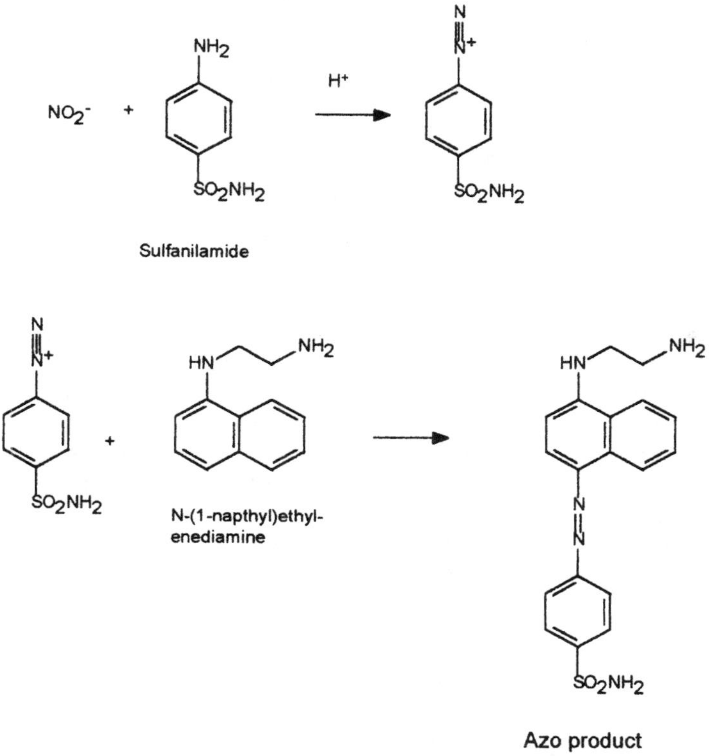

Scheme 1. The Griess reaction.

1.3. Formate-Nitrate Reductase

The conversion of nitrate to nitrite using formate-nitrate reductase is based upon the method of Taniguchi et al. *(4)*. Flavin mononucleotide (FMN) is added as a supplementary electron carrier and is reduced by the formate-dehydrogenase activity (formate:ferricytochrome b_1 oxidoreductase, EC 1.2.2.1) of this complex to $FMNH_2$. The $FMNH_2$ is then used to reduce nitrate to nitrite using nitrate-reductase activity (ferrocytochrome:nitrate oxidoreductase, EC 1.9.6.1). In contrast to NADPH and $NADP^+$, FMN does not interfere with the Griess reaction *(4)*. The nitrite formed in the reaction is then determined with the Griess reagent.

Formate Dehydrogenase

$$FMN + HCOO^- + H_2O \rightarrow FMNH_2 + CO_2 + OH^-$$

Nitrate Reductase

$$FMNH_2 + NO_3^- \rightarrow NO_2^- + FMN + H_2O$$

2. Materials

2.1. Determination of Nitrite Using the Griess Reaction

1. Griess solution A: Dissolve 0.1 g N-(1-naphthyl)ethylenediamine hydrochloride (Sigma N 5889) in 100 mL double-distilled water. Store at 0–4°C protected from light.
2. Griess solution B: Dissolve 1 g of sulfanilamide (Sigma S 9251) in 100 mL of 5% (v/v) orthophosphoric acid. Store at 0–4°C protected from light.
3. Working Griess reagent: Prepare by mixing one part Griess solution A with one part Griess solution B. Make immediately before use.
4. 10 m*M* nitrate standard: Dissolve 85.0 mg of sodium nitrate in 100 mL of double-distilled water. Store at 0–4°C protected from light. Prepare standard solutions from this stock ranging between 0-100 μ*M* nitrate.
5. 10 m*M* nitrite standard: Dissolve 69.0 mg of sodium nitrate in 100 mL of double-distilled water. Store at 0–4°C protected from light. Prepare standard solutions from this stock ranging between 0–100 μ*M* nitrite.

2.2. Preparation of Reagents for the NADPH-Dependent Nitrate-Reductase Assay

1. 14 m*M* sodium phosphate buffer, pH 7.4: Dissolve 101 mg of $Na_2HPO_4{\cdot}2H_2O$ and 20.8 mg of $NaH_2PO_4{\cdot}2H_2O$ in 50 mL of double-distilled water. Store at 0–4°C.
2. 10 μM nicotinamide adenine dinucleotide, reduced form: Dissolve 1.7 mg of NADPH (Sigma N 6504) in 2 mL of double-distilled water. Dilute 100-fold to obtain the working solution. Prepare fresh each d.
3. Conversion buffer sufficient for 125 samples: Dissolve 3.8 mg of glucose-6-phosphate (Sigma G 7250) in the 14 m*M* phosphate buffer. Add two units of glucose-6-phosphate dehydrogenase (Sigma G 4134, final conc. 400 U/l) and one unit of NADPH-dependent nitrate reductase (from *Aspergillus* species, Sigma N 7265, final conc. 200 U/l). Prepare fresh each d just prior to use.
3. Nitrate and nitrite standards as in **Subheading 2.1.**

These reagents are also available either individually or as a Total Nitrate plus Nitrite kit from Cayman Chemical Co. (Ann Arbor, MI). The kit has all the components required for the assay and produces rapid and reliable results.

2.3. Preparation of Reagents for Converting Nitrate to Nitrite Using the Formate Nitrate Reductase Assay

1. 0.5 *M* phosphate buffer, 0.8 *M* sodium formate, 0.4 m*M* FMN buffer, pH 6.0: Dissolve 0.61 g $Na_2HPO_4{\cdot}2H_2O$, 3.36 g $NaH_2PO_4{\cdot}2H_2O$, 2.72 g of sodium for-

mate (Sigma F 6502), and 10 mg FMN (FMN-Na·$2H_2O$, Sigma F 8399) in 50 mL of double-distilled water. pH to 6.0. Store at 0–4°C protected from light.
2. Formate-nitrate reductase (approx 2–4 kU/l): Suspend the lyophilized powder to a final concentration of 2 mg protein/mL in double-distilled water at 4°C. Prepare immediately before use from lyophilized powder (*see* **Subheading 3.4.** for the preparation of the enzyme).
3. Nitrate and nitrite standards as in **Subheading 2.1.**

2.4. Preparation of Reagents for the Formate-Nitrate Reductase Activity

1. Nitrate-reductase positive, nitrite-reductase deficient strain of *E. coli*, strain JCB387 *pcnB* Δ*nrf* (*see* **Note 1**).
2. Lennox L Broth media: Add 10 g LB broth base (Sigma L 3022), 5 g KNO_3, 5 g of sodium formate, 0.5 g K_2HPO_4 and 367 μL of glycerol to 450 mL of double-distilled water. Adjust the pH of the medium to 7.2 with KOH and make up to a final volume of 500 mL. Autoclave at 15 psi for 20 min at 121°C.
3. Kanamycin solution: Dissolve 50 mg of kanamycin (Sigma K 4000) in 1.0 mL of 0.9% (w/v) NaCl.

3. Methods

3.1. Determination of Nitrite in Samples

1. Pipet 100 μL of sample or nitrite standard into a 96-well microtiter plate (*see* **Note 2**). If less than 100 μL of sample is used, the volume should be made up to 100 μL with water or the medium in which the samples were prepared in.
2. Add 100 μL of Griess working solution and incubate at room temperature for 15 min to allow color development (*see* **Note 3**).
3. Measure the absorbance of the samples and standards at 540 nm (*see* **Note 4**).
4. Calculate the concentration of nitrite in the samples by comparison with the nitrite-standard curve. If only nitrite is to be measured, the standard curve is linear up to 250 μ*M*.

3.2. Determination of Total Nitrate Plus Nitrite Using the NADPH-Dependent Nitrate-Reductase Assay

1. Pipet 50 μL of sample, nitrite or nitrate standards into a 96-well microtiter plate (*see* **Notes 5–8** for sample preparation).
2. Pipet 40 μL of the conversion buffer into each well. For the determination of nitrite alone, prepare a conversion-assay buffer without the NADPH-dependent nitrate reductase and pipet 40 μL of this into a parallel set of wells containing the samples.
3. Pipet 10 μL of the NADPH solution into each well (*see* **Note 9**).
4. Shake the plate and incubate for 45 min at room temperature to convert the nitrate into nitrite.
5. Determine the total nitrite in the samples using the Griess reagent as in **Subheading 3.1.**

6. Compare the absorbance values for the nitrate- and nitrite standard curves. Typically, the conversion of nitrate to nitrite is greater than 95% using standards ranging between 0–100 μ*M* nitrate.
7. Calculate the total concentration of nitrite in the samples from the nitrite-standard curve. This gives the amount of nitrite stoichiometrically converted from nitrate, plus the nitrite originally present in the medium.
8. Determine the nitrate concentration by subtracting the nitrite concentration measured in the absence of the enzyme from the total nitrite, and correct this value if necessary to allow for the percentage conversion.

3.3. Determination of Total Nitrate Plus Nitrite Using the Formate Nitrate-Reductase Assay

1. Dilute one part of the phosphate-conversion buffer with 1.8 parts of double-distilled water before use (*see* **Note 10**).
2. Pipet 50 μL of sample (*see* **Notes 5–8** for sample preparation), nitrate and nitrite standards into the wells of a 96-well microtiter plate and add 40 μL of the conversion buffer, followed by 10 μL of the diluted formate-nitrate reductase (*see* **Note 11**).
3. Nitrite values in samples can be determined by adding the conversion buffer with 10 μL of water replacing the enzyme. There is no conversion of nitrate in the absence of the enzyme.
4. Incubate at 37°C for 60 min to convert the nitrate to nitrite.
5. Determine the total nitrite in the samples using the Griess reagent as in **Subheading 3.1.** and calculate the nitrate and nitrite concentration in the original medium as described in **Subheading 3.2., steps 6–8.**

3.4 Preparation of Formate-Nitrate Reductase

1. Grow *E. coli*, strain JCB387 *pcnB* Δ*nrf*, (*see* **Note 1**) anaerobically in 5 mL of media consisting of 10 g of LB media containing 5 μL of the stock-kanamycin solution for 8 h.
2. Aseptically transfer 0.5 mL of the culture to 500 mL of LB media and grow overnight.
3. Collect the cells by centrifugation at 2000*g* for 20 min at 4°C.
4. Resuspend the cells in 500 mL of distilled water and wash the cells by centrifugation at 2000*g* for 20 min at 4°C to remove any nitrate or nitrite from the medium.
5. Repeat the washing procedure a further two times.
6. Freeze the cell pellet for between 2–16 h at –20°C to destroy any formate-nitrite reductase activity.
7. Resuspend the cell pellet (approx 2 g) in 10 mL (5 volumes) of ice-cold distilled water and disrupt the cells by sonication for 5 min with cooling (10–20 kilocycles).
8. Centrifuge the sonicate at 2000*g* for 20 min at 4°C to collect undamaged cells.

9. Remove the supernatant and centrifuge at 20000*g* for 40 min to collect the membrane fraction containing the membrane-bound formate-nitrate reductase. Wash the pellet twice by resuspending in distilled water, followed by centrifugation at 20,000*g* for 40 min at 4°C.
10. The cell pellet obtained in **step 8** can be resuspended in distilled water and further sonicated and centrifuged as described in **steps 7–9** to increase the yield if necessary.
11. The final membrane fraction is resuspended in 1 mL of distilled water and dialyzed against 1 L of distilled water at 4°C overnight.
12. Adjust the protein content of the dialyzate to 10 mg/mL. This can be used directly in the nitrate-reductase assay; however, for long-term storage, it is better to lyophilize small aliquots and store at –70°C.

4. Notes

1. Other strains such as Yamagutchi (IFO 12433), strain B, or other *E. coli* strains can be used provided they contain significant amounts of formate-nitrate reductase *(5)*. Do not add the kanamycin solution when using these strains. Normal strains of *E. coli* contain some formate-nitrite reductase. Though most of this is lost during the freezing and extensive washing of the membranes, potential interference by any residual activity can be determined by comparing nitrite-standard curves measured in the presence and absence of the enzyme following incubation as described in **Subheading 3.3.**
2. A number of compounds which may be present in the sample inhibit the diazotization of nitrite and interfere with the Griess reaction. This includes reducing agents such as cysteine, glutathione, ascorbic acid, NADH, and NADPH *(5)*. The presence of interfering compounds in different sample types can be determined by adding known amounts of nitrite into the sample as internal standards.
3. A higher absorbance value can be obtained if the components of the Griess working solution are added separately *(3)*; however, when processing a large number of samples, it is more convenient to add the premixed Griess reagent. If an increased sensitivity is required, then 50 μL of the N-(1-naphthyl)ethylenediamine hydrochloride solution should be added, followed immediately by 50 μL of the sulfanilamide solution to enhance the color development.
4. Primary filters anywhere in the range 530–570 nm can be used for the assay.
5. Samples should be stored at –20°C and analyzed as quickly as possible. If contamination by bacteria or molds are likely to be a problem, the samples should be sterilized by filtration to prevent any possible metabolism of the nitrate or nitrite. Protein precipitation is not normally required in these assays; however, if this procedure is necessary, then the precipitation method should be carefully chosen to prevent any interference with the nitrate-reductase enzymes.
6. Urine samples: The concentration of nitrate plus nitrite in urine samples is typically between 200 μM and 2000 μ*M* and, therefore, urine samples require dilution by a factor of approx 10. Dilute in water for the formate-nitrate reductase

(FNR) assay and assay buffer for the NADPH-dependent nitrate-reductase assay. Dilutions of urine between 0.5-10% (v/v) do not affect the conversion with either assay, although with the NADPH-dependent nitrate reductase assay it may be necessary to prolong the incubation time to achieve maximal conversion.

7. Culture media: A number of tissue-culture media formulations contain high levels of nitrate and, therefore, these should be avoided and replaced by low nitrate containing media. Some types of media may interfere with the color development (though not generally the conversion) and produce slight deviations from linearity in the Griess reaction, particularly if they contain high levels of serum. This can be overcome by preparing standards by dilution of the stock nitrate and nitrite solutions in the media to be used and using a curve-fit program rather than linear regression to calculate the unknown concentrations from the standard curves. With the NADPH-dependent nitrate-reductase assay, it may also be necessary to prolong the incubation time to achieve maximal conversion and the exact time course for the conversion with any particular type of sample should be determined.
8. Plasma samples: Nitrate and nitrite levels can be measured without deproteinization of the plasma; however, it is best to dilute the plasma to 20% with either water for the FNR assay or assay buffer for the NADPH-dependent nitrate-reductase assay. If the plasma contains a large amount of particulate matter, it should be centrifuged or filtered prior to the assay to prevent interference. Although dilute plasma samples do not affect the percentage conversion, plasma does interfere with the quantitation of nitrite in the Griess reaction, with the maximum absorbance being reduced by approx 10%. This can be overcome by comparison of standards prepared in water and diluted serum. If the percentage conversion of nitrate to nitrite is less than 95%, extend the incubation period with the conversion enzyme.
9. The NADPH must be added separately for the reaction to occur *(3)*.
10. If the sensitivity of the assay is a problem, then the initial dilution of the conversion buffer can be omitted and the volume of the sample can be increased up to 75 µL, in which case add 15 µL of the undiluted conversion buffer.
11. When measuring large numbers of samples, the enzyme can be added to the conversion buffer immediately before use, in which case add 50 µL of this mix to each well. Ensure that the enzyme remains in suspension.

References

1. Gross, S. S. and Wolin, M. S. (1995) Nitric oxide: Pathophysiological mechanisms. *Ann. Rev. Physiol.* **57,** 737–769.
2. Wu, G. Y. and Brosnan, J. T. (1992) Macrophages can convert citrulline into arginine. *Biochem. J.* **281,** 45–48.
3. Verdon, C. P., Burton, B. A., and Prior, R. L. (1995) Sample pretreatment with nitrate reductase and glucose-6-phosphate dehydrogenase quantitatively reduces nitrate while avoiding interference by $NADP^+$ when the Griess Reaction is used to assay nitrite. *Anal. Biochem.* **224,** 502–508.

4. Taniguchi, S., Takahashi, K., and Noji, S. (1985) Nitrate, in *Methods of Enzymatic Analysis* (vol. 7) (Bergmeyer, J. and Graβl, M., eds.), VCH Verlagsgesellschaft mbH, Weiheim, Germany, pp. 578–585.
5. Davison, W. and Woof, C. (1978) Comparison of different forms of cadmium as reducing agents for the batch determination of nitrate. *Analyst* **103,** 403–406.
6. Green, L. C., Wagner, D. A., Glogowski, J., Skipper, P. L., Wishnock, J. S., and Tannenbaum, S. R. (1982) Analysis of nitrate, nitrite and [^{15}N]nitrate in biological fluids. *Anal. Biochem.* **126,** 131–138.

10

Determination of NOS Activity Using Cyclic-GMP Formation

Suzanne G. Laychock

1. Introduction

Nitric oxide (NO) produced from L-arginine during the activation of nitric oxide synthase(s) (NOS) in cells has pleiotropic effects and can behave as an intracellular modulator of cellular activity, as well as an intercellular messenger which diffuses from the originator cell to affect other "target" cells in a paracrine manner. NO effects on cells include adenine diphosphate (ADP)-ribosylation and nitrosation/nitration of proteins, and the activation of soluble guanylyl cyclase (GC-S) to form guanosine 3',5'-cyclic monophosphate (cGMP) *(1)*. GC-S contains heme as a prosthetic group, and NO apparently binds to the heme moiety and induces a conformational change in the enzyme which either dis-inhibits or activates the catalytic site *(2)*. NO activation of GC-S has been demonstrated for numerous tissues. NO's ability to activate GC-S and increase cGMP levels has been used to advantage to determine the production and levels of NO in originator cells as well as in target cells employed in the capacity of a bioassay. Whereas the half-life of NO is very short and the gas is difficult to measure during short periods of stimulation, in the presence of a phosphodiesterase inhibitor, the levels of cGMP produced in response to GC-S activation are stable and quantitative. Variations on the assay methods for NO-stimulated cGMP production include incubating the cytosolic fraction of cells with purified GC-S for quantitation of cGMP *(3)*, prelabeling cells with [^{3}H]guanosine for quantitation of [^{3}H]cGMP *(4)*, and adding cells or the supernatant fractions from cells stimulated for NOS activity to NO target cells, such as RFL-6 cells, which are rich in GC-S and respond in a

From: *Methods in Molecular Biology, Vol. 100. Nitric Oxide Protocols*
Edited by: Michael A. Titheradge © Humana Press Inc., Totowa, NJ

sensitive manner to exogenously supplied NO, for quantitation of cGMP production *(5)*.

The quantitation of cGMP commonly employs radioimmunoassay (RIA) methodology *(6,7)*. cGMP in tissue homogenates or released into cell/tissue incubation medium is readily quantitated by RIA. RIA is a sensitive competitive-binding assay which relies upon antibodies specific for cGMP. The acetylation of cyclic nucleotides improves the sensitivity of the RIA and allows for the measurement of low levels of cGMP at femtomole concentrations.

The methodology involved in cyclic nucleotide determination commonly includes the incubation of cells or tissue with a phosphodiesterase inhibitor, such as 3-isobutyl-1-methylxanthine (IBMX), to allow for accumulation of the nucleotides. The addition of acid to the tissue samples followed by several freeze–thaw cycles assists in cell-membrane fracturing and release of the endogenous cyclic nucleotides, though sonication may achieve similar results. cGMP can be separated and purified from other nucleotides using ion-exchange chromatography. The recovered cGMP can be concentrated by lyophilization prior to RIA. If the levels of cGMP are sufficiently high and the RIA antibody highly specific for cGMP, the tissue extracts may be assayed directly by RIA without prior purification, as long as the samples are neutralized prior to acetylation and RIA. Experimental samples of cGMP, and cGMP standards for generation of the standard curve, are acetylated using triethylamine and acetic anhydride just prior to analysis in the RIA. The RIA utilizes [^{125}I]cGMP succinyl methyl ester which does not require acetylation. The percentage of [^{125}I]cGMP bound to the antibody in the RIA is calculated to generate a standard curve for determination of cGMP levels in experimental tissue samples (unknowns).

2. Materials

2.1. Tissue Preparation and Chromatographic Separation

1. Dowex AG 1-X8 (formate form; 200–400 mesh) (Bio-Rad Laboratories, Hercules, CA). Keep refrigerated. An aqueous slurry of Dowex (1:1, v/v), with fines decanted following the initial suspension and wash, is used to load chromatography columns. Used Dowex can be stored, refrigerated, and regenerated for later use.
2. Chromatography columns. Plastic chromatography columns (0.8 × 20 cm) (Kontes, Inc.,Vineland, NJ), with removable tips which can be fitted with a fritted disk or glass wool-plug to retain the Dowex, and a funnel top for sample and eluent addition, are convenient, re-usable, and inexpensive.
3. [^{3}H]cGMP; diluted in water.
4. Hydrochloric acid (HCl; 1 *M*).

2.2. Radioimmunoassay

1. Triethylamine (99%).
2. Acetic anhydride (99+%).
3. 10 × 75 mm borosilicate-glass tubes.
4. Rabbit antibody to thyroglobulin-linked 2'-O-succinyl cGMP (antibody may be purchased commercially or as part of a RIA kit).
5. Guanosine 3',5'-cyclic phosphoric acid, 2'-O-succinyl 3-[^{125}]iodotyrosine methyl ester (2200 Ci/mmol).
6. cGMP (free acid).
7. Sodium acetate buffers: 50 m*M* and 250 m*M*, pH 4.75; 50 m*M*, pH 6.2. Adjust pH of these buffers with 1 *M* acetic acid. Store refrigerated.
8. 0.1% γ-globulin (human, HG-II) (Sigma Chemical Co., St. Louis, MO) prepared in 50 m*M* sodium-acetate buffer, pH 4.75. Store frozen.
9. Human plasma (expired plasma from local blood bank) containing 0.15% ethylenediamine tetraacetic acid, disodium salt (EDTA) (*see* **Note 1**). Store frozen.
10. 12% polyethylene glycol ("Carbowax" PEG 6000) in 50 m*M* sodium-acetate buffer, pH 6.2. Store refrigerated.
11. Refrigerated centrifuge capable of 1500*g*. A capacity of approx 146 tubes (10 × 75 mm).
12. Gamma counter.
13. Digital manual pipettors. Continuously adjustable volume pipettors between 10–100 µL are convenient.

3. Methods

3.1. Tissue Preparation

1. Intact cells (2–5 × 10^6) or tissue can be washed and incubated at 37°C in oxygenated buffered medium containing IBMX (0.2 m*M*) for 10 min. It is convenient to use plastic microfuge tubes (1.5–1.9 mL) for the incubations. Then, centrifuge the cells/tissue at low speed, aspirate, discard the medium, and replace it with fresh medium containing IBMX in the presence or absence of stimuli for NO production in primary cells, or the addition of other NO-producing cell types deficient in GC-S, or treated-cell supernatant fractions if NO levels are to be monitored. Reserve duplicate samples for zero-time controls, and add stopping reagent at time zero (*see* below). Incubate the remaining experimental samples for 5 min, or longer (dependent upon the concentration of NO produced), to obtain measureable levels of cGMP.
2. For determination of total cGMP produced, stop the experimental samples by adding 0.1 *M* HCl (final concentration) (*see* **Note 2**). Total-sample volumes should be kept small, i.e., 100–200 µL. Intracellular levels of cGMP can be determined if the incubation medium is removed prior to the addition of acid to stop the reaction. cGMP released to the medium can be determined by assay of the medium following centrifugation to remove cells/tissue. Acidification will

aid in deproteinizing the sample prior to RIA and may help to reduce blank effects.

3. Prepare buffer only samples as blanks, which are identical to experimental samples, and are incubated and processed identically, except for the absence of cells/tissue.
4. Add $[^3H]$cGMP (20,000 cpm) to several blank samples in order to determine percent recovery of cGMP through the purification process if chromatographic separation methods are used (*see* **Note 3**). Reserve an aliquot of $[^3H]$cGMP for determination of total cpm added to each sample.
5. Freeze samples at –20°C or lower for storage.
6. Freeze–thaw samples three times and/or sonicate to disrupt tissue, and centrifuge (14,000 rpm for 5 min in a microfuge) to precipitate acid-insoluble protein and particulates. The protein in the precipitate is quantitated for normalization of cGMP levels among different samples. The supernatants are used for direct quantitation of cGMP by RIA, or for chromatographic purification of cyclic nucleotides (*see* **Note 4**).

3.2. Column Separation of Cyclic Nucleotides

1. Add 1.5 mL of Dowex (in a slurry of 1:1 deionized (d) H_2O) to plastic disposable columns fitted with fritted disks or with a plug of glass wool to retain the resin. The amount of Dowex required will depend on the levels of cyclic nucleotides in the samples. All columns should have equal volumes of Dowex. Air must be removed from the disks or plugs to ensure a rapid and constant rate of flow through the columns; run 1 mL of dH_2O through the columns before adding the Dowex.
2. Wash the Dowex onto the column bed with 1–5 mL of dH_2O, and avoid air bubbles in the resin bed. Allow water to run through the columns, but do not allow the columns to dry prior to adding experimental samples.
3. Add experimental samples gently to the surface of the prepared resin bed to avoid turbulence.
4. Immediately after each sample has run into the resin bed, add 4 mL of dH_2O and discard the eluate. This step will aid in removing any residual protein from the column.
5. cAMP will be eluted with 10 mL of 2 *M* formic acid, and this eluate can be discarded.
6. cGMP will be eluted with 10 mL of 5 *M* formic acid, and this fraction is collected in plastic vials and frozen at –70°C.
7. The frozen cGMP-containing eluates are then lyophilized to dryness and resuspended in dH_2O. An aliquot is removed for determination of $[^3H]$cGMP recovery by liquid-scintillation counting. Compare the amount of $[^3H]$cGMP recovered in the total volume of eluted sample, with the total cpm initially added to each sample. cGMP recovery can be expected to average about 75%. If recovery is low, the cGMP may be retained on the column (acid-washed Dowex will retain cyclic nucleotides longer than dH_2O washed Dowex), and 12 mL of 5 *M* formic acid may be necessary to elute the cGMP.

Table 1
Typical Standard Curve for cGMP Radioimmunoassay[a]

Tube #	cGMP standards/ unknowns (50 µL)	$[^{125}I]$cGMP (c.p.m.)	cGMP antibody	avg. cpm bound	% Bound	unknown cGMP (fmol)
1,2	(acetylated dH_2O)	10,000	—	100	(nonspecific binding)	
3,4	(acetylated dH_2O)	10,000	20 µL	4000	[total binding (100%)]	
5,6	1.25 fmol	10,000	20 µL	3500	87%	
7,8	5 fmol	10,000	20 µL	2440	60%	
9,10	10 fmol	10,000	20 µL	1933	47%	
11,12	20 fmol	10,000	20 µL	1465	35%	
13,14	50 fmol	10,000	20 µL	958	22%	
15,16	100 fmol	10,000	20 µL	646	14%	
17,18	200 fmol	10,000	20 µL	451	9%	
19,20	Blank sample	10,000	20 µL	3300	82%	1.65
21,22	unknown # 1	10,000	20 µL	980	22.5%	47
23,24	unknown # 2	10,000	20 µL	1950	47%	10

[a]Buffer contents of the tubes are not listed. All tubes contain equal final volumes.

3.3. Radioimmunoassay

1. cGMP (free acid) for standard preparation should be diluted to final concentrations of 1×10^{-9} *M* and 4×10^{-9} *M* for use in preparing the standard-curve samples for RIA (*see* **Note 5**). cGMP standards can be stored at –20°C.
2. Experimental samples should be diluted with dH_2O such that the estimated concentration of cGMP is between 10–100 fmol per 50 µL.
3. Acetylate in 10 × 75 mm borosilicate-glass tubes:
   a. Between 0.25–0.5 mL aliquots of unknowns;
   b. 0.025–4 n*M* cGMP standards in a volume equal to that of the unknowns;
   c. Equal volumes of water for use in volume equalization in RIA-assay tubes lacking standards or unknowns. Acetylation is carried out by the addition of 10 µL of triethylamine and 5 µL of acetic anhydride in rapid succession, followed by vigorous vortexing, at room temperature (*see* **Note 4**). The acetylated samples may be assayed immediately, or they may be frozen for 10 min at –70°C and thawed to improve the reproducibility of pipeting; these samples may also be stored frozen until the assay has been analyzed.
4. Label 10 × 75-mm glass tubes for the number of samples in the RIA, and immerse at least one inch of the tubes in an ice-water bath. A high-capacity polypropylene tube rack is convenient for this purpose.
5. Into the numbered tubes, pipet 50 µL in duplicate for each acetylated-cGMP standard. The most accurate range being between 10–100 fmol cGMP per 50 µL. Also, pipet up to 50 µL of each experimental (unknowns) and blank sample into tubes, in duplicate (*see* **Table 1**). If less than 50 µL of sample is assayed, the tube

volumes should be made equal by addition of acetylated water. All tubes must contain equal, total-final volumes.

6. Include duplicate samples of 50 μL of acetylated water in numbered tubes designated for "nonspecific binding" (tubes 1 and 2), and in duplicate tubes for "total binding" (tubes 3 and 4) (*see* **Table 1**).
7. Add 10 μL of 250 m*M* sodium acetate buffer, pH 4.75, to each assay tube. This may be combined with the [^{125}I]cGMP-addition step (*see* **step 8**) in order to reduce pipeting effort.
8. Dilute [^{125}I]cGMP with 50 m*M* sodium acetate buffer, pH 4.75, to approx 10,000 cpm per 40 μL. Add 40 μL of the diluted radioactivity to each tube.
9. To make a working solution of antibody, dilute the antibody with 0.1% γ-globulin in 50 m*M* sodium acetate, pH 4.75, such that the total binding in the assay will be 20–40% of the amount of [^{125}I]cGMP added per tube. Invert the tubes to mix the diluted antibody; vigorous vortexing may adversely affect antibody-binding activity. Dilutions will depend upon the initial titer of the antibody obtained. Add 20 μL of the antibody-working solution to all tubes except the nonspecific-binding tubes. To the nonspecific-binding tubes, add 20 μL of the 0.1% γ-globulin solution. Vortex each tube gently immediately following antibody addition (avoid foaming).
10. Store the tubes at 4°C overnight, covered with Parafilm.
11. Keeping the tubes in an ice-water bath, add 20 μL of cold human plasma containing 0.15% EDTA to each tube, followed immediately by 1 mL of ice cold 12% PEG in 50 m*M* sodium acetate buffer, pH 6.2, and vortex each tube immediately after the additions to achieve thorough mixing.
12. Allow the tubes to remain immersed in the ice bath for 1 h.
13. Centrifuge the samples at 5°C for 30 min at 2000*g*.
14. A Pasteur pipet attached to a vacuum aspirator is used to aspirate and discard the supernatant fractions without disturbing the pellets. The supernatant contains the unbound [125^{125}I]cGMP and should be collected in a trap bottle for disposal as radioactive waste.
15. An additional 0.5 mL of the cold 12% PEG is added down the sides of the tubes without disturbing the pellets. The centrifugation step and supernatant aspiration steps are repeated.
16. The radioactivity in the pellet in each tube is counted in a gamma counter.

3.4. Calculations

1. Subtract the averaged-nonspecific binding (tubes 1 and 2) (which should be less than 500 cpm) from all other samples (*see* **Table 1**).
2. Prepare a standard curve of the percent [^{125}I]cGMP bound by dividing the radioactivity in each standard curve tube by the "total bound" radioactivity (average of tubes 3 and 4) (*see* **Table 1**). The percentages bound are plotted against the cGMP-standard concentrations on semilogarithmic graph paper.
3. Determine the amount of cGMP in each experimental (unknown) sample and the blank samples by determining the percent bound, and reading the fmol quantity

of cGMP corresponding with the percent-bound value on the standard curve (*see* **Table 1**).

4. Correct the samples for percent recovery of [^{3}H]cGMP.
5. Subtract blank values or zero-time values from other experimental samples.
6. Correct the fmol cGMP values to the total volume of the resuspension following lyophilization to determine the total amount of cGMP in the original cell/tissue sample.

4. Notes

1. Protective gloves should be worn when handling blood products. If particulate matter is observed in the thawed plasma, filter the plasma through two layers of wetted gauze prior to addition to the RIA.
2. Trichloroacetic acid (TCA; 10% final concentration) or perchloric acid (0.5 *M*) *(7)* can also be used to stop the cell/tissue incubation. The acidified samples can be applied to Dowex columns. If the TCA-treated samples are to be used without chromatographic purification, water-saturated ether is used to wash the samples (three ether washes with the addition of four times the volume of the experimental sample, and vigorous vortexing) to remove the acid and neutralize the samples. The ether (top) layer is removed following each wash by vacuum aspiration in a ventilated hood.
3. [^{3}H]cGMP can be added to each sample to determine recovery as long as the mass of the radiolabeled cGMP is negligible compared to the tissue levels of cGMP being measured in the RIA.
4. Chromatographic separation of cyclic nucleotides may not be required if the concentration of cGMP in the samples is high and cAMP does not compete significantly in the binding assay. In this case, acidify incubation medium/cells and assay directly. If the sample is acid (i.e., 0.1 *M* HCl), add additional triethylamine to neutralize the pH (i.e., 3:1 with acetic anhydride). Alternatively, an aliquot may be neutralized with 250 m*M* sodium acetate prior to acetylation.
5. cGMP (free acid) is solubilized in a small volume of dilute NaOH and suspended in dH_2O to an approximate concentration of 4 m*M*. The final concentration should be determined from the molar-extinction coefficient (ε 11350, pH 7.0) by spectrophotometric absorption at 256.5 nm.

References

1. Kerwin, Jr., J. F., Lancaster, J. R., and Feldman, P. L. (1995) Nitric oxide: a new paradigm for second messengers. *J. Med. Chem.* **38,** 4342–4362.
2. Schmidt, H. H. H. W., Lohmann, S. M., and Walter, U. (1993) The nitric oxide and cGMP signal transduction system: regulation and mechanism of action. *Biochim. Biophys. Acta* **1178,** 153–175.
3. Mulsch, A. and Busse, R. (1991) Nitric oxide synthase in native and cultured endothelial cells—calcium/calmodulin and tetrahydrobiopterin are cofactors. *J. Cardiovasc. Pharmacol.* **17,** S52–S56.

4. Pou, S., Pou, W. S., Rosen, G. M., and El-Fakahany, E. E. (1991) N-Hydroxylamine is not an intermediate in the conversion of L-arginine to an activator of soluble guanylate cyclase in neuroblastoma N1E-115 cells. *Biochem. J.* **273,** 547–552.
5. Ishii, K., Chang, B., Kerwin, Jr., J. F., Huang, Z.-J., and Murad, F. (1990) N-Omega-nitro-L-arginine: a potent inhibitor of endothelium-derived relaxing factor formation. *Eur. J. Pharmacol.* **176,** 219–223.
6. Steiner, A. L., Kipnis, D. M., Utiger, R., and Parker, C. (1969) Radioimmunoassay for the measurement of adenosine 3',5'-cyclic phosphate. *Proc. Natl. Acad. Sci. USA* **64,** 367–373.
7. Domino, S. E., Tubb, D. J., and Garbers, D. L. (1991) Assay of guanylyl cyclase catalytic activity. *Methods Enzymol.* **195,** 345–355.

Barry W. Allen

11

Determination of NO with a Clark-Type Electrode

Kurt Schmidt and Bernd Mayer

1. Introduction

Reliable methods for the specific detection and quantitative determination of nitric oxide (NO) release from tissues are a prerequisite for a better understanding of the complex biological functions of this widespread cellular messenger. Although several techniques are available for the sensitive detection of NO, most of these methods either involve sophisticated and expensive equipment (e.g., electron paramagnetic resonance [EPR] spectroscopy and chemiluminescence) or lack the required specificity (UV/VIS spectroscopy). In the past few years, electrochemical methods have been developed that are highly selective for NO and exhibit a sensitivity in the nanomolar range. Within electrochemical sensors, Clark-type electrodes are most widely used, as they are commercially available and easy to handle. The principle of these sensors is that NO diffuses through a gas-permeable membrane and a thin film of electrolyte, followed by oxidation on the working electrode. This oxidation creates a current that is proportional to the concentration of NO outside the membrane. According to the manufacturer, the electrode is insensitive to O_2, N_2, CO, and CO_2 but not to NO_2. However, this interference may be only a problem for gas-phase measurements, as in solution NO_2 is highly unstable and quickly degrades to nitrite and nitrate.

According to our experience, the detection limit of the electrode is ~20 n*M* with a linear response up to 10 μ*M*. The sensitivity thresholds described for detection of NO by chemiluminescence or EPR are about 10-fold lower *(1)*, but both methods require sophisticated and expensive equipment. Clark-type electrodes are also apparently less sensitive than porphyrinic-based NO sensors *(2)*, but the latter are not commercially available. A commonly used method for detection of NO is the oxyhemoglobin assay. It is based on the NO-induced conversion of oxyhemoglobin to methemoglobin which can

From: *Methods in Molecular Biology, Vol. 100. Nitric Oxide Protocols*
Edited by: Michael A. Titheradge © Humana Press Inc., Totowa, NJ

be monitored photometrically at 401 nm *(3)* or 577 nm *(4)*. The sensitivity of this method is comparable to that achieved with Clark-type sensors, but the assay lacks the required specificity. Conversion of oxyhemoglobin to methemoglobin is also mediated by superoxide *(5)* and peroxynitrite induces spectral changes identical with those elicited by NO *(6)*. None of these species, however, interfere with the electrochemical detection of NO *(7)*.

Although Clark-type NO sensors offer many advantages over other NO-detection systems, the usefulness of this electrode may be limited by its rather long response time (~2–3 s). For comparison, porphyrinic-based NO sensors *(8)* have been reported to response within 10 ms, and the reaction of NO with oxyhemoglobin is also very fast *(9)*. This may be of concern if transient changes of NO in a flow system are to be measured or rapid reactions of NO with other molecules are to be studied. In spite of this limitation, Clark-type NO sensors provide a unique tool for selective and sensitive detection of NO in many experimental situations.

In the present chapter, we report on our 3-yr experience with a Clark-type NO sensor. A detailed description of our methodological approach is given and selected data are presented that demonstrate the usefulness of the sensor.

2. Materials

1. Calibration solution: 0.1 *M* KI in 0.1 *M* H_2SO_4. Prepare fresh each day just prior to use.
2. Nitrite-standard solutions: 0.1 or 1 m*M* KNO_2 in distilled water. Prepare fresh each day.
3. DEA/NO stock solution: 2 m*M* DEA/NO (2,2-diethyl-1-nitroso-oxyhydrazine sodium salt, NCI Chemical Carcinogen Repository, Kansas City, MO) in 0.01 *M* NaOH. Prepare fresh each day (*see* **Note 1**).
4. SIN-1 stock solution: 100 m*M* SIN-1 (3-(4-morpholinyl)-sydnonimine-hydrochloride, Hoechst Marion Roussel, Frankfurt am Main, Germany) in 50 m*M* sodium phosphate buffer, pH 5.0. Prepare fresh each day (*see* **Note 2**).
5. Incubation buffer A (for measuring NO release from DEA/NO): 100 m*M* sodium phosphate buffer, pH 7.4
6. Incubation buffer B (for measuring NO release from SIN-1): 100 m*M* sodium phosphate buffer, pH 8.0.
7. Incubation buffer C [for measuring NO release from purified neuronal NO synthase (NOS)]: 50 m*M* triethanolamine/HCl buffer, pH 7.0, containing 0.5 m*M* $CaCl_2$, 1 m*M* L-arginine, 0.3 m*M* nicotinamide adenine dinucleotide phosphate (NADPH), 5 μ*M* flavin adenine dinucleotide (FAD), and 5 μ*M* flavin mononucleotide (FMN).
8. Clark-type electrode (Iso-NO, World Precision Instruments, Berlin, Germany) connected to an Apple Macintosh computer via an analog to digital converter (MacLab, ADInstruments Ldt, Hastings, UK). For data acquisition, the Chart™

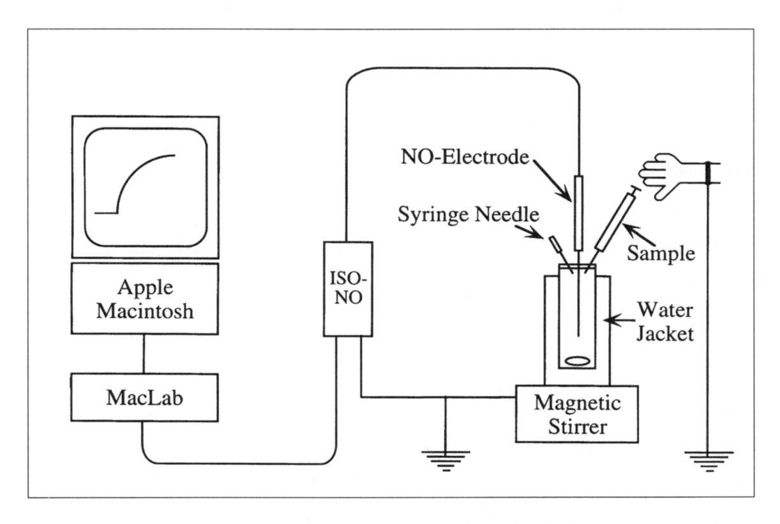

Fig. 1. Experimental setup for electrochemical detection of NO.

MacLab application program (Version 3.4., ADInstruments Ldt., Hastings, UK) was used.

3. Methods

3.1. Experimental Setup

The following procedure describes the setup and handling of the electrode (a scheme of the experimental setup is shown in **Fig. 1**):

1. Connect the NO meter to the data-acquisition system.
2. Place a water jacket on top of magnetic stirrer and make a grounding connection between the stainless-steel plate of the stirrer and the NO meter (*see* **Note 3**).
3. Insert a small stirring bar into the sample vial (1.8 mL) and fill it with calibration solution or incubation buffer. Close the vial with a rubber cap, insert it into the water jacket, and fix it with a clamp. Ensure that the solution is stirring (*see* **Note 4**).
4. Insert a syringe needle through the rubber cap. This needle serves as outlet when the sensor is inserted into the vial.
5. Perforate the rubber cap with a needle and insert the tip of the sensor through this hole into the vial (*see* **Note 5**).
6. Wait until the output current becomes stable and adjust baseline to zero.
7. Slowly inject solutions to be tested (usually 1–20 μL) through the cap into the vial.
8. Record output current by collecting data at sampling rates of 0.3–10 Hz (depending on the resolution required).

After each experiment, immerse the tip of the sensor for at least 5 min into distilled water to clean the membrane (*see* **Note 6**). When not in use, the sensor should be also immersed into distilled water.

3.2. Calibration Procedure

Calibration of the electrode is performed by chemical generation of NO. This procedure is based on the reaction of nitrite with iodide in sulfuric acid (*see* **Note 7**).

$$2\ KNO_2 + 2\ KI + 2\ H_2SO_4 \rightarrow 2\ NO + I_2 + 2\ H_2O + 2\ K_2SO_4.$$

The calibration procedure is as follows:

1. Deoxygenate the calibration solution by bubbling for 1 min with helium or argon and fill the sample vial with the gassed solution (1.8 mL).
2. Close the vial with a rubber cap and insert the sensor as described in **Subheading 3.1.**
3. Select the setting "20.00 nA" on the NO meter. Wait until the background current becomes stable and adjust the baseline to zero.
4. Inject slowly 7.2 µL of a 0.1 m*M* solution of KNO_2 into the vial. This will produce an NO concentration of 0.4 µ*M* and should result in a current of ~0.1 nA (*see* **Note 8**). Wait until the current becomes stable and inject another 7.2 µL of the standard solution. Repeat this procedure three times. The resulting curve should be similar to that shown in **Fig. 2A**.
5. Construct a calibration curve (**Fig. 2B**) and calculate the calibration factor (= sensitivity) from the slope of the linear-curve fit. This value can be then used as conversion factor in the Chart™ application program (*see* **Note 9**).

Alternatively, the electrode can be calibrated by using the slope function of the NO meter. This method has the advantage that NO concentrations can be directly read from display of the NO meter, but the disadvantage that only an one-point calibration can be performed. The procedure is as follows:

1. Perform **steps 1** and **2** as previously described.
2. Select the setting "20.00 µ*M*" on the NO meter. Wait until the background current becomes stable and adjust baseline to zero.
3. Inject slowly 3.6 µL of a 1 m*M* solution of KNO_2 into the vial. This will yield on NO concentration of 2 µ*M*. As soon as the reading on the display becomes stable, adjust the slope so that the correct concentration (2.00 µ*M*) is displayed on the meter.
4. Verify the calibration by injection of another 3.6 µL of the standard solution. The reading on the display should now be in the range of 3.8–4.2 µ*M* (*see* **Note 10**). If the values for theoretical and measured NO concentration differ by more than 5%, repeat the calibration procedure.

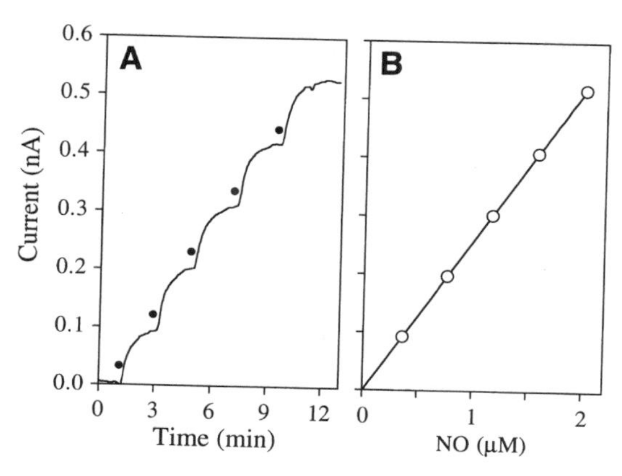

Fig. 2. Calibration of the NO electrode. **(A)** At the indicated time points (•), 7.2 μL of a 0.1 m*M* solution of KNO_2 were added to 1.8 mL of a helium gassed solution of 0.1 *M* KI in 0.1 *M* H_2SO_4 (25°C) yielding NO concentrations of 0.4, 0.8, 1.2, 1.6, and 2.0 μ*M*. **(B)** Correlation between NO concentration and output current of the electrode (3.85 nM/pA).

After calibration of the electrode, use the protocol described in **Subheading 3.1.** to monitor release of NO from donor compounds (*see* **Note 11**), formation of NO by NOS (*see* **Note 12**) or release of NO from intact cells (*see* **Note 13**).

4. Notes

1. NONOates are stable under alkaline conditions but decomposes rapidly at physiological pH *(10)*. We, therefore, prepared a stock solution of DEA/NO in 0.01 *M* NaOH and monitored NO release at pH 7.4 (Incubation buffer A).
2. SIN-1 is stable under acidic conditions and decomposes at alkaline pH *(11)*. Thus, a stock solution of SIN-1 was prepared in 50 m*M* phosphate buffer, pH 5.0, and NO release was monitored at pH 8.0 (Incubation buffer B).
3. Because the obtained current is usually in the pA range, successful measurements may critically depend on efficient grounding to protect the sensor from electrical interferences. According to our experience, grounding of stirrer and NO meter is usually sufficient when NO concentrations > 100 n*M* are to be measured. However, grounding of the operator is recommended for maximal sensitivity of the method.
4. For reliable measurements, stirring of the samples is essential. As shown in **Fig. 3**, addition of nitrite (final concentration 1 μ*M*) to a stirred calibration solution resulted in a current of ~0.26 nA, but the value declined to ~0.14 nA when the stirrer was switched off. Upon restirring of the sample, the original output current was obtained again. The reason for this phenomenon is unclear but electrical interferences can be excluded as basal currents were identical in stirred and nonstirred samples.

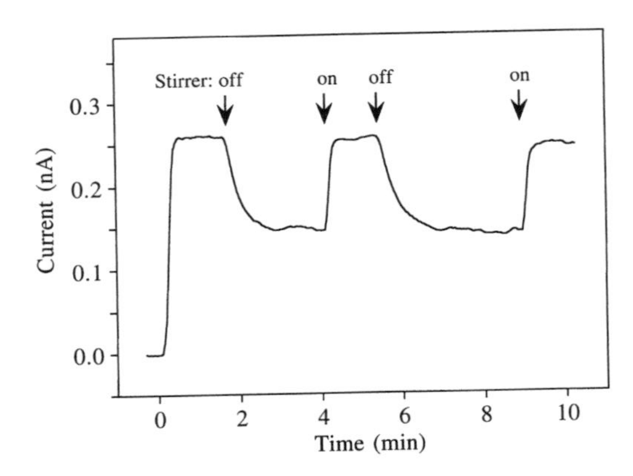

Fig. 3. Effect of stirring on output current. At time point zero, 1.8 μL of a 1 m*M* solution of KNO_2 was added to 1.8 mL of a helium gassed solution of 0.1 *M* KI in 0.1 *M* H_2SO_4. Stirring of the solution was stopped and started as indicated.

5. Be very careful when handling the electrode to avoid damage of the membrane covering the tip of the electrode. This membrane prevents the working electrode from ions, which could be oxidized and produce a current. The integrity of the membrane can be easily checked by ascertaining that the current remains low and stable when the tip of the sensor is immersed in a strong saline solution.
6. Aside from avoiding damage, a major concern is to keep the membrane clean. Immersing the tip in distilled water after each experiment should be sufficient, but soaking the membrane in base solution may be necessary if experiments have been performed in protein-rich solutions. Although unusually low sensitivity or slow response of the sensor are good indications for a dirty membrane, they may also result from a no-longer-functioning electrolyte. According to our experience, refilling of the sensor with new electrolyte becomes necessary within 3–4 wk. Although it is recommended by the manufacturer to replace the (expensive) membrane sleeve each time it is removed from the electrode, we observed that the sleeves can be reused several times without affecting sensitivity or selectivity of the sensor.
7. Quantitative reduction of NO_2^- to NO under acidic conditions allows a reliable calibration of the electrode as demonstrated by the excellent agreement with the values predicted by kinetic modeling (*see* Chapter 27, this volume). Alternatively, calibration can be performed with aqueous-NO standards prepared from a saturated-NO solution. Because preparation of accurate NO standards is time-consuming and requires meticulous exclusion of O_2, calibration of the electrode with nitrate may be, at least from a practical point of view, the better alternative. The use of NO donors for calibrating the electrode is not recommended, because decomposition rate and stoichiometry of NO release may vary depending on the incubation conditions.

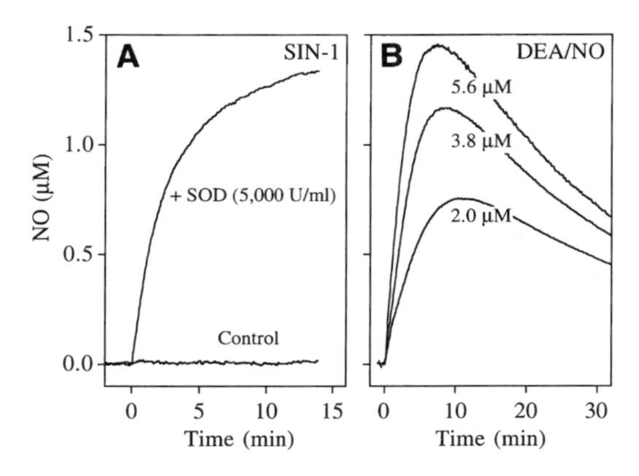

Fig. 4. Detection of NO generated by donor compounds. **(A)** At time point zero, 18 μL of a 100 m*M* solution of SIN-1 in 50 m*M* phosphate buffer (pH 5.0) was added to 1.8 mL of Incubation buffer B (25°C) yielding an initial donor concentration of 1 m*M*. Experiments were performed in the absence (control) and presence of 5,000 U/mL SOD. **(B)** At time point zero, 1.8, 3.4, or 5 μL of a 2 mM solution of DEA/NO in 0.01 NaOH were added to 1.8 mL of Incubation buffer A (25°C) yielding initial donor concentrations of 2, 3.8, and 5.6 μ*M*, respectively.

8. Note that the output signal from the NO meter is 0.1 mV/pA. Thus, a current of 0.1 nA will create a voltage of 10 mV. If the NO meter is used in the μ*M* setting, the output signal is 0.1 V/μ*M*.
9. Although the sensitivity of the sensor usually remains unchanged within a working session, day to day variation are in the range of 0.5 nM/pA, making it necessary to calibrate the electrode daily. Furthermore, we observed that the sensitivity of the sensor varies with temperature (~2 and ~4 nM/pA at 25° and 37°C, respectively), demonstrating that recalibration is necessary when the temperature is changed.
10. According to our experience, an error of ±5% is within the experimental uncertainty of this one-point calibration. If greater accuracy is desired, a calibration curve should be constructed.
11. Note that the decomposition of certain NO donors (e.g., SIN-1 or other sydnonimines) results in the simultaneously release of NO and superoxide which both rapidly combine to peroxynitrite. As a consequence, release of NO from these compounds is not detectable unless sufficient SOD is present to outcompete the peroxynitrite reaction (*see* **Fig. 4A**). In contrast to SIN-1, NONOates (such as DEA/NO) do not produce superoxide and thus release of NO can be monitored in the absence of SOD (*see* **Fig. 4B**).
12. Detection of NO produced by neuronal (n) NOS is also critically depending on the presence of SOD. As shown in **Fig. 5A**, no detectable signal of the electrode is obtained by incubation of NOS with L-arginine and cofactors in the absence or presence of exogenously added tetrahydrobiopterin (BH_4). However, subsequent

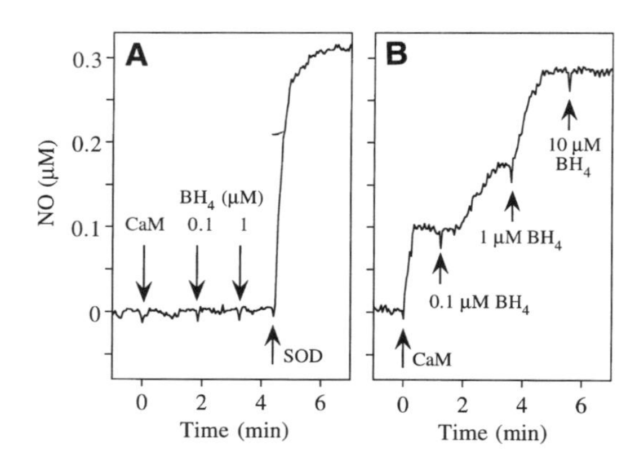

Fig. 5. Detection of enzymatically formed NO. Experiments were performed in the absence **(A)** and presence of 5000 U/mL SOD **(B)** at ambient temperature. Vials were filled with 50 m*M* triethanolamine/HCl buffer, pH 7.0, containing 0.5 m*M* $CaCl_2$, 1 m*M* L-arginine, 0.3 m*M* NADPH, 5 μ*M* FAD, 5 μ*M* FMN, and 5 μg of purified brain NOS. Where indicated, CaM (10 μg/mL), BH_4 (0.1 to 10 μ*M*), or SOD (5000 U/mL) were added (final concentrations in parenthesis).

addition of 5000 U/mL of SOD results in the generation of NO up to a steady-state concentration of ~0.3 μ*M* after about 1 min. In the presence of SOD, addition of calmodulin (CaM) already produces a signal corresponding to ~0.1 μ*M* NO, and this is further increased to ~0.3 μ*M* in a concentration-dependent manner by added BH_4 (*see* **Fig. 5B**). We have previously demonstrated that purified brain NOS, not saturated with BH_4, generates hydrogen peroxide ***(12)***, a reaction that involves production of superoxide as intermediate ***(13)***. Furthermore, we found that BH_4 induces oxidation of NO ***(7)*** apparently owing to superoxide generation in the course of BH_4 autoxidation ***(14)***. Thus, at low concentrations of BH_4, NOS generates superoxide owing to uncoupled oxygen activation, whereas saturating concentrations of the pteridine, which prevent the enzymatic-oxygen reduction, produce superoxide nonenzymatically. In each case, the amounts of generated superoxide are apparently sufficient for complete inactivation of NO.

13. We have made several attempts to monitor NO release from cultured aortic-endothelial cells but independent of the incubation conditions (±SOD, ±L-arginine, ±bradykinin, ±Ca-ionophore) or the experimental setup (putting cells directly into the incubation vial, putting microbeads with attached cells into the incubation vial, or putting the electrode as closely as possible to a monolayer of cells growing in a Petrie dish) we were not able to detect an NO signal so far. Interestingly, NOS activity in these cells was detectable by other methods (conversion of ^{3}H-arginine to ^{3}H-citrulline, conversion of oxyhemoglobin to methemoglobin, increase in endothelial cGMP production) indicating that the amounts of NO released were below the detection limit of the electrode.

References

1. Archer, S. (1993) Measurement of nitroc oxide in biological models. *FASEB J.* **7,** 349–360.
2. Malinski, T. and Taha, Z. (1992) Nitric oxide release from a single cell measured in situ by a porphyrinic-based microsensor. *Nature (London)* **358,** 676–678.
3. Feelisch, M. and Noack, E. A. (1987) Correlation between nitric oxide formation during degradation of organic nitrates and activation of guanylate cyclase. *Eur. J. Pharmacol.* **139,** 19–30.
4. Murphy, M. E., Piper, H. M., Watanabe, H., and Sies, H. (1991) Nitric oxide production by cultured aortic endothelial cells in response to thiol depletion and replenishment. *J. Biol. Chem.* **266,** 19,378–19,383.
5. Winterbourn, C. C., McGrath, B. M., and Carrell, R. W. (1976) Reactions involving superoxide and normal and unstable haemoglobins. *Biochem. J.* **155,** 493–502.
6. Schmidt, K., Klatt, P., and Mayer, B. (1994) Reaction of peroxynitrite with oxyhaemoglobin: Interference with photometrical determination of nitric oxide. *Biochem. J.* **301,** 645–647.
7. Mayer, B., Klatt, P., Werner, E. R., and Schmidt, K. (1995) Kinetics and mechanism of tetrahydrobiopterin-induced oxidation of nitric oxide. *J. Biol. Chem.* **270,** 655–659.
8. Kiechle, F. L. and Malinski, T. (1993) Nitric oxide—biochemistry, pathophysiology, and detection. *Am. J. Clin. Pathol.* **100,** 567–575.
9. Kelm, M., Feelisch, M., Spahr, R., Piper, H. M., Noack, E., and Schrader, J. (1988) Quantitative and kinetic characterization of nitroc oxide and EDRF released from cultured endothelial cells. *Biochem. Biophys. Res. Cummun.* **154,** 236–244.
10. Maragos, C. M., Morley, D., Wink, D. A., Dunams, T. M., Saavedra, J. E., Hoffman, A., Bove, A. A., Isaac, L., Hrabie, J. A., and Keefer, L. K. (1991) Complexes of NO with nucleophiles as agents for the controlled biological release of nitric oxide. Vasorelaxant effects. *J. Med. Chem.* **34,** 3242–3247.
11. Noack, E. and Feelisch, M. (1989) Molecular aspects underlying the vasodilator action of molsidomine. *J. Carciovasc. Pharmacol.* **14 (Suppl. 11),** S1-S5.
12. Heinzel, B., John, M., Klatt, P., Böhme, E., and Mayer, B. (1992) Ca^{2+}/calmodulin-dependent formation of hydrogen peroxide by brain nitric oxide synthase. *Biochem. J.* **281,** 627–630.
13. Pou, S., Pou, W. S., Bredt, D. S., Snyder, S. H., and Rosen, G. M. (1992) Generation of superoxide by purified brain nitric oxide synthase. *J. Biol. Chem.* **267,** 24,173–24,176.
14. Davis, M. D. and Kaufman, S. (1993) Products of the tyrosine-dependent oxidation of tetrahydrobiopterin by rat liver phenylalanine hydroxylase. *Arch. Biochem. Biophys.* **304,** 9–16.

12

The Measurement of NO in Biological Systems Using Chemiluminescence

Evangelos D. Michelakis and Stephen L. Archer

1. Introduction

With an extremely short half life and susceptibility to postsynthetic oxidation, it has been a challenge to measure nitric oxide (NO). This chapter discusses the chemiluminescence assay for NO and its intermediate and end products of oxidation—NO_2^- and NO_3^-. This is one of the most simple, sensitive, and accurate assays available. It will emphasize several practical aspects of the assay and give examples of recent applications, such as the study of breath NO.

1.1. History

Processes in which excited state molecules release light are termed luminescent and can be produced from a chemical (chemiluminescence) or a biological source (bioluminescence) *(1)*. The kinetics of the gas-phase reaction of NO with O_3 were studied in 1954 by Johnston and Crosby *(2)*, but it wasn't until 1970 that Fontijn et al. used the amount of light emitted by this reaction to calculate the amount of NO present, developing a chemiluminescence assay *(3)* (**Table 1**). In order to measure NO-oxidation products (NO_2^-, NO_3^-), however, these have to be reduced to NO before the assay, because they are not volatile and they don't react with O_3. In 1980, Cox first measured NO_2^- and NO_3^- in aqueous solutions using potassium iodide (KI) and ammonium sulfate and molybdate as reducing agents *(4)*. Braman later used vanadium (iii) as a stronger reducing agent, enabling the measurement of NO from NO_2^- and NO_3^- in the same sample *(5)*. Zafiriou and McFarland modified the assay in 1980, essentially to its current form *(6)*. In 1987, Ignarro et al. *(7)* and Palmer et al.

From: *Methods in Molecular Biology, Vol. 100. Nitric Oxide Protocols*
Edited by: Michael A. Titheradge © Humana Press Inc., Totowa, NJ

Table 1
History of Development of the NO Chemiluminescence Assay (EDRF: Endothelium Derived Relaxing Factor)

Systematic description of the kinetics of the $NO + O_3$ reaction	1954, **ref.** ***2***
The chemiluminescence reaction between NO and O_3 is described.	1964, **ref.** ***15***
Use of the chemiluminescence reaction with O_3 to measure total NO in atmospheric pollutants	1970, **ref.** ***3***
The chemiluminescence assay takes its current form and its specificity and sensitivity studied systematically.	1980, **ref.** ***6***
NO oxides are measured in liquids. KI is used as the reducing agent for NO_2^- and ferrous ammonium sulfate and molybdate for NO_3^-.	1980, **ref.** ***4***
Vanadium is described as a more powerful reducing agent for the study of NO oxides in one step.	1987, **ref.** ***5***
Using chemiluminescence it is shown for the first time that EDRF is NO.	1987, **ref.** ***7*** 1987, **ref.** ***8***

(8) used the chemiluminescence assay to show that endothelium derived relaxing factor (EDRF) was NO, and this stimulated NO research.

1.2. NO Chemistry

NO is a radical and, although not particularly unstable in low concentrations and in the absence of O_2 *(9)*, it is very unstable in biological models, primarily owing to the presence of O_2 and O_2 radicals. Tanks of 10 ppm NO have remained stable for mo although theoretically NO can very slowly self-react in a thermodynamic-disproportionation reaction ($4NO \rightarrow N_2O_3 + N_2O$) *(10)*.

NO tends to exist as a gas and is poorly soluble in water. It has a very high-partition coefficient of >20 (molecules per unit volume gas/molecules per unit volume solution) and does not undergo any kind of hydration reaction. This hydrophobic property is very important for the physiologic functions of NO and permits the small molecule to pass freely across cell boundaries. The fate of NO, in the presence of O_2, depends on the phase in which the reaction occurs (gas or liquid, with or without the presence of hemoproteins).

In the gas phase, for example, NO reacts with O_2 to give NO_2^- and N_2O_3 *(11)*:

$$2NO + O_2 \rightarrow 2NO_2^-, NO + NO_2^- \rightarrow N_2O_3.$$

The autoxidation of NO in the liquid phase (in the absence of hemoproteins) has recently been studied in detail *(12)*. The product was found to be 100% NO_2^- with NO_3^- below the limit of detectability, corresponding to the stoichiometry of the equation:

$$4NO + O_2 + 2H_2O \rightarrow 4HNO_2 \rightarrow 4NO_2^- + 4H^+$$

In biological systems, the fate of NO is altered by the presence of oxidizing agents such as oxyhemoproteins and superoxide anion, which favor the production of NO_3^- *(12)* and peroxynitrite ($ONOO^-$) *(13,14)*, respectively. Furthermore, NO_3^-, which is the dominant species in vivo, is also formed from NO_2^- in the presence of strong oxidizing agents.

1.3. The Ozone–NO Reaction

The NO-chemiluminescence assay is based on the fact that some of the NO_2 produced by the reaction of NO with O_3 is in the excited state *(15)*: NO + O3 → $NO_2^* + O_2 \rightarrow NO_2 + hv$. As the NO_2^*'s unstable electrons return to their original ground state, they dissipate energy (hv). This energy can either be used in reactions between NO_2 and other gas molecules, or released as a photon *(15)*. Light emission is linearly related to the NO content of the sample *(6,10)* and increases significantly when the pressure in the reaction chamber is reduced below 0.3 atm *(16)*. This is probably because the concentration of other gas molecules that can quench the emitted energy (by reacting with NO_2) is reduced. Furthermore, the light emission is strongly dependent on temperature *(16)*. The reaction of NO with O_3 is fast (rate constant at room temperature is 10^{-7} L mol^{-1} s^{-1}) *(15)* and this allows for on-line monitoring of NO levels. In contrast to the reaction of O_3 with NO, which has a very low energy of activation (10.5 KJ), other nitrogen oxides such as NO_2 have much higher energies and react much more slowly *(2)*. Thus, although NO_2 is present, it will not affect the chemiluminescence reaction between the O_3 and NO. The same is true for other substances (pollutants) like CO_2, C_2H_4, CO, NH_3, SO_2, or H_2O as shown by Fontijin et al. in 1970 *(3)*. The high speed and the specificity make this reaction very attractive for use in an assay to measure NO.

2. Materials

2.1. The NO Chemiluminescence Analyzer

Like all luminometers *(17)*, the basic structure of the chemiluminescence analyzer is simple: it consists of a reaction chamber, a light detector which collects the emitted light and transforms it to an electrical signal, and a recorder (**Fig. 1**).

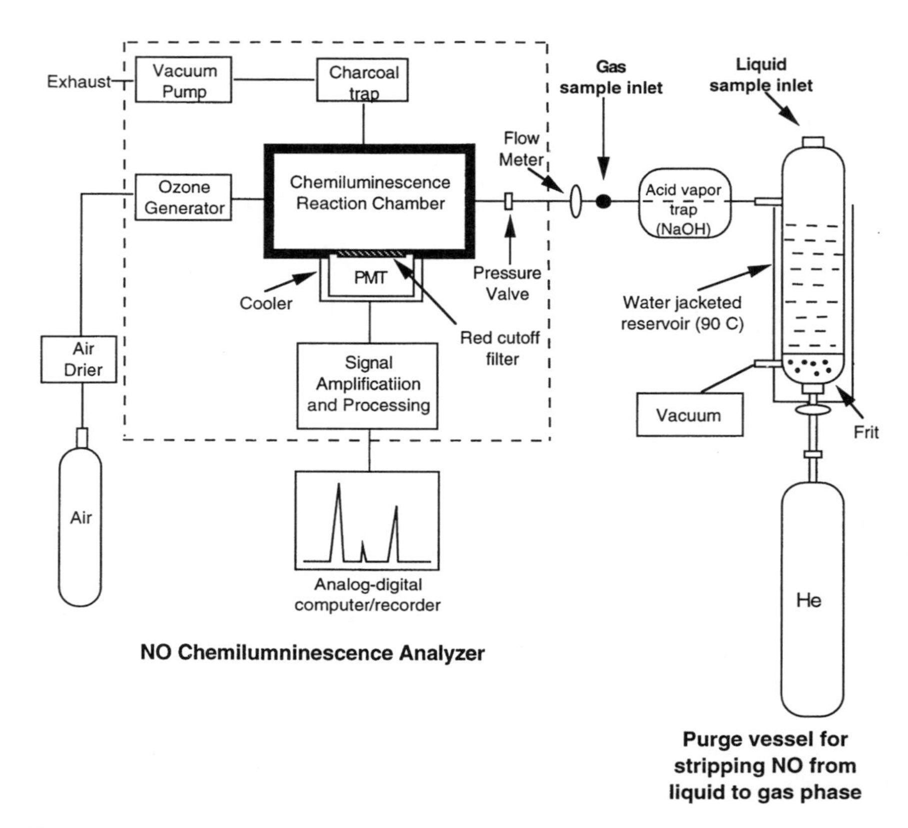

Fig. 1. A schematic diagram of the NO chemiluminescence analyzer and a purge vessel for stripping NO from the liquid to gas phase (*see* text for details).

The reaction chamber, where the chemiluminescence reaction takes place, has two inputs: the samples to be analyzed are injected (sample inlet) from one, and the O_3, which is produced to the attached O_3 generator, is supplied from the other. The pressure in the chamber is constantly kept low (1–15 mmHg) with a vacuum pump (*see* **Note 1**). The flow rate from the sample inlet into the reaction chamber is controlled by a needle valve; it is adjusted so that authentic NO gives rapidly rising peaks with no slurring of the signal.

O_3 is produced by a high-voltage electrostatic discharge. Compressed, dried air is supplied to the O_3 generator at a constant rate ***(18)***.

The effluent from the chamber (O_3, NO, NO_2) is externally vented. NO_2 is first passed through a charcoal filter, which eliminates most of this species.

A photomultiplier is the light detector of choice because it is mostly sensitive to low levels of light at the red end of the spectrum and NO_2^- emits weak red and infrared light (*see* **Note 2**).

A red cut-off filter is interposed between the reaction chamber and the photomultiplier (*see* **Note 3**).

Because most internal coolers keep the analyzer's temperature at a fixed level below the ambient temperature, the room temperature should be kept cool and relatively constant (*see* **Note 4**).

Most analyzers give an analog signal. This can be output to a recorder, an analog/digital converter, allowing visualization of the signal contours (*see* **Note 5**).

A suitable analyzer is the Sievers 270 or 280 (Sievers, Boulder, CO) chemiluminescence analyzer.

2.2. Chemicals and Tubing

1. Omnisolve (NO_2^- free water) (Curtin Matheson Scientific, Houston, TX).
2. 1 *M* hydrochloric acid (Baker Chemical Co, Phillipsburg, NJ).
3. 1 *M* KI, (MW:166): To prepare a 1 *M* solution, dissolve 16.6 g in 100 mL of Omnisolve (Sigma Chemical Co., St. Louis, MO).
4. 0.1 *M* vanadium (iii) chloride, (MW: 157.3): To prepare a 0.1 *M* solution dissolve 1.57 g vanadium (iii) in 100 mL of 2 *M* HCl. To prepare the 2 *M* HCl solution, put 7.2 g of HCl in 100 mL of Omnisolve (Aldrich Chemical Co., Milwaukee, WI).
5. 1 m*M* $NaNO_2$ (MW: 69): Dissolve 0.69 mg of $NaNO_2$ in 10 mL of Omnisolve water (Sigma Chemical Co., St Louis, MO). Prepare fresh. Protect from light.
6. 1 m*M* $NaNO_3$. (MW: 84.99) : Dissolve 0.85 mg in 10 mL of Omnisolve water (Sigma Chemical Co., St. Louis, MO). Prepare fresh. Protect from light.
7. Dow Corning FG-10 Antifoam emulsion (Dow Corning Corp., Midland, MI).
8. Septa #2-0668 thermogreen, LB-1 half-hole type (Superlco Inc., Bellefonte, PA).
9. Chemware FEP 1.2 L gas-collecting bag (Curtin Matheson Scientific Inc., Houston, TX).
10. Masterflex 6409-25, tygon tubing (Cole-Parmer Instruments, Chicago, IL).

3. Methods

The NO chemiluminescence reaction is always performed in the gas phase (*see* **Note 6**). If liquid specimens are measured, the NO is stripped in the gas phase prior to entry into the analyzer. Furthermore, nitrogen oxides must first be reduced to NO before they can be measured. The fact that the assay is exclusively in gas phase enhances its specificity; to be detected, a substance must be volatile and react with O_3-emitting light.

3.1. NO Measurement in Gas Samples (Breath NO)

3.1.1. Calibration for Gas NO

1. Connect a source of "NO free" air (for example, "medical air," *see* **Note 7**) to the gas inlet port of the NO analyzer.

2. Allow the pressurized gas to flow through the analyzer for 5 min to ensure equilibration.
3. Read the analog/digital recording; this value is taken as "0 ppb."
4. Connect a (commercially available) pressurized gas tank with a known amount of NO (100–1000 ppb, for example, 800 ppb) to the gas inlet of the analyzer. The pressurized tank will ensure constant flow of the gas to the analyzer.
5. After 5 min of flow through the machine, accept the recorded value as "800 ppb."
6. The analyzer is now calibrated for gases and will display values in ppb (all the analyzers have software to guide one step by step through the calibration process).

3.1.2. Gas Sampling-Bag Acquisition

1. Exhale breath air (or collect from a respirator in intubated patients, or aspirated during bronchoscopy) into a Chemware FEP 1.2-L gas-collecting bag, using a Tygon 25 tubing (*see* **Note 8**). The Tygon 25 tubing is connected to the metallic valve of the gas-collecting bag.
2. Immediately after the sample is collected, seal off the ambient air from the gas-collecting bag using a septum (SEPTA #2-0668 thermogreen LB-1 Half Hole Type). The sample has to be studied within one h *(26)*.
3. Connect the Tygon 25 tube to the gas-sample inlet of the NO analyzer. Twist open the metallic valve of the gas-collecting bag. Using the vacuum created by the analyzer's pump, the sample will be suctioned into the analyzer at approx 40 mL/s.
4. Read the analog/digital recording; this is the NO content of the gas sample.
5. Repeat the process 2–3 times.

3.1.3. Direct Acquisition of Breath by Cannulation

1. Place a Tygon tube (5–8 cm long, internal diameter 5.5 mm, external diameter 8 mm) approx 1 cm inside the mouth, left or right nostril.
2. Connect the tube to the gas inlet of the NO analyzer. Using the vacuum of the analyzer's pump, air will be drawn in the analyzer (at approx 10 mL/s).
3. Record the signal for 30–40 s of normal breathing (or breath holding following a deep inspiration, or while the subjects are asked to read; *see* **Fig. 2** and **Note 9**).

We recently used a chemiluminescence-NO analyzer (Sievers NOA 280, Boulder CO) to measure NO in air exhaled from the nose, mouth, trachea, and distal airway in healthy volunteers and patients undergoing bronchoscopy *(26)*. The results of this study, on the origin of breath NO in humans, are summarized in **Note 9**.

3.2. NO Measurement in Liquid Samples

3.2.1. NO_2^- Measurement

1. a. Prepare fresh and light protected 1 m*M* NO_2^- standards and make serial dilutions over the range of interest (i.e., 10^{-9}–10^{-3} m*M*).

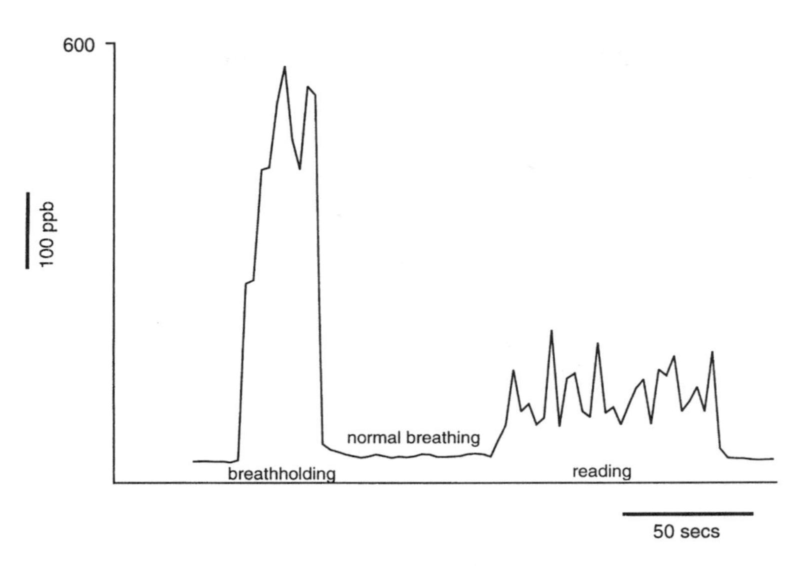

Fig. 2. Breath holding rapidly increases the nasal NO signal measured by direct aspiration from the nose (*see* also **Note 9**).

b. Add 3 mL of KI (1 *M*) and 3 mL of HCl (1 *M*) into the purge vessel and bubble with He for 5 min.
c. Add the samples with the known amount of NO_2^- (100 μL at a time) into the purge vessel. The recorded value (in mV, for maximal peak height of the signal or mV· s, for area under curve) is the amount of signal that a known amount of NO_2^- produces (*see* **Note 10**).
d. Prepare a calibration (standard) curve. On the X axis, plot the concentration of the standard, e.g., 10^{-9}, 10^{-8},10^{-7} *M*, etc. On the Y axis, plot the measured value (mV or mV·s) (*see* **Fig. 3**).

2. Collect 10 mL of the liquid sample in a gas-tight syringe (*see* **Note 11**).
3. Add 3 mL of KI (1 *M*) and 3 mL of HCl (1 *M*) into the purge vessel (*see* **Note 12**), under constant bubbling with He. Bubble the KI and HCl for 5 min prior to the addition of samples.
4. If the sample is proteinaceous, add 300 μL of antifoaming agent (Corning FG-10) to the purge vessel (*see* **Note 13**).
5. A large increase in the signal (spike) will initially appear (the etiology of which is unclear, *see* **Note 14**), but in approx 5 min the signal will return to its baseline. At this time, inject 100 μL of the sample into the purge vessel, under constant bubbling with He (*see* **Notes 15–16**).
6. Record the analog/digital reading. The NO_2^- content of the sample is then determined by reference to the standard curve (*see* **Note 17**).
7. After the signal returns to its baseline, repeat the sample injection 2–3 times.
8. After injection of approx 5 samples (100 μL of each), the KI, HCl, and antifoaming contents of the purge vessel have to be replaced with fresh solutions (*see* **Note 18**).

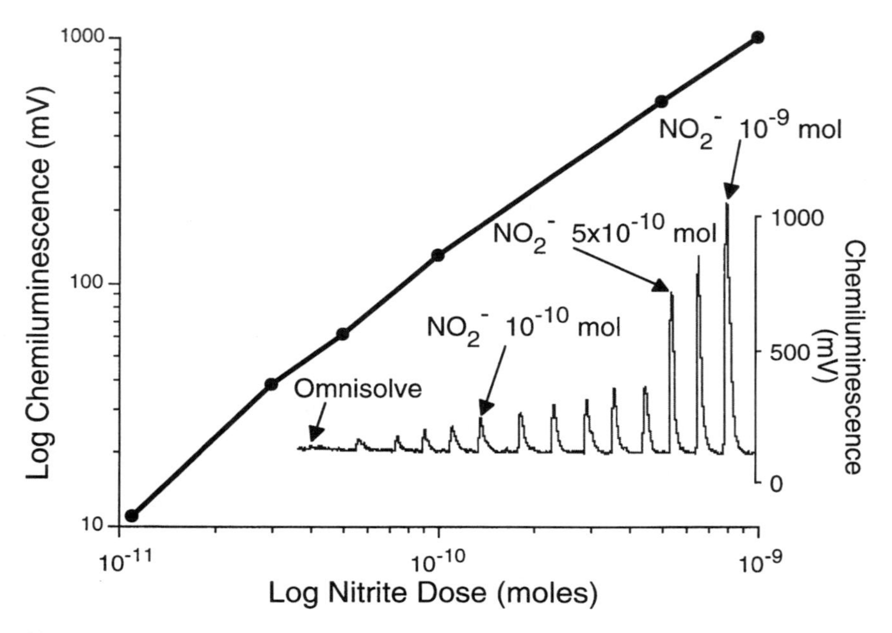

Fig. 3. Calibration curve created by incremental doses of NO_2^- , reduced by KI + HCl. The raw data are shown in the inset and illustrate the usual morphology of chemiluminescence spikes created at integration time 2 s. The relationship between nitrite dose and chemiluminescence (peak height of the signal, in mV) is a logarithmic function. (Used with permission from **ref. *10***.)

3.2.2. NO_3^- Measurement

1. Prepare NO_3^- standard-calibration curve as in the NO_2^- assay, using 1 m*M* $NaNO_3$.
2. Heat the purge vessel to 90°C.
3. Add 3 mL of vanadium (iii) (0.1 *M*) and 3 mL of HCl (1 *M*) to the purge vessel (*see* **Note 19**).
4. If the sample is proteinaceous, add 300 µL of antifoaming agent (Corning FG-10) to the purge vessel.
5. Inject 100 µL of the sample in the purge vessel.
6. Record the analog/digital reading. This is the total NO_2^- + NO_3^- content of the sample (*see* **Note 20**).
7. After the signal is back to the baseline, repeat the sample injection 2–3 times.
8. After injection of approx 5 samples (100 µL each) the KI, HCl, vanadium (iii), and antifoaming contents of the purge vessel have to be replaced with fresh solutions (*see* **Note 18**).
9. The above assay can be performed at room temperature as another means of measuring NO_2^-. The value obtained is then subtracted from the value obtained at 90°C to obtain the NO_3^- content (*see* **Note 19** and **Fig. 4**).

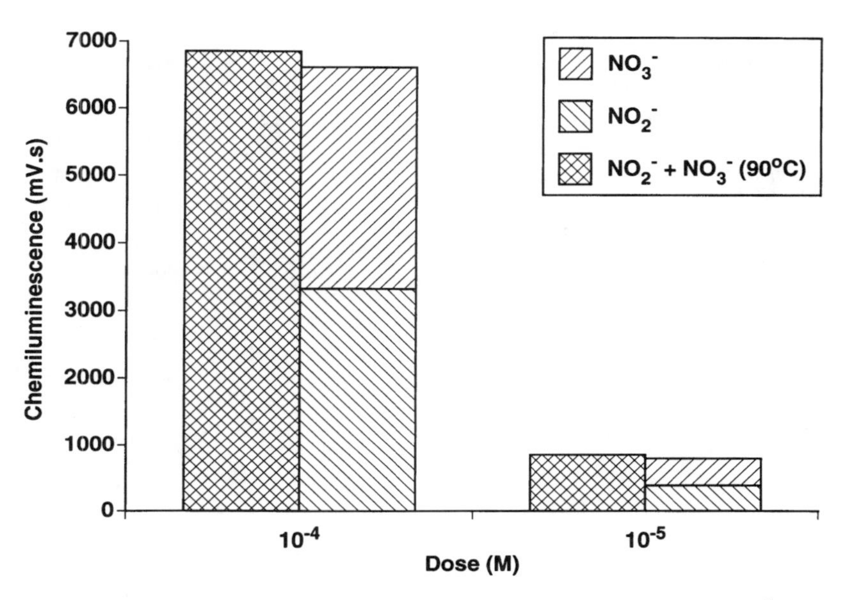

Fig. 4. Known amounts of NO_2^- and NO_3^- standards (0.1 cc of 10^{-4} or 10^{-5} *M* solutions) were measured either separately or after being mixed together. NO_2^- alone was measured using "cold" (25°C) vanadium (iii) as a reducing agent. NO_3^- alone as well as the NO_2^- + NO_3^- mixture were measured using "hot" (90°C) vanadium (iii). NO_3^- can be measured accurately by subtracting the NO_2^- value (obtained while the vanadium is cold) from the NO_2^- + NO_3^- total (measured with hot vanadium).

3.3. Maintenance

Routine measurements of the chemiluminescence signal resulting from a fixed dose of NO standard (e.g., 50 mL of 10 ppm NO) are the simplest way to know if the analyzer needs maintenance. A troubleshooting algorithm based on our own experience is shown in **Fig. 5**.

Using the above method, the NO-chemiluminescence assay is one of the most sensitive (*see* **Note 21**) and specific (*see* **Note 22**) NO assays. Because the NO + O_3 reaction takes place in the gas phase, the assay is ideal for measurements of gas NO. It has proved to be very useful in the new and exciting field of breath NO. The method, however, can be easily modified to permit measurements of NO in liquid samples. Furthermore, the NO nitrite and nitrate can be measured with better sensitivity than can be achieved by the Griess reaction *(10)*.

4. Notes

1. The decrease in pressure results in a higher light signal by reducing the number of gas molecules that absorb some of the NO_2's emitted energy. Furthermore,

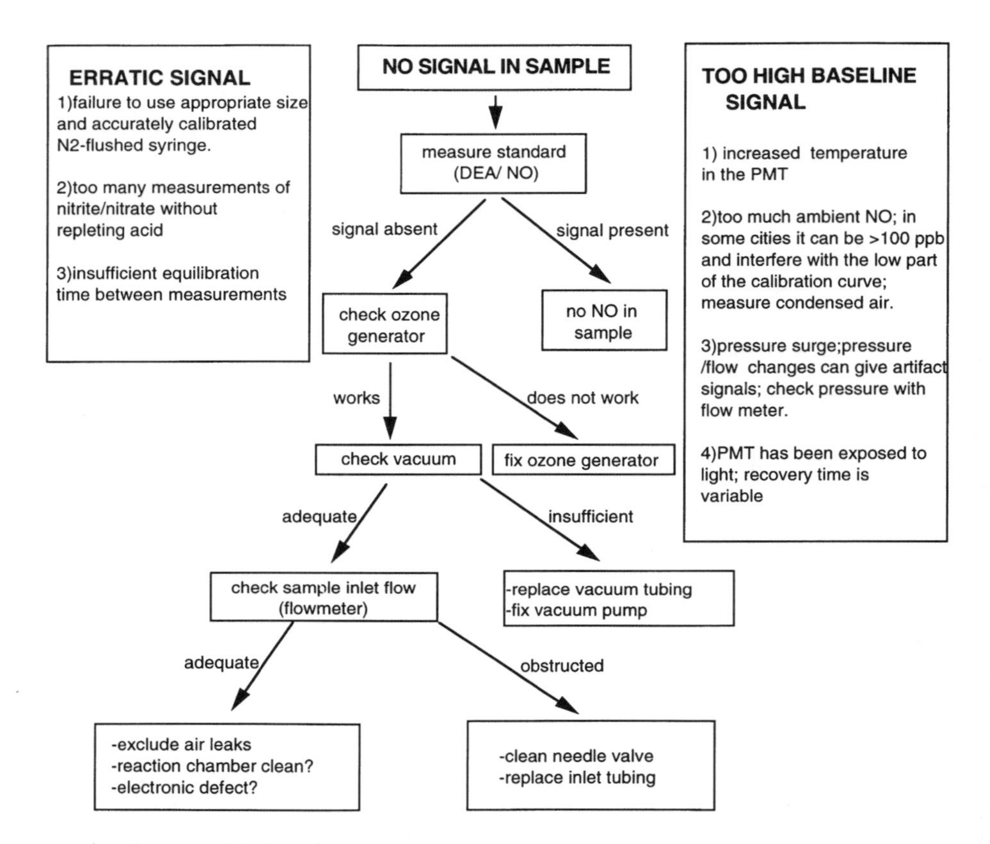

Fig. 5. A troubleshooting flow diagram for the NO-chemiluminescence assay.

the vacuum constantly removes O_2, stabilizing any NO present, and facilitates the flow of samples from the inlet port. Gorimar et al. ***(16)*** showed that at a stable temperature, the light output of the NO + O_3 reaction varies directly as the sample-flow rate and inversely as the pressure and is independent of the chamber volume and O_3 concentration (assuming stable supply of O_3 in large excess to the NO).

2. Although the peak intensity of the emitted light is at ~1100 nm (640–3000 nm range) and photomultipliers are sensitive to wavelengths below 900 nm, the amount of light emitted in the 640–900 range is still enough for very sensitive measurements ***(16)***. The sensitivity of the photomultiplier is very high. For each photon that strikes the photosensitive surface, there are millions of electrons detected in the anode at the other end of the tube ***(17)***.

Several factors can increase the background noise (dark current) of the photomultiplier. Exposure to large amounts of NO leads to enormous light emissions and increases dark current, which can take h to stabilize. For this reason, when large amounts of NO are expected in the sample, dilutions are recommended. If a light leak occurs, exposing the photomultiplier to ambient light, the recovery period is often more than a d. This may be expected when the photomultiplier tube is changed or when the chamber is opened for service ***(17)***.

3. The photomultiplier might also detect light with wavelengths below 640 nm, like the emission from alkenes and sulfur-containing species or from the poorly understood interaction between O_3 and the reaction-chamber wall. For this reason, a red cut-off filter is interposed between the reaction chamber and the photomultiplier ***(19)***.
4. An important determinant of dark current is temperature. For every 10°C increase in the temperature, the dark current doubles ***(17)***.
5. This is preferable to relying solely on the analyzer's light-emitting diode display. A good analog/digital converter program gives information about the signal, which helps to identify artifacts and permits data analysis. The analog/digital converter software and computer are usually purchased from MacLab (Milford, MA).
6. Confusion sometimes arises with the use of the terms "ppm," parts per million or "ppb," parts per billion. In liquids, ppm is a measure of weight, whereas in gases, ppm is a measure of volume. In a fluid with a density of 1 g/mL, like water, 1 ppm means 1 mg/L. In contrast, ppm for a gas is the number of parts of gas per million parts of contaminated air by volume. Because at 25°C and 760 mmHg, 1 mol of any gas has a volume of 22.4 L, the NO content of a gas can be easily translated from ppm to moles.
7. Ambient air should not be used in the calibration process. In some cities, such as Paris, the ambient NO level is as high as 120 ppb. Medical air usually contains less than 5 ppb NO. The key points of the calibration are known amounts of NO (e.g., 0 and 800 ppb) and constant flow through calibration. It takes 10 min to complete.
8. We use 1200 mL gas-sampling bags, made of fluoroethylene propylene (which does not react with NO, unlike rubber which contains sulfhydryl groups) and sealed with a metal valve (Curtis Matheson, Houston, TX).
9. In the absence of O_2 and reactive O_2 species, gas NO in low concentrations is relatively stable. NO has been shown to be present in human breath ***(20,21)***. Because NO is produced in large amounts in inflammation, it could be useful in monitoring of patients with lung inflammation or asthma. Breath-NO production has been attributed by different authors to the pulmonary-vascular epithelium ***(21)***, the epithelium of the small airways ***(22,23)***, and the nose ***(24)*** or paranasal sinuses. It has also been suggested that NO in the nose or paranasal sinuses might be of bacterial origin ***(25)***.

   Figure 2 shows typical data obtained using this method to examine NO in breath. Samples were acquired by having the volunteers exhale directly in the collecting bag or directly by a cannula from nose or mouth during rest breathing. Breath NO levels were greater in the nose (25 ± 2 ppb) than in the mouth (6 ± 1 ppb), trachea (3 ± 1 ppb), or distal airway (1 ± 2 ppb) ***(26)***. Beclomethasone (a topical steroid administered by nasal inhaler) but not antibiotics, decreased nasal NO without affecting mouth NO levels ***(26)***. These results suggest that, in normal human beings who do not have asthma, NO is synthesized in the nose or paranasal sinuses by a steroid-inhibitable NO synthase (NOS), presumably at least some of which is inducible (i) NOS, as iNOS is the only isoform that is inhibited by steroids. Interestingly, the nasal-NO levels measured by cannulation, increased sharply during 30 s of breath holding (**Fig. 2**). During breath holding, the lower-airway air (which contains much less NO than the nose) does not dilute the high local-nasal NO, the production of which remains unaffected. During the first few s, the NO measured reflects the locally produced NO in the nose and paranasal sinuses. At the end of the breath-holding period, however, the NO signal starts to plateau, because the air removed from the nasal cavity by the cannula suction is replaced by ambient air. When the volunteers were reading (and so breathing predominantly through their mouth), the nasal-cavity air was only partially diluted by the ambient- and lower-airway air, because speaking partially closes the soft palate. Therefore, the NO signal in their nasal cavity was intermediate between the signal during rest breathing and during breath holding (**Fig. 2**) ***(26)***.

The area of breath/airway NO research is new and expanding. The NO-chemiluminescence assay is fast, sensitive [detection threshold is 1 ppb, 1 ppb ***(26)***] and gives reproducible results in breath NO. Whereas our results indicate that the breath NO in normal volunteers originates in the nasal cavity, this is not entirely true in situations with significant lung inflammation. For example, in asthma the amount of NO originating from the lower airways is significantly increased ***(27–29)***. Therefore, this assay might prove to be helpful in the clinical assesment of exacerbations of asthma. More studies are needed to determine if breath NO could serve as a marker in other inflammatory or noninflammatory pulmonary or systemic diseases like cirrhosis or congestive heart failure.

10. The correlation between the amount of NO administered and chemiluminescence is more linear when peak height of the signal (mV), rather than area under the curve (mV·s), is used, owing in part to ambiguity in defining where the signal ends for the purposes of integration in the area under the curve method. However, when measuring NO_2^- or NO_3^-, the peak heights of the signal can vary (because conversion to NO can be slow), whereas the total amount of NO is accurately reflected by the area under the curve. In general, both techniques appear to yield comparable results ***(10)***.
11. It is unlikely that free NO, as such, exists in blood or protein-containing body fluids. However, in vitro, in perfused-organs experiments, various strategies can be used to decrease the amount of O_2 and oxidants in samples to increase NO detection. These strategies include lowering O_2 tension and the addition of superoxide dismutase (SOD) in order to stabilize the small amount of NO present ***(30)***. One has to be alert to the fact that hypoxia can affect NO levels either indirectly, through modulation of the NOS activity ***(31)***, or directly by affecting the partitioning of NO ***(10)***. The effect of hypoxia in an vitro system is to favor the survival of NO in the liquid phase and decrease the percentage of NO oxidized to NO_2^- ***(31)***. SOD has been used to permit detection of NO produced by arterial rings ***(30)***.

12. In case of NO-oxides measurements, an acid–vapor trap is used to prevent damage to the analyzer by the acids, KI, or vanadium (iii) used in these assays.
13. NO stripping of proteinaceous fluids can be a challenge, because these fluids tend to form bubbles very easily. These bubbles can be aspirated into the analyzer, coat the photomultiplier tube, and decrease the sensitivity of the assay. We use the defoaming agent FG-10 (Dow Corning, Midland, MI) at 0.1 mL per 1 mL of sample. FG-10 alone or in combination with acids or reducing agents does not give a chemiluminescence signal.
14. Acid and KI separately do not cause chemiluminescence, but together they cause a large and reproducible spike that disappears within one minute. The origin of this spike, which appears to depend on the presence of O_3, is unknown, but it may be a chlorine oxide or singlet O_2 ***(32)***.
15. NO has a high-partition coefficient of ~20. When a liquid sample is injected into a purge vessel, NO will rapidly escape from the liquid into the headspace; 20 times more NO will finally be in the gas than in the liquid phase. NO_2^- and NO_3^- are nonvolatile and remain in the liquid phase. NO alone can be aspirated from the headspace into the analyzer, and this accounts for the specificity of the assay in liquid samples. Although the escape of NO into the headspace is very fast (in equilibrium within 1 min), this can be accelerated by bubbling the liquid sample with a constant flow of an inert gas like helium (He). This is commonly referred to as stripping ***(33)***. A purge-vessel system to accomplish this is shown next to the NO analyzer in **Fig. 1**. He or another inert gas flows into the vessel through a porous frit at a constant rate. The optimal flow rate is above 8–10 mL/min and this can be judged by interposing a flowmeter in the circuit.
16. Many authors recommend that the valve between the purge vessel and the analyzer remains closed for the first 15–30 s of the sample stripping. Theoretically, opening of the valve releases NO as a bolus, improving the chemiluminescence signal (i.e., makes a sharp, spike-like signal contour). However, in our experience, opening and closing the valve results in significant changes to pressure and flow, creating confusing artifactual changes in the chemiluminescence signal ***(34)***. It should be emphasized that gas samples should be injected directly into the analyzer and not through the purge vessel. For gas samples, the purge vessel is unnecessary and some sample can be lost in transit.
17. In the presence of acid, NO_2^- is converted to the nitrosonium ion (NO^+): $NO_2^- + 2H^+ \rightarrow NO^+ + H_2O$. NO^+ reacts rapidly with nucleophiles like iodide (NaI or KI) or vanadium (iii), to form NO. For example:

$$NO^+ + I^- \rightarrow ONI\ ,\ 2ONI \rightarrow 2NO + I_2 \text{ or, in summary:}$$

$$NO_2^- + 2H^+ + e^- \rightarrow NO + H_2O$$

We and others have previously shown that the assay is less sensitive for NO_2^- than for NO gas ***(10)***; therefore, we cannot detect a stoichiometric reduction of NO_2^- to NO. This might reflect incomplete reduction of NO_2^- during the 30–60 s mixing with acid and KI and incomplete stripping of the NO as it is produced in the purge vessel. Stronger reducing conditions will convert NO_3^- to NO as well *(4)*: $NO_3^- + 4H^+ + 3e^- \rightarrow NO + 2H_2O$

18. It is best to initially add a large amount of acid and KI [or NaI or vanadium (iii)] and make multiple NO_2^- measurements, rather than adding acid and KI fresh prior to each measurement, as it was originally reported. This way one will not have to wait for the acid + KI signal to disappear before proceeding to the next measurement. In practice, addition of 3 mL of acetic acid or HCl (1 *M*) and 3 mL of stock KI solution (3 *M*) to the purge vessel is enough for 3–5 repeated injections of 0.1–0.2 mL samples.
19. In the original report by Cox ***(4)***, a mixture of ferrous ammonium sulfate (4%) and ammonium molybdate (2%) was used to reduce NO_3^- to NO; later, Braman introduced the use of vanadium (iii) ***(5)***. While at room temperature, vanadium (iii) reduces NO_2^- and not NO_3^-. By increasing the temperature up to 80–95°C, all the NO_3^- is reduced to NO as well. That way both oxides can be measured in one specimen (**Fig. 3**). Acidic vanadium (iii) is first heated without the sample to decontaminate it from NO-oxidation products. It is then cooled to room temperature, the sample is added, and NO from the reduction of NO_2^- is measured. Then, the mixture is heated again and NO from the reduction of NO_3^- is measured. Other protocols for the reduction of the sample, using columns of copper-plated cadmium filings ***(35)***, or bacterial ***(10,36)*** or commercial-nitrate reductase to convert NO_3^- to NO_2^- are also used.
20. As discussed earlier, the amount of free NO in biological fluids is very small, if any, and stripping alone would not reflect at all the amount of NO present because most of it would be already oxidized to NO_2^- or NO_3^-. In an acidic environment, and with strong reducing agents [such as KI, NaI, or vanadium (iii)], these oxides can be reduced back to NO ***(4,5,37)***, which would then escape to the headspace and be measured by the NO assay. In body fluids, under these reducing conditions, other bioactive-NO adducts, like nitrosothiols and nitrosoamines, are reduced to NO as well ***(38)***. Thus, measuring NO_2^- and NO_3^- may overestimate the amount of NO which was biologically available. Furthermore, NO_3^- is abundant in certain foods and beverages.

    Because almost all sources of water are contaminated with NO_2^- , the background-chemiluminescence signal from the control solution has to be subtracted from the sample's signal. To minimize this artifact, use of a relatively NO_2^--free water, such as Omnisolve (Curtin Matheson Scientific, Houston, TX), is advised.
21. The detection threshold for gas NO by the chemiluminescence assay is agreed by most authors to be 20–50 pmol ***(19,39)*** or 1 ppb in gas phase NO ***(26)***. In our authentic NO calibrations, we have found the NO-chemiluminescence relationship to be linear between 20 and 200 pmol, but somewhat nonlinear at higher concentrations ***(10)***.
22. The NO chemiluminescence assay is highly, but not entirely, specific for NO. The few other substances that can give a chemiluminescence signal (like sulfides, amines, H_2S) are either nonvolatile, having significantly smaller partition

coefficients than NO, or do not occur in biological systems. However, we have found that dimethyl sulfoxide (DMSO—a very commonly used solvent for drugs) at high doses can cause a chemiluminescence signal ***(19)***. We avoid this by using very small doses of DMSO (less than 0.1 mL/10 mL) and by interposing the red cut-off filter in front of the PMT, as we discussed earlier and shown in **Fig. 1**. NaOH can be used to remove hydrogen sulfide (*see* **Fig. 1**). $FeSO_4$ or reduced hemoglobin will remove NO, but also hydrogen sulfide. Reduced hemoglobin can be added to the sample in the purge vessel; it rapidly extinguishes the NO in less than 1 min ***(10)***.

Acknowledgments

All NO traces were performed in a Sievers 270 or 280 (Sievers, Boulder, CO) chemiluminescence analyzer. The 280 analyzer was provided as a gift by Sievers.

References

1. Van Dyke, K. (1985) Introduction, in: *Bioluminescence and Chemiluminescence: Instruments and Applications* (Van Dyke, K., ed.), CRC Press, Boca Raton, FL, pp. 1–7.
2. Johnston, H. S. and Crosby, H. J. (1954) Kinetics of the fast gas phase reaction between ozone and nitric oxide. *J. Chem. Phys.* **22,** 689–692.
3. Fontijn, A., Sabadell, A. J., and Ronco, R. J. (1970) Homogenous chemiluminescent measurement of nitric oxide with ozone. Implications for continuous selective monitoring of gaseous air pollutants. *Anal. Chem.* **42,** 575–579.
4. Cox, R. D. (1980) Determination of nitrate and nitrite at the parts per billion level by chemiluminescence. *Anal. Chem.* **52,** 332–335.
5. Braman, R. S. and Hendrix, S. A. (1989) Nanogram nitrite and nitrate determination in environmental and biological materials by vanadium (iii) reduction with chemiluminescence detection. *Anal. Chem.* **61,** 2715–2718.
6. Zafiriou, O. C. and McFarland, M. (1980) Determination of trace levels of nitric oxide in aqueous solution. *Anal. Chem.* **52,** 1662–1667.
7. Ignarro, L. J., Buga, G. M., Wood, K. S., Byrns, R. E., and Chaudhuri, G. (1987) Endothelium-derived relaxing factor produced and released from artery and vein is nitric oxide. *Proc. Natl. Acad. Sci. USA* **84,** 9265–9269.
8. Palmer, R. M. J., Ferrige, A. G., and Moncada, S. (1987) Nitric oxide release accounts for the biological activity of endothelium-derived relaxing factor. *Nature* **327,** 524–526.
9. Braker, W. and Mossman, A. L., (eds.) (1975) *The Matheson Unabridged Gas Data Book* (6th ed.) Matheson, East Rutherford, NJ, pp. 20–24.
10. Archer, S. L., Shultz, P. J., Warren, J. B., Hampl, V., and DeMaster, E. G. (1995) Preparation of standards and measurement of nitric oxide, nitroxyl, and related oxidation products. *Methods: A Companion to Methods Enzymol.* **7,** 21–34.
11. Henry, Y., Ducrocq, C., Drapier, J., Servent, D., Pellat, C., and Guissani, A. (1991) Nitric oxide, a biological effector. Electron paramagnetic resonance detection of nitrosyl-iron-protein complexes in whole cells. *Eur. Biophys. J.* **20,** 1–15.

12. Ignarro, L. J., Fukuto, J. M., Griscavage, J. M., Rogers, N. E., and Byrns, R. E. (1993) Oxidation of nitric oxide in aqueous solution to nitrite but not nitrate: comparison with enzymatically formed nitric oxide from L-arginine. *Proc. Natl. Acad. Sci. USA* **90,** 8103–8107.
13. Radi, R., Cosgrove, T. P., Beckman, J. S., and Freeman, B. A. (1993) Peroxinitrite-induced luminol chemiluminescence. *Biochem. J.* **290,** 51–57.
14. Beckman, J. S., Chen, J., Ischiropoulos, H., Crow, and J. P. (1994) Oxidative chemistry of peroxynitrite. Oxygen radicals in biological systems. *Methods Enzymol.* **233,** 229–240.
15. Clyne, M. A. A., Thrush, B. A., and Wayne, R. P. (1964) Kinetics of the chemiluminescent reaction between nitric oxide and ozone. *Trans. Faraday Soc.* **60,** 359–370.
16. Gorimar, T. S. (1985) Total nitrogen determination by chemiluminescence, in: *Bioluminescence and Chemiluminescence: Instruments and Applications.* (Van Dyke, K., ed.), CRC Press, Boca Raton, FL, pp. 77-93.
17. Turner, G. K. (1985) Measurement of light from chemical and biochemical reactions, In: *Bioluminescence and Chemiluminescence: Instruments and Applications.* (Van Dyke, K., ed.), CRC Press, Boca Raton, FL, II, 43-78.
18. Sievers nitric oxide analyzer NOA™ 270B (1992). Operation and service manual. Boulder, CO: Sievers Instruments.
19. Archer, S. (1993) Measurement of nitric oxide in biological models. *FASEB J.* **7,** 349–360.
20. Barnes, P. J. and Belvisi, M. G. (1993) Nitric oxide and lung disease. *Thorax.* **48,** 1034–1043.
21. Borland, C., Cox, Y., and Higenbottam, T. (1993) Measurement of exhaled nitric oxide in man. *Thorax* **48,** 1160–1162.
22. Persson, M. G., Wiklund, N. P., and Gustafsson L. E. (1993) Endogenous nitric oxide in single exhalations and the change during exercise. *Am. Rev. Respir. Dis.* **148,** 1210–1214.
23. Persson, M. G., Midtvedt, T., Leone, A. M., and Gustafsson, L. E. (1994) Ca^{2+}-dependent and Ca^{2+}-independent exhaled nitric oxide, presence in germ-free animals, and inhibition by arginine analogues. *Eur. J. Pharmacol.* **264,** 13–20.
24. Gerlach, H., Rossaint, R., Pappert, D., Knorr, M., and Falke, K. (1994) Autoinhalation of nitric oxide after endogenous synthesis in nasopharynx. *Lancet* **343,** 518–519.
25. Lundberg, J. O. N., Farkas-Szallasi, T., Weitzberg, E., et al. (1995). High nitric oxide production in human paranasal sinuses. *Nature Medicine* **1,** 370–373.
26. Dillon, W. C., Hampl, V., Shultz, P. J., Rubins, J. B., and Archer, S. L. (1996) Origins of breath nitric oxide in humans. *Chest* **110,** 930–938.
27. Kharitonov, S. A., Yates, D., Robbins, R. A., Logan-Sinclair, R., Shinebourne, E. A., and Barnes, P. J. (1994) Increased nitric oxide in exhaled air of asthmatic patients. *Lancet.* **343,** 133–135.
28. Persson, M. G., Zetterström, O., Agrenius, V., Ihre, E., and Gustafsson, L. E. (1994) Single-breath nitric oxide measurements in asthmatic patients and smokers. *Lancet* **343,** 146–147.

29. Alving, K., Weitzberg, E., and Lundberg, J. M. (1993) Increased amount of nitric oxide in exhaled air of asthmatics. *Eur. Respir. J.* **6,** 1368–1370.
30. Archer, S. L. and Cowan, N. J. (1991) Measurement of endothelial cytosolic calcium concentration and nitric oxide production reveals discrete mechanisms of endothelium-dependent pulmonary vasodilation. *Circ. Res.* **68,** 569–1581.
31. Archer, S. L., Freude, K. A., and Shultz, P. J. (1995) Effect of graded hypoxia on the induction and function of inducible nitric oxide synthase in rat mesangial cells. *Circ. Res.* **77,** 21–28.
32. Wallace, J., Springer, G., and Stedman, D. (1980) Photochemical ozone and nitric oxide formation in air-nitrogen dioxide mixtures containing sulfur dioxide or chlorine. *Atmosph. Environ.* **14,** 1147–1157.
33. Aoki, T. (1990) Continuous flow determination of nitrite with membrane separation/chemiluminescence detection. *Biomed. Chromatograp.* **4,** 128–130.
34. Hampl, V., Walters, C. L., and Archer, S. L. (1996) Determination of nitric oxide by the chemiluminescence reaction with ozone, in: *Methods in Nitric Oxide Research* (Feelisch, M. and Stamler, J. S., eds.) Wiley, Chichester, West Sussex, UK, pp. 310–318.
35. Green, L. C., de Luzuriaga, K. R., Wagner, D. A., et al. (1981) Nitrate biosynthesis in man. *Proc. Natl. Acad. Sci. USA* **78,** 7764–7768.
36. Granger, D. L., Hibbs, J. B., Perfect, J. R., and Durack, D. T. (1990) Metabolic fate of L-arginine in relation to microbiostatic capability of murine macrophages. *J. Clin. Invest.* **85,** 264–273.
37. Termin, A., Hoffman, M., and Bing, R. J. (1992) A simplified method for the determination of nitric oxide in biological solutions. *Life Sci.* **51,** 1621–1629.
38. Walters, C. L., Gillatt, P. N., Palmer, R. C., and Smith, P. L. R. (1987) A rapid method for the determination of nitrate and nitrite by chemiluminescence. *Food Add. Contam.* **4,** 133–140.
39. Brien, J., McLaughlin, B., Nakatsu, K., and Marks, G. (1991) Quantitation of nitric oxide formation from nitrovasodilator drugs by chemiluminescence analysis of headspace gas. *J. Pharmacol. Methods.* **25,** 19–27.

Barry W. Allen

13

Measurement of NO Using Electron Paramagnetic Resonance

S. Tsuyoshi Ohnishi

1. Introduction

1.1. Principle of Electron Paramagnetic Resonance (EPR)

Absorption spectroscopy is based upon the principle that the absorption of radiation by a molecule involves the transition of the energy level from its ground state to an excited state. If we denote the energy difference as ΔE and the frequency of radiation as ν, then the relationship is expressed as

$$\Delta E = \hbar\nu \quad (1)$$

where $\hbar$ is the Planck's constant. In optical-absorption spectroscopy, the absorption may be caused by π-electrons in proteins or conjugated double bonds. In infrared spectroscopy, the absorption may depend on bond angles and strength. In EPR, it is the magnetic interaction between the electron spin of a compound and the magnetic field applied by the instrument. In nuclear-magnetic resonance (NMR), the interaction between the nuclear spin and the applied field is detected. **Table 1** shows approximate frequencies and wavelengths typical to these spectroscopic techniques. The exact wavelength (λ) can be calculated from the equation, $c = \nu\lambda$, where c is the light velocity (3×10^{10} cm/s).

In EPR, several microwave frequencies are used. The wavelength of the instrument determines the size of the cavity resonator to be used, because the size is the same as the wavelength. Most commercial instruments use The X-band, which is approx 9.5 GHz (GHz = 10^9 Hz), and the size of the resonator is approx 3.2 cm.

From: *Methods in Molecular Biology, Vol. 100. Nitric Oxide Protocols*
Edited by: Michael A. Titheradge © Humana Press Inc., Totowa, NJ

Table 1
Approximate Frequencies and Wavelengths of Absorption Spectroscopy for Biological Materials

	Transition frequency, n (Hz)	Wavelength, λ (cm)
UV absorption	10^{16}	10^{-6} - 10^{-5}
Visible absorption	10^{15}	10^{-5} - 10^{-4}
Infrared	10^{14} - 10^{13}	10^{-4} - 10^{-2}
EPR (microwave)	10^{11} - 10^{9}	10^{-1} - 10^{1}
NMR (radio frequency)	10^{9} - 10^{8}	10^{1} - 10^{2}

The energy of interaction between a magnetic moment μ and the magnetic-field H is given by

$$E = \mu H. \tag{2}$$

The magnetic moment of electron spin is given by

$$\mu = -g\beta S \tag{3}$$

where g is a proportional constant called the g-value, $\beta = (e\hbar\nu/4\,\pi mC) = 0.92732 \times 10^{-20}$ erg/Gauss (the Bohr magneton), e is the electron charge, m is the electron mass, and $S = \pm 1/2$ (electron spin numbers).

Without an applied magnetic field, the energy levels of electrons with $S = 1/2$ and $S = -1/2$ are the same (**Fig. 1**, left line). When a magnetic field H is applied, this line splits into two levels, one higher than the other (**Fig. 1**, right lines). From **Eqs. 1–3**, the energy difference (ΔE) between two levels will be

$$\Delta E = g\beta H. \tag{4}$$

If a microwave with the frequency of ν is introduced to this system under the applied magnetic-field H, then the energy absorption occurs when the relationship

$$\hbar\nu = g\beta H. \tag{5}$$

is fulfilled.

For technical reasons, the microwave frequency is kept constant and the magnetic field is scanned. **Figure 2A** shows schematically that the absorption occurs at the place where the magnetic field (abscissa) fulfills the aforementioned relationship. In order to detect a small signal from the background of noises, EPR devices use a frequency modulation method that records the first derivative of the absorption. Thus, the recorded spectrum looks like that of **Fig. 2B**. For a free electron, the g value is 2.0023, and at microwave frequency of 9.5 GHz, the absorption takes place at 3400 Gauss (or 340 mT). Conven-

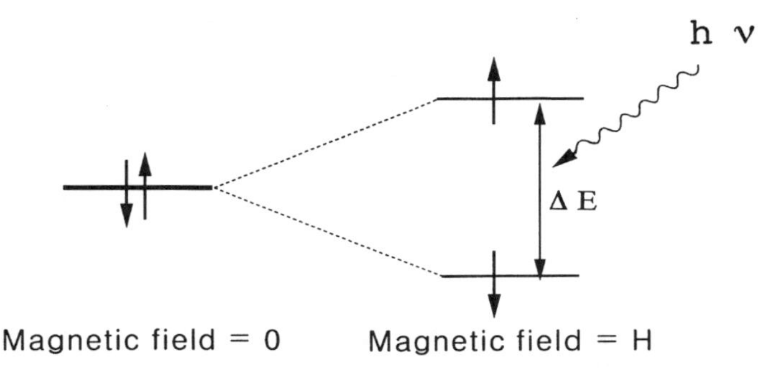

Fig. 1. Principle of EPR. The microwave absorption occurs when energy difference DE between two levels of spin states, which is created by the external magnetic field H, is equal to hv.

tionally, the abscissa of the absorption spectra is indicated by the strength of the magnetic field. However, in a precise discussion, it is expressed by the g-value.

1.2. Detection of Nitric Oxide (NO) by EPR

Nitric oxide (NO) has an unpaired electron (**Fig. 3**), which is essentially localized in the p-orbital of the nitrogen atom. Thus, NO is paramagnetic and can be detected using EPR. Because nitrogen has a nuclear-magnetic moment, the magnetic moment of the nucleus is superimposed onto the applied-field *H*, thus producing a hyperfine splitting. The number of splitting is given by $2I + 1$, where I is the nuclear-spin number. Because $I = 1$ for ^{14}N, there are three hyperfine absorption spectral lines (**Fig. 2C**). If nitrogen of NO is replaced by ^{15}N, then the total hyperfine splitting line will be two, because the nuclear spin number of ^{15}N is 1/2 (**Fig. 2D**).

Being a free radical, NO has a short half-life on the order of s. This problem can be overcome by the use of a "spin-trapping agent," which traps the unstable NO radicals to form a more stable radical compound. Iron is known to form such a compound. Deoxyhemoglobin or carbon monoxide-hemoglobin can be used as a "NO-trapping agent" (**Fig. 4A**) *(1–3)*. For the detection and quantitation of the tissue NO level, a smaller molecule may be preferred from the standpoint of effective tissue delivery. Vanin's group developed a method of injecting diethyldithiocarbamate (DETC) and iron separately to form the DETC-iron compound in vivo, which in turn traps NO radicals (**Fig. 4B**) *(4–5)*. Because DETC-iron is not water-soluble, Lai et al. developed a method to use a water-soluble carbamate-iron compound (**Fig. 4C**) *(6–8)*.

A water-insoluble trapping agent has a disadvantage in that DETC and iron have to be injected separately (if mixed, they precipitate). However, this short-

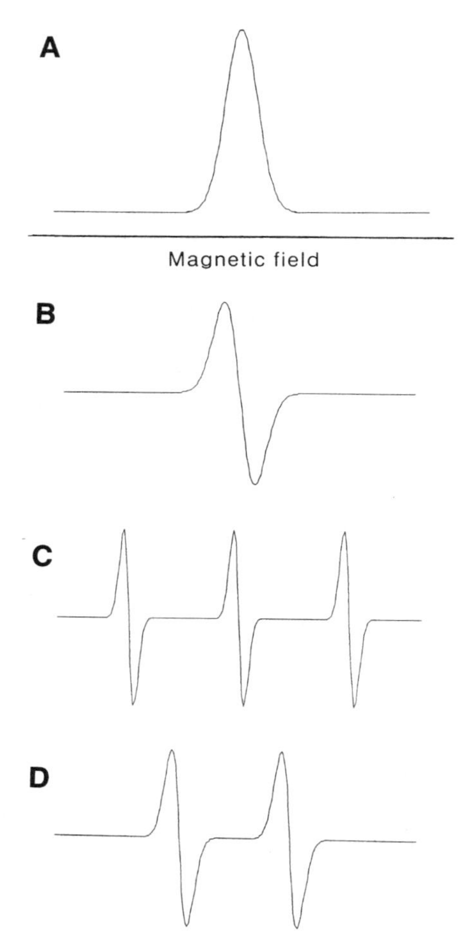

Fig. 2. **(A)** Absorption of microwave energy. **(B)** First derivative of the absorption. **(C)** Hyperfine splitting by ^{14}N. **(D)** Hyperfine splitting by ^{15}N. The abscissa represents the applied magnetic field H.

coming can also be an advantage because as it goes into cell membranes, it can detect NO in and around the membranes. On the contrary, a water-soluble trapping agent can be used to detect NO levels in blood or body fluid.

NO can be measured directly using a NO-sensitive electrode. Compared with the EPR method, the electrode method has an advantage in that it allows continuous recording, which is especially useful for in vitro measurements. However, for in vivo measurements, it measures NO production only around the micro-electrode tip, and it has the disadvantage in tissue studies in that the interpretation of the signals is not easy because the electrode may be influ-

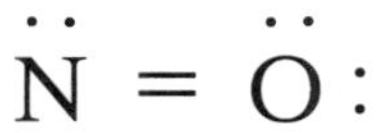

$$\cdot\ddot{N} = \ddot{O}:$$

Fig. 3. An unpaired electron in NO makes the molecule paramagnetic.

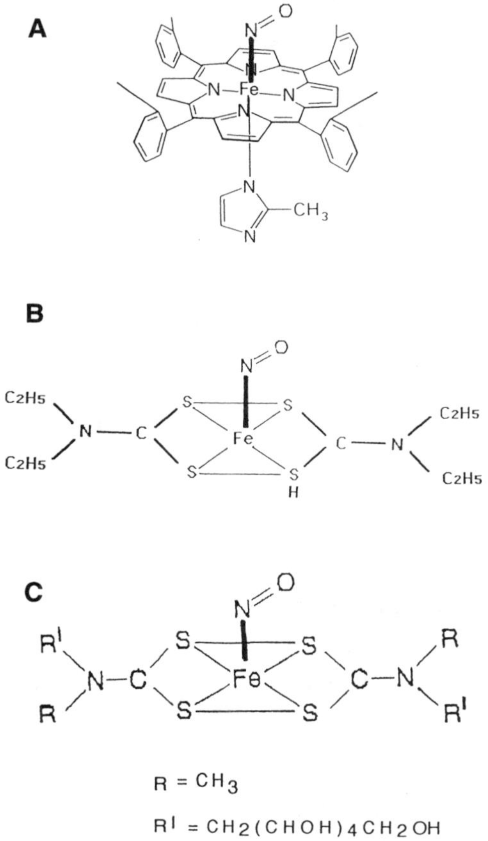

A

B

C

$R = CH_3$

$R^I = CH_2(CHOH)_4CH_2OH$

Fig. 4. **(A)** Iron in hemoglobin can trap NO. **(B)** A water-insoluble diethyldithiocarbamate-iron complex as a NO trap. **(C)** A water-soluble N-methyl D-glycamine dithiocarbamate-iron complex as a NO trap.

enced by temperature, ion compositions, pH, and other radicals. For this reason, the EPR method and the electrode method are considered to be complementary, but not competitive. The sensitivity of detection using EPR for the tissue-NO level is about 1 μ*M*.

The EPR technique has not been widely used as yet in the biomedical field, partly because: its usage has mostly been limited to the study of transition metal centers, and it is not suitable for unfrozen water-containing biological materials.

However, for the detection and analysis of free radicals in medical studies, EPR is a powerful technique. Because many pathologic conditions such as ischemia, hypoxia, ischemia-reperfusion injury, diabetes, cancer, multiple sclerosis, arteriosclerosis, arthritis, and others are related to the production of free radicals, the use of EPR should be made available for many more investigators in biomedical and clinical sciences.

In this chapter, the DETC-iron method will be discussed. We have used this method successfully for the assay of NO levels in the brain ***(9–13)***, spinal cord ***(14)***, lung ***(15,16),*** and the liver ***(17)***.

2. Materials

2.1. Instrumentation

Currently, Bruker (Germany) manufacture several lines of EPR instrumentation worldwide. Similarly JOEL (Japan) manufacture several highly sensitive EPR instruments, although their products are not marketed world-wide and they are not available in the United States.

The Varian Company (Palo Alto, CA) discontinued production more than a decade ago; however, their instruments for X-band EPR (Century Series E-4 or E-104, 4 in magnet; and E-9 and E-109, 9 in magnet) can still be found in many Chemistry or Physics Departments and are adequate for detecting NO using EPR.

2.2. Nitrogen Flow Attachment

The absorption of microwave energy by water increases with the increase of frequency. In X-band EPR, which employs 9.5 GHz, the absorption by tissue water poses a serious problem (*see* **Note 1**). To overcome this, either a thin flat cell or a thin capillary tube is used. For the sake of ease of handling with satisfactory sensitivity, it is the best to use a 4-mm (od) quartz tube and to freeze the specimen to eliminate the effect of absorption by water. Normally, a low temperature from –120°C to –196°C (the temperature of liquid nitrogen) is used. Lower temperatures actually favor the measurements, because radical species are more stable and because the interference from copper signals is less at lower temperatures (*see* **Fig. 5**).

An EPR instrument with a cryogenic attachment is desirable. However, if the instrument does not have such an attachment, then a dip-type dewar-flask (e.g., WG-816 manufactured by Wilmad Glass, Buena, NJ) can be used with the specimen kept in liquid nitrogen throughout the measurements (**Fig. 6**). This is simple to use, but it has a drawback in that nitrogen bubbles are constantly produced in the dewar-flask, which vibrates the specimen and makes the recording noisy. Because the signal of NO from the tissue is relatively small, the noise from the nitrogen bubbles can become a problem.

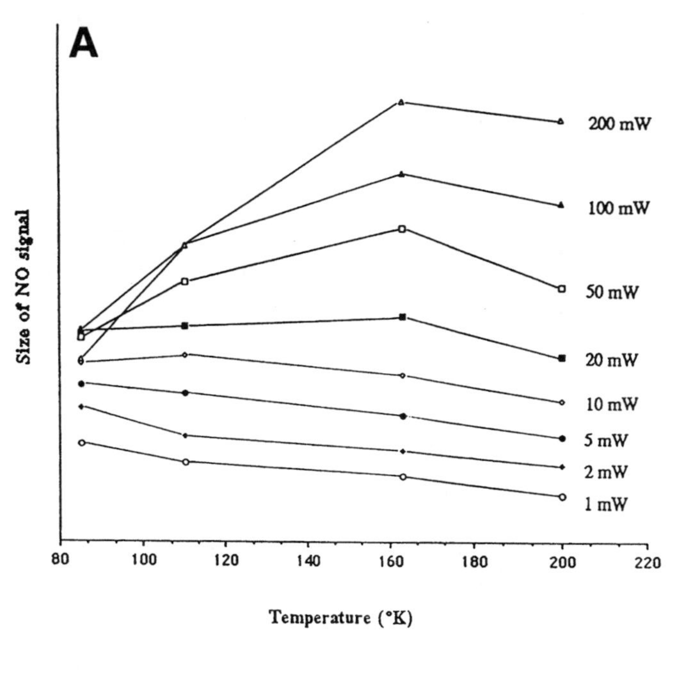

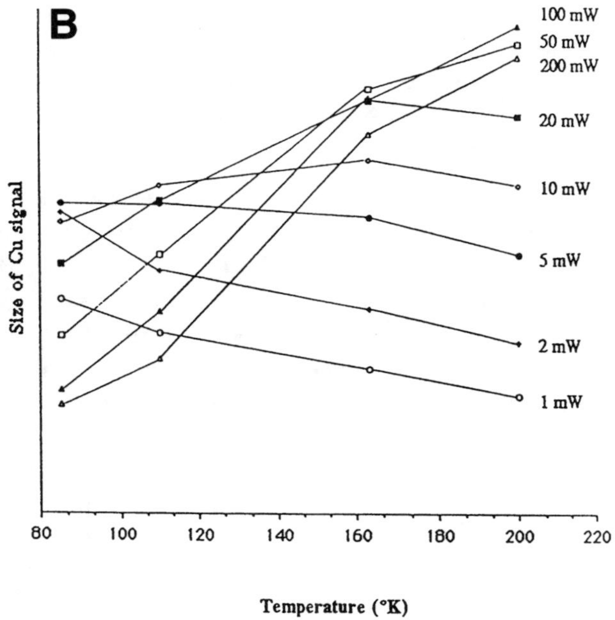

Fig. 5. Effects of temperature and microwave power on NO (**A**) and copper signals (**B**). (Used with permission from **ref.** ***13***.)

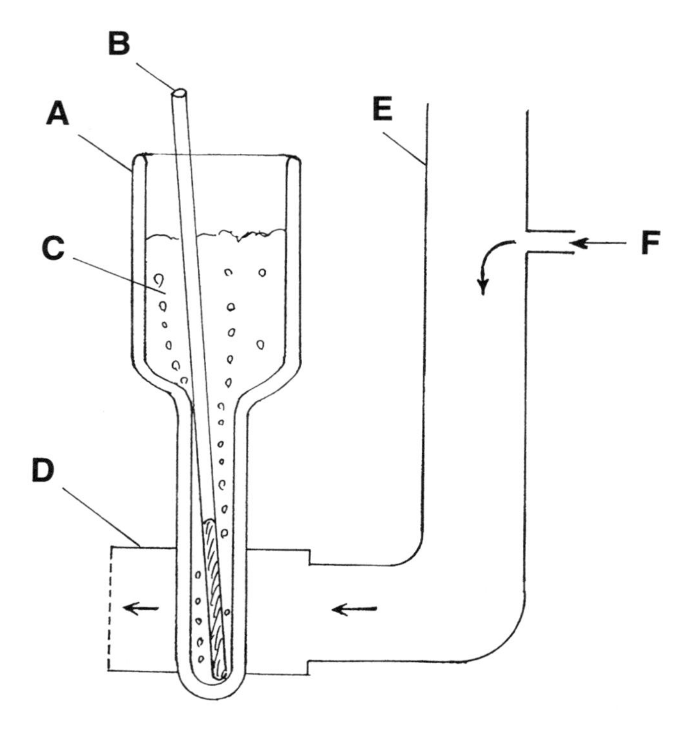

Fig. 6. A dewar-flask method for cooling specimens with liquid nitrogen. **A**, Dewar-flask; **B**, quartz tube with tissue specimens; **C**, liquid nitrogen; **D**, cavity resonator; **E**, microwave guide; **F**, nitrogen gas flow to purge possible water condensation.

For sensitive, precise measurements, an inexpensive nitrogen-flow apparatus can be built. **Figure 7** shows a schematic illustration of the entire system. The dewar-insert with sample support dimples (e.g., WG-821) and the transfer dewar (e.g., WG-760) can be purchased from Wilmad Glass. The copper tubing should have an od of 1/4 in. (6.3 mm) and a length of 25 ft (7.5 m), and should be made into a coil that is placed inside a large kitchen dewar-jar with id of about 6 in. (15 cm). The two gas-flow meters should have a capacity of about 15 L/min and are connected to a nitrogen-gas source (either nitrogen-gas cylinder or a gas outlet of a liquid-nitrogen container). Neither air nor carbon-dioxide gas can be used because they liquidify at liquid-nitrogen temperature. The gas flow inside the dewar-insert is to cool the specimen, and the flow connected to the waveguide (about 10 L/min) is to prevent frost formation outside of the dewar-insert and inside of the cavity resonator.

The specimen temperature is regulated by the flow velocity inside the dewar-insert, and can be measured by a pocket thermocouple thermometer (e.g., Cole-

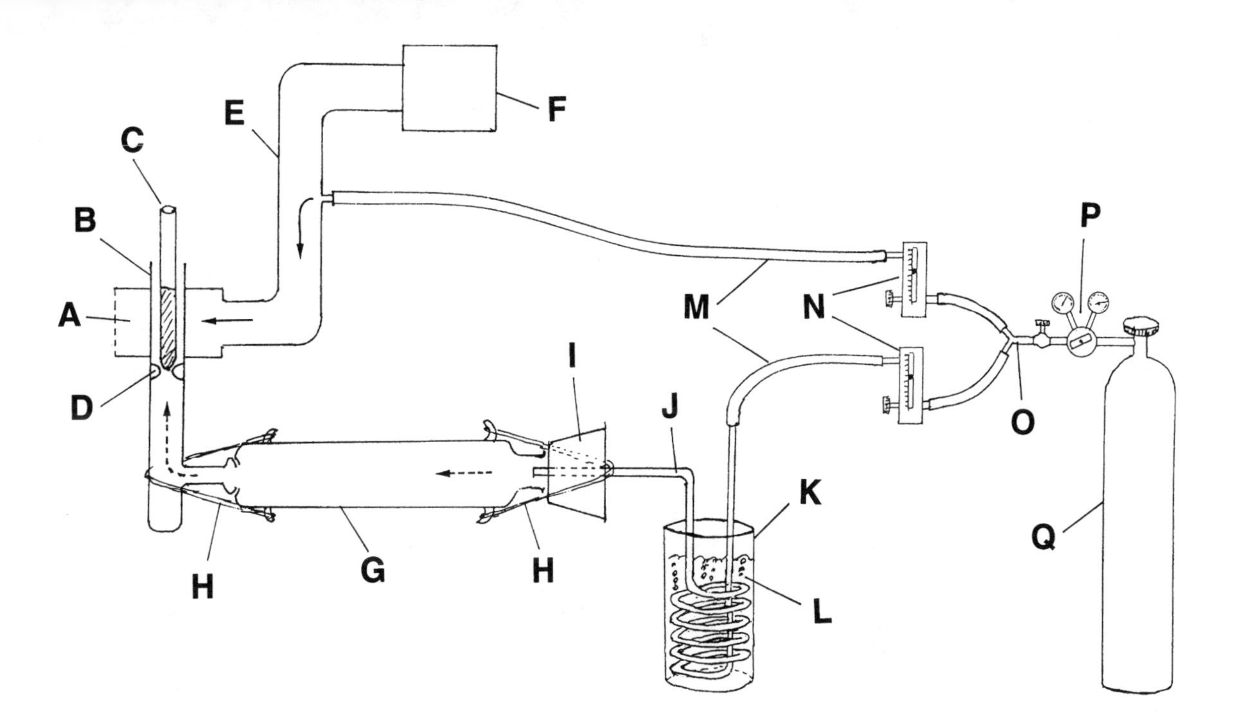

Fig. 7. Schematic illustration of cooling specimen by a nitrogen flow system. **A**, Cavity resonator; **B**, dewar-insert; **C**, quartz tube; **D**, dimples; **E**, micro-wave guide; **F**, microwave generator and detector; **G**, transfer dewar; **H**, rubber band; **I**, silicon rubber stopper; **J**, copper coil; **K**, kitchen dewar-jar; **L**, liquid nitrogen; **M**, tubings; **N**, flow meters; **O**, Y-coupling; **P**, double-stage regulator; **Q**, nitrogen gas cylinder (a gas outlet of liquid nitrogen container can also be used). Drawings do not show actual sizes.

Table 2
A Relationship Between the Nitrogen Flow Velocity and Specimen Temperature

Flow velocity (L/min)	Temperature, (°C)
5	–110
6.5	–133
8	–150
10	–170

Parmer Instrument Company, Vernon Hills, IL). **Table 2** shows the relationship between the flow rate and the temperature of my instrument.

With this simple set-up, the temperature can be regulated within ± 2°C during the experiment. As shown in **Fig. 5**, this degree of fluctuation of the temperature does not adversely affect the accuracy of EPR measurement of NO radicals.

2.3. Device for Anesthesia

1. Use of an anesthetic vaporizer: Connect the inlet of an isoflurane vaporizer (*see* **Note 2**) to a cylinder of compressed air via a flow meter and connect the outlet to a facial mask. This can be easily prepared by rolling up a piece of aluminum foil into a conical shape using two or three layers of aluminum foil (**Fig. 8A**).
2. If a vaporizer is unavailable, a plastic container or a glass bell-jar with a layer of paper towel at the bottom will suffice (**Fig. 8B**). Drop a few mL of isoflurane onto the paper and put the animal into the container until it falls asleep. Then, take out the animal and quickly administer trapping agents subcutaneously.

2.4. NO-Trapping Agents

2.4.1. For Rat Experiments

1. For DETC solution: Dissolve 200 mg of diethyldithiocarbamic acid (sodium salt) per mL of deionized water that had been bubbled with argon or nitrogen. Prepare fresh daily and keep frozen or in ice until immediately before administration (*see* **Note 3**).
2. For Fe-citrate solution: Dissolve 20 mg of $FeSO_4$ and 100 mg of sodium citrate per mL of deionized water that had been bubbled with argon or nitrogen. Prepare fresh daily and keep frozen or in ice until immediately before administration.

2.4.2. For Mice Experiments

1. For DETC solution: Dissolve 100 mg of diethyldithiocarbamic acid (sodium salt) per mL of deionized water that had been bubbled with argon or nitrogen. Prepare fresh daily and keep frozen or in ice until immediately before administration.

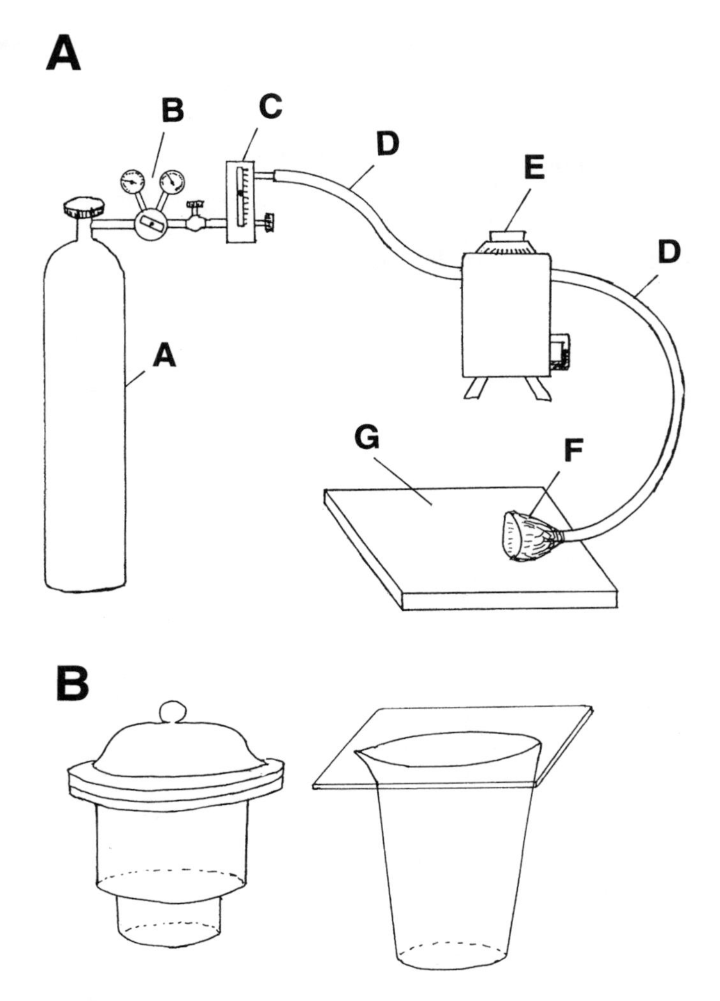

Fig. 8. **(A)** A method of anesthetizing animals using a vaporizer. **A**, Compressed air cylinder; **B**, double-stage regulator; **C**, flow meter; **D**, tubings; **E**, vaporizer; **F**, facial mask made from aluminum foils; **G**, surgical table (made from polystyrene foam plate). **(B)** A method of anesthetizing animals using a container.

2. For Fe-citrate solution: Dissolve 10 mg of $FeSO_4$ and 50 mg of sodium citrate per mL of deionized water that had been bubbled with argon or nitrogen. Prepare fresh daily and keep frozen or in ice until immediately before administration.

2.5. NO Generating Agents

1. For 10 m*M* DEA/NO: Dissolve 1,1-diethyl-2-nitroso-hydrazine (diethylamine-NONOate from Cayman Chemicals Company, Ann Arbor, MI) in nitrogen-bubbled 10 m*M* NaOH to give a final concentration of 10 m*M*.

2. For sodium-phosphate buffer: Prepare a solution containing 140 m*M* NaCl and 3 m*M* sodium-phosphate buffer, bubble with argon or nitrogen, and adjust the pH to 7.4.

2.6. Quartz EPR Tubes for Tissue-NO Study

Use 4-mm (od) quartz tubes, precision grade (714-PQ-8, Wilmad Glass).

2.7. Quartz EPR Tubes for Spinal-Cord NO Study

1. Use 3-mm (od) quartz tubes, precision grade (705-PQ-6.25, Wilmad Glass).
2. Use a quartz tube whose id is slightly larger than 3 mm so that it can hold the precision 3-mm tubes in it. (This is optional. *See* **Subheading 3.3.2.**)

2.8. Others

1. Two 1 mL disposable plastic syringes with 27 gage and 1/2 in-long needle; one for DETC and the other for Fe-citrate administration.
2. 5 mL disposable plastic syringe for extruding tissue samples.
3. An IV catheter (*see* **Note 4**).
4. A small dewar-jar to keep liquid nitrogen for freezing specimens in quartz tubes.
5. Petrie dishes in which to place tissue specimens. It is recommended that these are kept on crushed ice to keep the specimens cool.
6. Isoflurane (*see* **Notes 2** and **5**).
7. Surgical tools.
8. Surgical table (a polystyrene foam plate).
9. A cardio-perfusion device that consists of a reservoir, which is held about 1 m above the surgical desk, and an IV catheter attached with a 18-gage needle.

2.9. Device Needed for Spinal-Cord Study

Obtain a small vacuum pump with a compressor outlet and connect a silicon tube to the outlet. The compression pressure should reach at least 30 lb/in^2 (2 atm pressure). Connect a plastic pipet-tip (for a 1 mL pneumatic pipet) to the outlet of the silicon tubing. Remove any protruberances from the outside of the tip with a razor blade and sand it down until it becomes smooth to make a good airtight fitting with the silicon tubing. The connection should be secured by winding a few turns of a wire around the tubing. Cut the pipet tip to make the id of the tip-opening about 2.5 mm (*see* **Subheading 3.2.3.2.**).

3. Methods

3.1. Injection of NO-Trapping Agents to Animals

3.1.1. Injection into Rats

1. Anesthetize the animals by inhalation of isoflurane (*see* **Note 6**).
2. Administer 2 mL/kg body weight of the DETC solution for rats subcutaneously on one side of the lower abdomen.

3. Administer 2 mL/kg of the Fe-citrate solution for rats subcutaneously on the other side of the lower abdomen (*see* **Note 7**). Important: Keep enough distance between the two sites so that both chemicals do not mix *in situ* and precipitate.

3.1.2. Injection into Mice

1. Anesthetize animals by inhalation of isoflurane (*see* **Note 6**).
2. Administer 4 mL/kg body weight of the DETC solution for mice subcutaneously on one side of the lower abdomen.
3. Administer 4 mL/kg of the Fe-citrate solution for mice subcutaneously on the other side of the lower abdomen (*see* **Note 7**). Important: Subcutaneous administration in the lower abdomen gives the most reproducible results.

3.2. Methods of Transferring Tissue Samples into a Quartz Tube

Thirty min after the injection of NO-trapping agents, the animal is euthenized to remove organs to measure tissue levels of NO. In case of ischemia or hypoxia experiments, ischemia or hypoxia should be produced 30 min after the injection of trapping agents, and tissues removed after the desired time periods.

3.2.1. Internal Organs, i.e., Lung, Liver, and Heart

1. Remove the tissue quickly.
2. Mince with a pair of scissors.
3. Transfer into a 5 mL syringe.
4. Push the contents out through an iv catheter tubing to transfer into a 4-mm (od) quartz tube (**Fig. 9A**).
5. Freeze at liquid-nitrogen temperature (*see* **Note 8**).

3.2.2. Brain

3.2.2.1. For Rats

1. For the rat brain, the volume of the brain is far greater than the volume of tissue needed for the EPR measurement (a height of about 25 mm). Because the NO level in the brain is not uniform, reproducibility is lost if the whole brain is put into the syringe and extruded out into a quartz tube. There are two ways in which this problem can be overcome: dissect an appropriate section of the brain of interest, or homogenize the brain inside the syringe with a thin spatula and make the specimen uniform before extruding out into a quartz tube.
2. Conduct the **steps 1–5** as described above, except for the **step 2**, as mincing is not necessary.

3.2.2.2. For Mice

The whole brain gives a height of only about 25 mm when it is transferred into a 4-mm quartz tube. This is just about the height of the X-band cavity resonator. If the iv-catheter method is used (**Fig. 9A**), some tissue may be lost

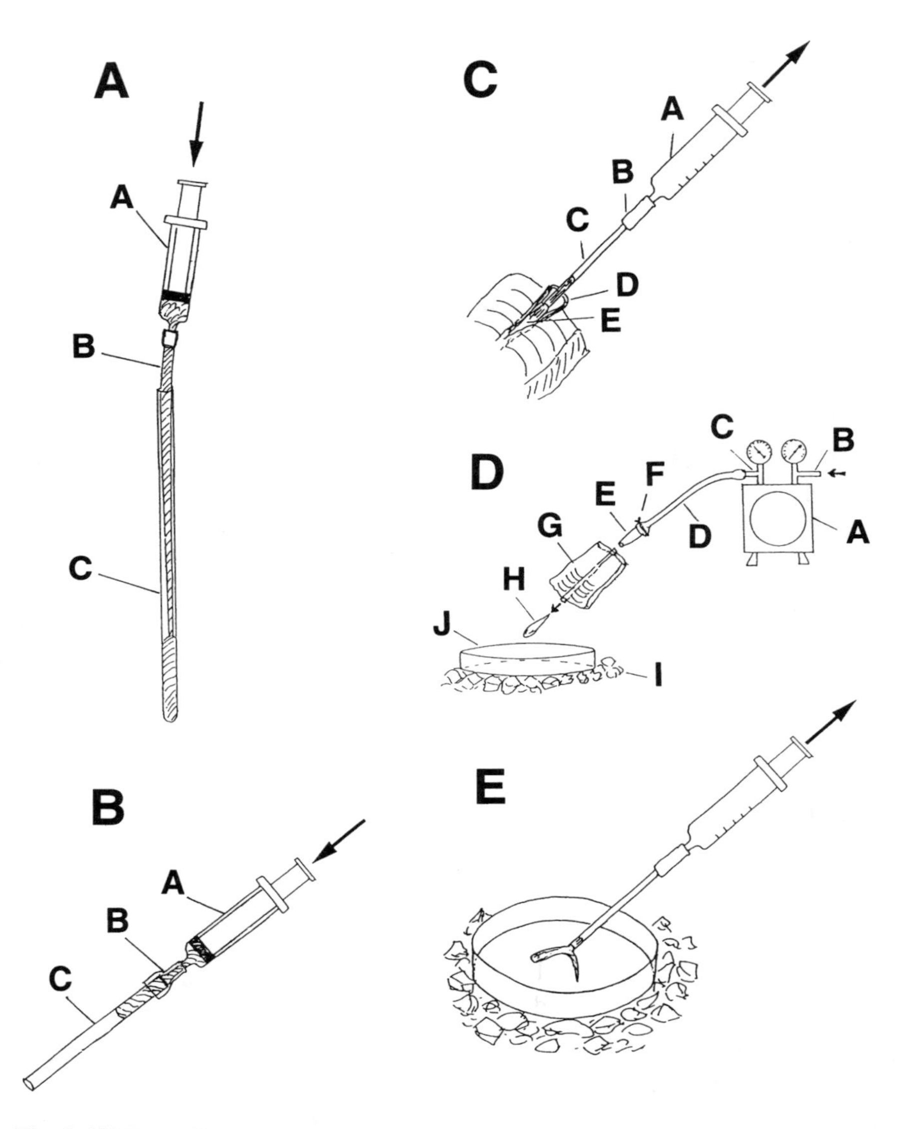

Fig. 9. **(A)** Extruding tissue specimens into a sealed 4 mm quartz tube. **A**, 5 mL syringe; **B**, iv catheter; **C**, sealed 4 mm quartz tube. **(B)** Transferring mouse brain tissue into an open-ended 4 mm quartz tube. **A**, 3-mL syringe; **B**, silicone tubing; **C**, 4 mm open-ended quartz tube. **(C)** Aspirating rat spinal cord directly from the spinal column. **A**, 5-mL syringe; **B**, silicone tubing; **C**, 3-mm quartz tube; **D**, rat spinal column; **E**, rat spinal cord. **(D)** A device for blowing out mouse spinal cord. **A**, vacuum pump; **B**, inlet; **C**, outlet; **D**, silicone tubing; **E**, 1-mL pipet tip with the tip cut to a 2.5 mm opening; **F**, tying wire; **G**, spinal column with muscle and bones; **H**, extruded spinal cord; **I**, crashed ice; and **J**, Petrie dish. **(E)** Aspirating the removed mouse spinal cord into a 3-mm quartz tube.

because it sticks to the catheter wall. In this case, use a 4-mm open-ended tube and transfer the tissue specimen directly from one end of the tube as shown in **Fig. 9B**.

3.2.3. Spinal Cord

Because the spinal cord is very sticky, the procedures described previously cannot be used. The following methods are recommended.

3.2.3.1. Rat Spinal Cord

1. Cut a 3-mm (od) quartz tube to a length of about 50 mm.
2. Aspirate the spinal cord into the tube by applying a negative pressure by sucking from a 5-mL syringe (**Fig. 9C**).

3.2.3.2. Mouse Spinal Cord

1. Take out a block of spinal column that includes the thoracic and lumbar regions together with muscle and ribs.
2. Apply compressed air from the lumbar side, at least 30 lb/in^2 (or 2-atm pressure) through the pipet-tip.
3. Pneumatically extrude the spinal cord onto an ice-cooled Petrie dish (**Fig. 9D**).
4. Aspirate the extruded cord into the 3-mm (od) tube (with the both ends open) from the middle portion of the spinal cord (*see* **Fig. 9E**).
5. Freeze the tube by immersing it into liquid nitrogen (*see* **Note 9**).

3.3. Measurements

3.3.1. Positioning of the 4-mm Quartz Tube

Adjust the position of the dewar-insert in the cavity resonator so that the tissue specimen is properly located inside the resonator.

3.3.2. Positioning of the 3-mm Quartz Tube

A 3-mm quartz tube cannot be held by the dimples of the dewar-insert. Three methods are applicable to hold and locate the 3-mm tube properly:

1. A precision-grade 3-mm (od) quartz tube (e.g., Wilmad Glass 705-PQ-6.25) must be used; however, these will not normally fit inside a 4-mm precision tube. Therefore, select the largest-diameter quartz tube from a number of nonprecision 4-mm (od) quartz tubes that are sufficiently large enough to hold the 3-mm (od) tube (**Fig. 10A**). The holder-tube should be washed clean from time to time. It is good practice to occasionally run the measurement with only this holder-tube to confirm that it has no EPR-active contamination.
2. Cut a 4-mm (od) tubing to a length of 3 or 4 mm. Mount it on top of the dimples in the dewar-insert to hold the 3-mm tube (**Fig. 10B**).
3. Use 3-mm (od) tubing with a length sufficient to stick out from the top of the dewar-insert. Then attach a piece of labeling tape on the tube so that it hangs in the dewar with the tissue specimen at the position of the cavity (**Fig. 10C**).

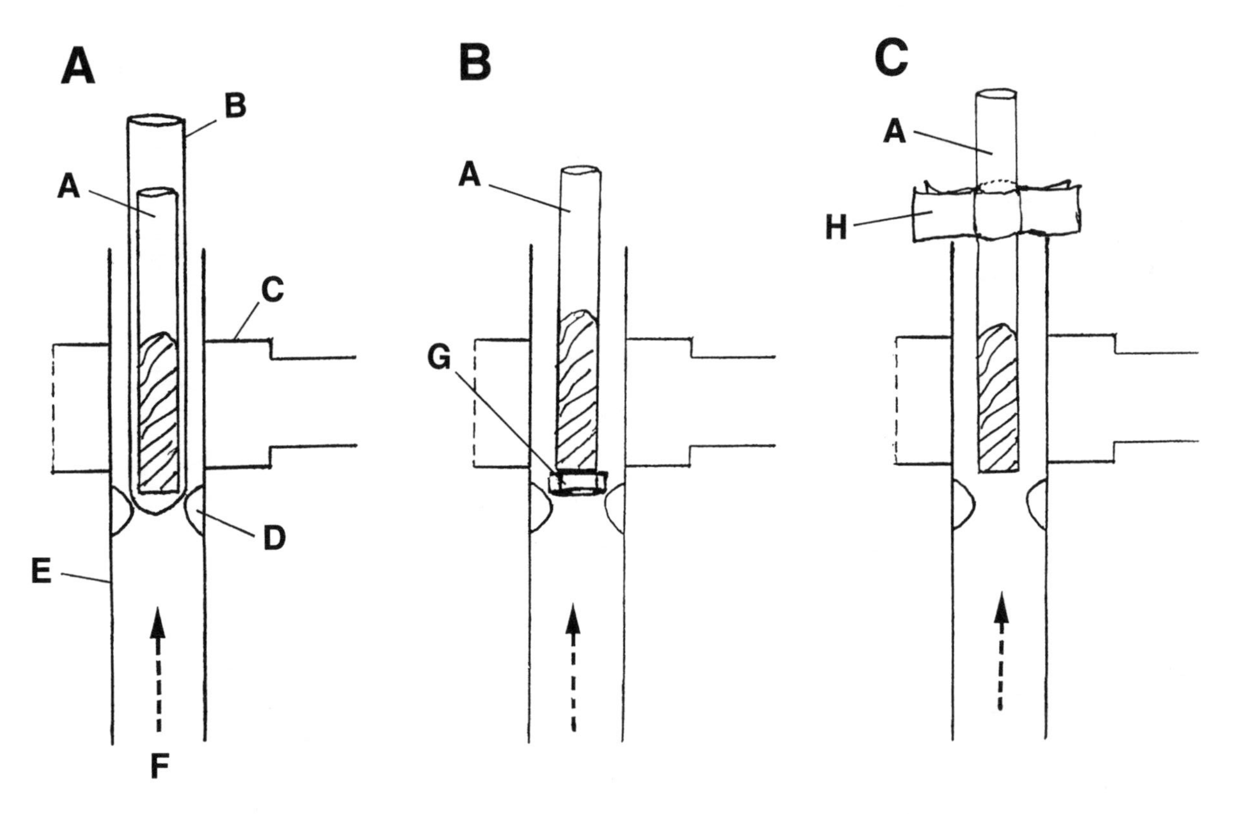

Fig. 10. **(A)** Supporting a 3 mm precision tube at an appropriate place by inserting into a nonprecision tube which has the i.d. somewhat larger than 3 mm. **(B)** Inserting a small piece of precision 4-mm tube on top of the dimples to support a 3 mm tube. **(C)** Supporting a 3 mm tube by a labeling tape. **A**, precision 3-mm tube; **B**, nonprecision 4-mm tube; **C**, cavity resonator; **D**, dimples; **E**, dewar-insert; **F**, nitrogen flow; **G**, a tiny piece of precision 4-mm tube; **H**, labeling tape.

Important: The quartz tube should be secured with a thin rubberband that pulls the tube against the dewar-insert so that it does not move or rattle in the flow of nitrogen gas.

3.3.3. EPR Settings

1. First, run an EPR standard (weak pitch or manganese oxide) to calibrate the instrument.
2. Select the position of g = 2.00 (where the standard has the free-radical peak) at the center of the recording paper.
3. Use a scan range of 400 Gauss for checking the NO signals.
4. Choose a scan speed of 4–16 min for full-scale.
5. Choose the time-constant in such a way that spectra is recorded properly.
6. Adjust the microwave power to 20 mW. This seems to give the best sensitivity/noise ratio without serious power saturation (for more detail, *see* **ref. *13***).
7. Select a scan range to 100 Gauss and change the position of g = 2.00 to the right edge of the recording paper for accurate measurements.
8. Use computer software for signal averaging and repeat the scanning 2–3 times to take the average to increase the reliability of the EPR measurements (*see* **Note 10**).

3.3.4. Negative Control

1. Prepare specimens from an animal into which only DETC was injected. Because this has no iron, only the DETC-Copper signal appears and no NO signal should be observed (**Fig. 11A**). If Fe-citrate is also injected, a physiological level of NO would be observed (**Fig. 11B**).
2. Adjust the gain to make the height of the largest peak, denoted as C (the copper signal, which indicates the tissue concentration of DETC), to about one-half to two-thirds of the recording paper (or computer screen). This should normally be done at a gain of 5×10^3 to 2×10^4 (*see* **Notes 11** and **12**).

3.3.5. Positive Control

When EPR measurements of NO are made for the first time, a positive control should be run to determine the shape and positions of NO signals. Two methods are available.

3.3.5.1. Use of Lipopolysaccharide (LPS)

1. Inject an animal with LPS (1–5 mg/kg i.p.) and wait for 6 h.
2. Inject DETC and Fe-citrate.
3. Collect the tissue in which you are interested 30 min later.
4. Transfer the tissue into a quartz tube and freeze it.
5. Run EPR measurements. You should see a large increase in NO equivalent to the production of 10–30 μ*M* (**Fig. 11C**) *(16,17)*.

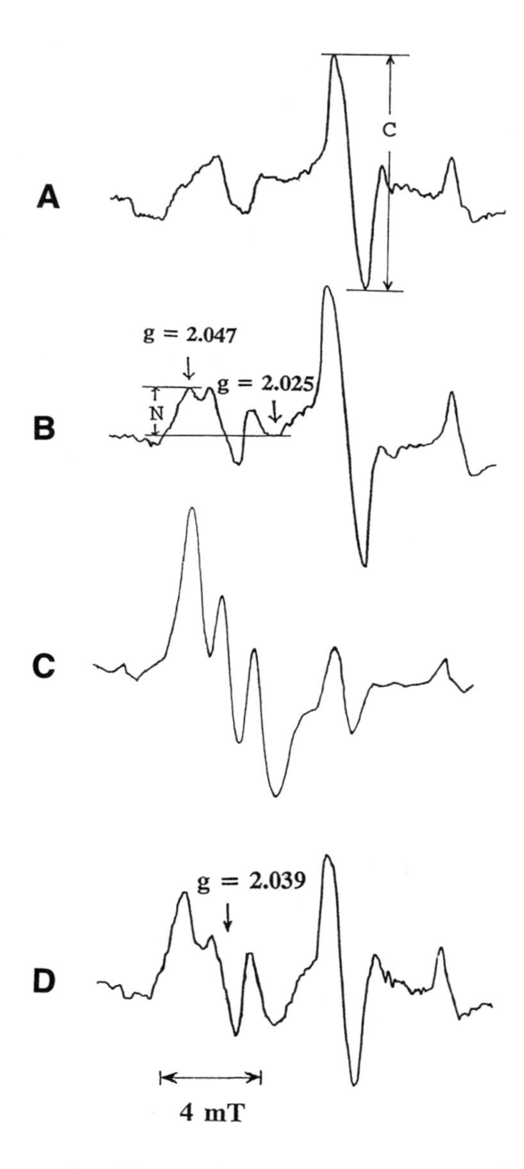

Fig. 11. **(A)** Copper signal from animals administered with only DETC. **(B)** A small NO signal (as indicated by N) observed in the normal rat brain. **(C)** Spin-trapped NO signal from the brain of an LPS-injected rat. **(D)** Spin-trapped NO signal from the rat brain induced by hypotention.

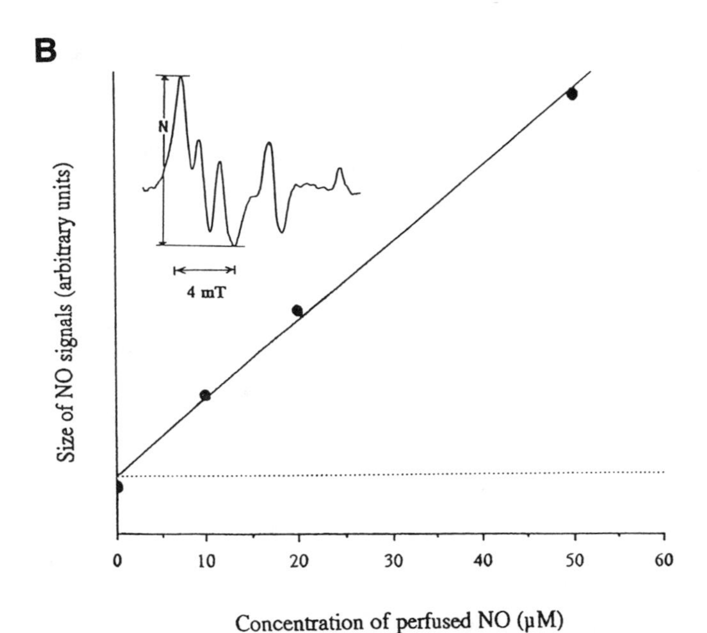

Fig. 12. **(A)** A NO generating compound, DEA/NO. **(B)** An example of calibration using DEA/NO perfusion. (Used with permission from **ref.** ***12***.)

3.3.5.2. Use of Hypotention-Induced NO Production

1. Anesthetize the animal.
2. Inject DETC and Fe-citrate.
3. After 30 min anesthetize the animal again.
4. Open the chest and remove the heart.
5. Collect the tissue of interest 10 min later and freeze.
6. Perform EPR measurement (*see* **Note 12**). The level of NO should be between 5–10 μ*M* (**Fig. 11D**).

3.4. Calibration

Calibration can be performed using a NO-generating compound, 1,1-diethyl-2-nitroso-hydrazine (called DEA/NO, which was originally synthesized by Keefer et al. [**Fig. 12A**]) ***(18)***. The concentration of NO in the solution is 1.5 times of the concentration of DEA/NO at pH 7.4 ***(18)***.

1. Anesthetize the animal.
2. Dilute the stock solution of 10 m*M* DEA/NO with a solution of nitrogen-bubbled solution containing 140 m*M* NaCl and 3 m*M* sodium-phosphate buffer, pH 7.4, to generate a known level of NO (1.5 times that of the concentration of DEA/NO).
3. Adjust the hydrostatic pressure of the perfusate to about 1 *M*.
4. Perfuse the mixture immediately into the left ventricle.
5. Cut open the femoral artery to let the mixture freely flow out.
6. Collect the tissue 5 min later and transfer into a quartz tube and freeze.
7. Perform EPR measurements.
8. Calibrate the instrument by measuring the height of the peak. **Figure 12B** shows an example of such a calibration where the height "N" represents the tissue level of NO (*see* **Notes 13** and **14** for the rationale underlying the calibration method) *(12,13)*. By measuring the height of N, the production of NO in a particular tissue can be calibrated.

Important: When working with a small animal, it may not be possible to collect sufficient tissue to fill up the quartz tube to the size of the cavity (height of about 25 mm). This will produce a smaller EPR signal than that with a full tube; however, as long as the tissue level of NO is around 10 μ*M* or less (*see* **Note 15**), the signal can be normalized using the height of the copper signal (the signal C in **Fig. 10**). For example, if the size of the "C" signal using insufficent material is one-half of the signal when sufficient tissue is used, then the "N" signal should be multiplied by a factor of two to correct for the lack of tissue. In other words, the ratio of N/C represents the tissue level of NO as long as it is less than 10 μ*M*. The EPR method of measuring NO would have many more applications (*see* **Notes 16** and **17**).

4. Notes

1. Biological specimens contain water. Therefore, the use of the X-band EPR is limited mostly to frozen specimens because water absorbs microwaves. The use of a longer wavelength, namely the use of the S-band (3 GHz) and L-band (1 GHz) EPR has been actively studied. Because the radiation energy is proportional to the frequency, the sensitivity of EPR instruments is also proportional to the frequency. Therefore, the S-band EPR is about three times less sensitive than the X-band EPR. A geometrical factor also poses a problem. Because the size of the cavity resonator of the S-band (10 cm) is about three times larger than that of the X-band (3.2 cm), if the specimen has the same size, then the ratio of specimen/cavity is about three times smaller in the S-band than in the X-band. It makes the efficiency of resonance absorption about 1/3 in the S-band than in the X-band. Altogether, the sensitivity in the S-band is about 1/10 of that in the X-band, if the same specimen is measured with two instruments. On the other hand, the S-band EPR has a definite advantage in that it can hold a larger sample and that the energy loss by water is much less. Using this system, three-

dimensional imaging EPR, based upon the similar principle to that in NMR imaging *(19)*, is possible and has been used to generate a three-dimensional image of NO production in ischemic-hypoxic rat brain *(21)*. Similarly, the Lai's group has developed a water-soluble spin trapping agent, N-methyl D-glucamine dithiocarbamate (MGD)-iron complex *(6,7)*. Following injection of a mouse with this complex, they have measured the NO level in the blood stream of a live mouse on-line by putting the tail of the mouse into the cavity of S-band EPR *(7)*.

2. Ideally, when isoflurane is used, a vaporizer designed for isoflurane should be used. However, if this is not available, you may substitute it with a vaporizer for halothane or enflurane. Vaporizers for these two inhalational anesthetics may not be actively used in the operating room today and may be obtained from the store room of Anesthesiology Departments. Diethyl ether is not recommended, because it is a fire hazard, it has a relatively high cardio-depressing effect, and it remains in the tissue.
3. DETC is an irritant. Wear an eye-goggle and a mask.
4. If you cannot find appropriate catheter tubing which fits into 4-mm (od) quartz tube (which has approx 3-mm id), obtain i.v. catheters from a hospital supplier (which has an od of about 3.5 mm). Cut the catheter to a length of about 20 cm, and stretch it in boiling water. When the tube is cooled in tap water, the diameter of the catheter should be reduced sufficiently to fit the inner diameter (3 mm) of the quartz tube (**Fig. 8A**).
5. In purchasing an inhalational anesthetic agent, the one manufactured for veterinary use is less expensive than that for human use.
6. To attain the best reproducibility, animals are anesthetized under inhalational anesthetics before subcutaneous injection of trapping agents. We use isoflurane (veterinary use), because its cardiodepressing effect is relatively small, and because the animal wakes up quickly when the anesthetic is taken away. Induce anesthesia at about the 4% (w/v) concentration of anesthetic gas. When the animal is asleep, reduce the gas concentration to about 1% (w/v) and administer the trapping agents.
7. Originally, Vanin's group used intraperitoneal injections of 500 mg DETC/kg body weight and 40 mg $FeSO_4$ + 200 mg sodium citrate/kg body weight for animal experiments. We found that 500 mg DETC/kg produced a severe hypotension, which in turn produced NO (which is a vasodilator) to increase blood flow to counteract the situation *(9–13)*. We found that the subcutaneous injections of both 400 DETC/kg body weight and 40 mg $FeSO_4$ + 200 mg sodium citrate/kg body weight produced satisfactory results for both rats and mice *(13)*.
8. If you immerse a sealed tube directly into liquid nitrogen and keep it until the tissue is frozen, occasionally the tube will crack because the contents expands upon freezing. In order to prevent this, two methods can be applied:
   a. Immerse the tube for an instant, and quickly take it out for few seconds. Then, repeat the procedure several times.
   b. Pour a small amount of ethyl alcohol in a polypropylene test tube and cool it with dry-ice. Then, dip the quartz tube into it for freezing.

Assumption 1: DETC-Fe complex is absorbed onto the endothelial cells.

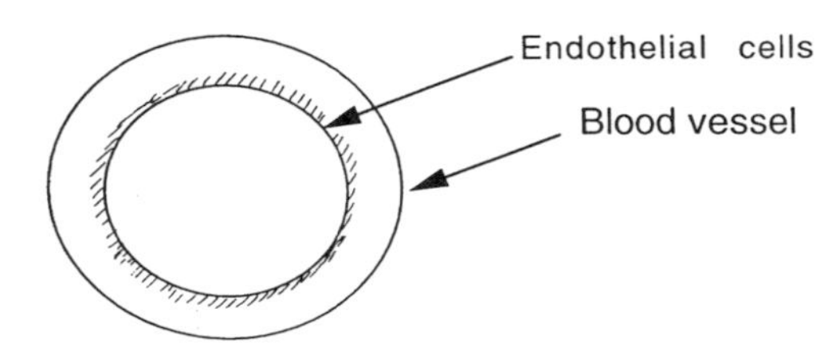

Assumption 2:

Absorbed DETC-Fe reacts equally with tissue NO and perfusate NO.

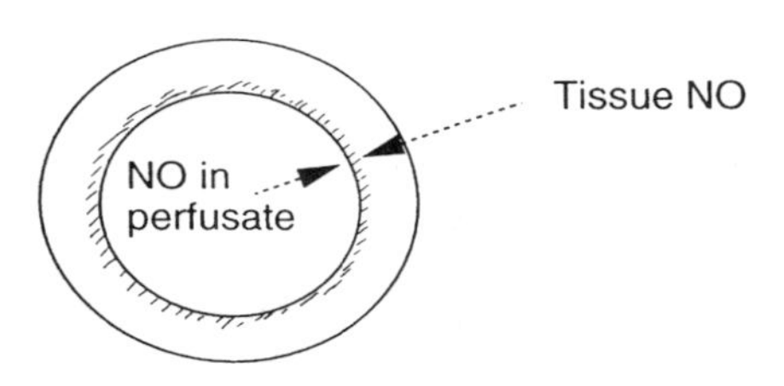

Fig. 13. Assumptions used for the calibration of tissue NO concentration.

9. Because the tube has an open end, and because the amount of the tissue is small, it does not crack when the contents freeze up.
10. If the instrument is too old to have such a system, you can still scan 2–3 times and take the average manually from these recordings. Examples of measurements of actual tissue-NO levels can be seen in the literature, e.g., in the brains ***(9–13)***, spinal cord ***(14)***, the lung ***(15,16)***, and the liver ***(17)***.
11. The size is different with different instruments. If you do not have enough signal height, check the concentration of DETC that you have prepared. If DETC becomes old, or oxidized, it will lose the copper signal.
12. Because of hypotension caused by the heart removal, the tissue produces NO to cause vasodilation (in an attempt to have more blood flow).
13. Detailed characterization of the method is published elsewhere ***(13)***. The points can be summarized as follows:
    a. Time course: The signal heights of the copper signal reach the maximum at 30 min after injection, and then slowly decrease. The height at 60 min is approx 60% of that at 30 min.

b. The NO signal "N" is relatively temperature insensitive between –110°C and –180°C as shown in **Fig. 12**.
c. The copper signal "C" increase with temperature. Therefore, the temperature range of –130°C to –170°C is preferred.
d. Origin of the copper signals are identified as the copper atoms in superoxide dismutase (SOD). Although the copper signals interfere with the NO signal, it can be used as a marker of the tissue concentration of DETC.

14. The calibration method described in this chapter is based upon following assumptions (**Fig. 13**):
    a. The DETC-iron complex is absorbed onto the inner membrane of the blood vessel.
    b. The membrane-bound complex is able to equally bind NO that is produced outside of the blood vessel or supplied from inside the blood vessel.
    c. The formation of DETC-iron-NO adduct is proportional to the local concentration of NO.
15. When the NO levels exceed 10 µ*M*, the copper signal "C" starts decreasing as the NO signal "N" increases.
16. For in vitro EPR measurements, the use of a water-soluble carbamate may be more advantageous, because the DETC-iron complex is not water soluble. The water-soluble MGD-iron complex has been used to measure the time course of NO production by isolated macrophages adhered onto the inner surface of a EPR quartz flat cells ***(6–8)***. Similarly, Kuppusamy et al. perfused the beating heart (an in vitro Langendorf model) with the MGD-iron complex and measured the concentration of NO in that beating heart. Using S-band EPR and 3-D imaging EPR ***(19)***, they have performed 3-D imaging of the NO production of the beating heart ***(20)***.
17. The development of EPR techniques that we can use for disease-related study and diagnosis is important. With the development of more sensitive spin-trapping agents, EPR will become more versatile and can be more effectively applied to the detection and quantitation of NO. Especially, a recent development of 3-D EPR imaging technique may make EPR a powerful tool in this field. The lowest limit of detecting tissue NO is about 1 µ*M* in X-band EPR. In the case of 3-D imaging using S-band EPR, it is still above 10 µ*M* ***(21)***. Therefore, there is an urgent need for finding more sensitive spin-trapping agents. If a trapping agent that is ten times more sensitive than the current agents is found, then the EPR method of detecting NO will become a very powerful technique.

Acknowledgment

The author thanks Drs. Teiji Tominaga and Shinya Sato from the Institute of Brain Diseases, Tohoku University, Dr. Tomoko Ohnishi from the University of Pennsylvania, and Drs. H. Koprowski and D.C. Hooper from Thomas Jefferson University for their collaboration.

References

1. Lancaster, J. R. II and Hibbs, J. B. II, (1990) EPR demonstration of iron-nitrosyl complex formation by cytotoxic activated macrophages. *Proc. Natl. Acad. Sci. USA* **87,** 1223–1227.
2. Kosaka, H., Watanabe, M., Yoshihara, H., Harada, N., and Shiga, T. (1992) Detection of nitric oxide production in lipopolysaccharide-treated rats by ESR using carbon monoxide hemoglobin. *Biochem. Biophys. Res. Comm.* **184,** 1119–1124.
3. Lin, R. F., Lin, T.-S., Tilton, R. G., and Cross, A. H. (1993) Nitric oxide localized to spinal cords of mice with experimental allergic encephalomyelitis: An electron paramagnetic resonance study. *J. Exp. Med.* **178,** 643–648.
4. Kubrina, L. N., Caldwell, W. S., Mordvintcev, P. I., Malenkova, I. V., and Vanin, A. F. (1992) EPR evidence for nitric oxide production from guanidino nitrogens of L-arginine in animal tissues in vivo. *Biochim. Biophys. Acta.* **1099,** 233–237
5. Voevodskaya, N. V. and Vanin, A. (1992) Gamma-irradiation potentiates L-arginine-dependent nitric oxide formation in mice. *Biochem. Biophys. Res. Comm.* **186,** 1423–1428
6. Komarov, A., Mattson, D., Jones, M. M., Singh, P. K., and Lai, C.-S. (1993) In vivo spin trapping of nitric oxide in mice. *Biochim. Biophys. Res. Commun.* **195,** 1191–1198
7. Lai, C.-S. and Komarov, A. (1994) Spin trapping of nitric oxide produced in vivo in septic shock mice. *FEBS Lett.* **345,** 120–124.
8. Kotake, Y., Tanigawa, T., Tanigawa, M., Ueno, I., Allen, D. R., and Lai, C.-S. (1996) Continuous monitoring of cellular nitric oxide generation by spin trapping with an iron-dithiocarbamate complex. *Biochim Biophys. Acta.* **1289,** 362–368.
9. Sato, S., Tominaga, T., Ohnishi, T., and Ohnishi, S. T. (1993) A spin trapping study of nitric oxide formation during bilateral carotid occlusion in the rat. *Biochim. Biophys. Acta.* **1181,** 195–197.
10. Tominaga, T., Sato, S., Ohnishi, T., and Ohnishi, S. T. (1993) Potentiation of nitric oxide formation following bilateral carotid occlusion and focal cerebral ischemia in the rat: In vivo detection of the nitric oxide radical by electron paramagnetic resonance spin trapping. *Brain Res.* **614,** 342–346.
11. Sato, S., Tominaga, T., Ohnishi, T., and Ohnishi, S. T. (1994) EPR study on nitric oxide production during brain focal ischemia and reperfusion in the rat. *Brain Res.* **647,** 91–96.
12. Tominaga, T., Sato, S., Ohnishi, T., and Ohnishi, S. T. (1994) EPR detection of nitric oxide produced during forebrain ischemia in the rat. *J. Cereb. Blood Flow Metab.* **14,** 715–722.
13. Sato, S., Tominaga, T., Ohnishi, T., and Ohnishi, S. T. (1995) Trapping of nitric oxide radicals in brain ischemia, in *Membrane-linked Diseases* (vol. 4) *CNS Trauma: Experimental Techniques*, (Ohnishi, S. T. and Ohnishi, T., eds.), CRC Press, Boca, Raton, FL, pp. 453–468.
14. Hooper, D. C., Ohnishi, T. S., Kean, R., Numagami, Y., Dietzschold, B., and Koprowski, H. (1995) Local nitric oxide production in viral and autoimmune diseases of the central nervous system. *Proc. Natl. Acad. Sci. USA* **92,** 5312–5316.

15. Ischiropoulos, H., Beers, M., Ohnishi, S. T., Fisher, D., Garner, S. E., and Thom, S. R. (1996) Nitric oxide production and perivascular tyrosine nitration in brain after carbon monoxide poisoning in the rat. *J. Clin. Invest.* **97,** 2260–2267.
16. Wizemann, T. M., Gardner, C. R., Quniones, S., Durham, S. K., Goller, N. L., Ohnishi, S. T., and Laskin, D. L. (1994) Production of nitric oxide and peroxynitrite in the lung following acute endotoxemia. *J. Leukocyte Biol.* **56,** 759–768.
17. Laskin, D. L, Rodriguez del Valle, M., Heck, D. E., Hwang, S.-M., Ohnishi, S. T., Durham, S. K., Goller, N. L., and Laskin, J. D. (1995) Hepatic nitric oxide production following acute endotoxemia in rats is mediated by increased inducible nitric oxide synthase gene expression. *Hepatology* **22,** 223–234.
18. Maragos, C. M., Morley, D., Wink, D. A., Dunams, T. M., Saavedra, J. E., Hoffman, A., Bove, A. A., Isaac, L., Hrabie, J. A., and Keefer, L. K. (1991) Complexes of NO with nucleophiles as agents for the controlled biological release of nitric oxide. *J. Medicinal Chem.* **34,** 3242–3247.
19. Kuppusamy, P., Chzhan, M., and Zweier, J. L. (1995) Development and optimization of 3-D spatial EPR imaging for biological organs and tissues. *J. Magn. Resonance B,* **105,** 122–130.
20. Kuppusamy, P., Chzhan, M., Wang, P., and Zweier, J. L. (1996) Three-dimensional gated EPR imaging of the beating heart: time-resolved measurements of free radical distribution during the cardiac contractile cycle. *Magn. Reson. Med.* **35,** 323–328.
21. Kuppusamy, P., Ohnishi, S. T., Numagami, Y., Ohnishi, T., and Zweier, J. L. (1994) Three-dimensional imaging of nitric oxide production in the rat brain exposed to ischemia-hypoxia. *J. Cereb. Blood Flow Metab.* **15,** 899–903.

14

Measurement of eNOS and iNOS mRNA Expression Using Reverse Transcription Polymerase Chain Reaction

Norbert Reiling, Artur J. Ulmer, and Sunna Hauschildt

1. Introduction

Numerous cell populations are capable of producing nitric oxide (NO) from L-arginine. The nitric oxide synthases (NOS) catalyzing the formation of NO have been grouped into two broad categories: the constitutive Ca^{2+}/calmodulin (CaM)-dependent endothelial NOS (eNOS) and neuronal NOS (nNOS) and an inducible Ca^{2+}/CaM-independent NOS (iNOS). If NO is produced in insufficient amounts, or if it reacts with other compounds, it often escapes detection. Because the failure to detect NO does not necessarily mean that cells are incapable of producing NO, other methods must be used to approach this question. One simple, adequate, and very powerful method is the analysis of NOS-mRNA expression by reverse transcription polymerase chain reaction (RT-PCR). This method allows the detection of minor amounts of different NOS-mRNA isoforms and points to the possible existence of the corresponding enzymes. Once translated, NOSs are only active in the presence of numerous cofactors. If the cofactors are missing, one will still find NOS-mRNA expression but no NO produced—indicating the potential of the cell to produce NO under appropriate conditions *(1,2)*. The following protocol enables the detection of the Ca^{2+}/CaM dependent eNOS and the iNOS in a limited number of cells by the use of RT-PCR using isolated mRNA. If the availability of tissue is not limited, then the protocol can easily be adapted to use total RNA in the RT reaction.

From: *Methods in Molecular Biology, Vol. 100. Nitric Oxide Protocols*
Edited by: Michael A. Titheradge © Humana Press Inc., Totowa, NJ

2. Materials

2.1. Poly A⁺-mRNA Preparation

1. Diethyl pyrocarbonate (DEPC) treated water (*see* **Notes 1** and **2**) or RNase-free water (United States Biochemical, Cleveland, OH).
2. Phosphate-buffered saline (PBS): 0.8 m*M* Na_2HPO_4, 0.2 m*M* KH_2PO_4, 137 m*M* NaCl, pH 7.4. Store at 4°C.
3. Oligo-$(dT)_{25}$- coated magnetic beads; magnetic particle concentrator (MPC), (Dynabeads, Dynal, Oslo, Norway).
4. RNase inhibitor (RNA guard (31200 U/mL); Pharmacia Biotech (Freiburg, Germany).
5. Lysis buffer: 10 m*M* Tris-HCl, pH 7.5, 0.14 m*M* NaCl, 5 m*M* KCl, 1% Nonidet P40. Store at 4°C.
6. Washing buffer I: 10 m*M* Tris-HCl, 0.15 m*M* LiCl, 1 m*M* ethylenediaminetertra-acetic acid (EDTA), pH 7.5, 0.2% sodium dodecyl sulfate (SDS). Store at 4°C.
7. Washing buffer II: 10 m*M* Tris-HCl, 0.15 m*M* LiCl, 1 m*M* EDTA, pH 7.5. Store at 4°C.
8. Elution buffer: 2 m*M* EDTA, pH 7.5. Store at 4°C.
9. 2X Binding buffer: 20 m*M* Tris-HCl, pH 7.5, 1 *M* LiCl, 2 m*M* EDTA, 0.7% SDS. Store at 20°C (*see* **Note 3**).
10. Cooled microcentrifuge (4°C).

2.2. Reverse Transcription

1. Oligo-dT_{20} (0.5μg/μL) in DEPC-treated water. Store at –20°C.
2. RNase inhibitor [RNA guard (31200 U/mL); Pharmacia Biotech]. Store at –20°C.
3. 10 m*M* Deoxynucleotide triphosphates (dNTP) (dATP, dCTP, dGTP, dTTP diluted in DEPC-treated water) (Ultrapure dNTP Set, Pharmacia Biotech). Store at –20°C.
4. Reverse transcriptase [Superscript (200 U/μL); Gibco-BRL, Eggenstein, Germany]. The enzyme is delivered together with 5X first-strand buffer (250 m*M* Tris-HCl, pH 8.3, 375 m*M* KCl, 15 m*M* $MgCl_2$) and 0.1 *M* dithiothreitol (DTT). Store at –20°C.
5. Two water baths (70°C and 37°C).

2.3. PCR

1. 10 m*M* dNTP (*see* **Subheading 2.2.3.**).
2. 10X Tricine buffer *(3)*: 300 m*M* N-tris [hydroxymethyl]-methylglycine, pH 8.4, 20 m*M* $MgCl_2$, 50 m*M* mercaptoethanol, 0.1% pork gelatin, 1% thesit (polyoxyethylene-9-laurylether). (All chemicals from Sigma, Deisenhofen, Germany.) Store at –20°C.
3. *Taq* DNA polymerase (Gibco-BRL, 5 U/μL). Store at –20°C.
4. Mineral oil (Sigma).
5. DNA-thermal cycler 480 (Perkin-Elmer Cetus, Norwalk, CT).

2.4. Agarose Gel Electrophoresis

1. 50X TAE buffer: 2 *M* Tris, 0.5 M EDTA, adjusted to pH 8.0 with acetic acid.
2. Agarose (Ultrapure, Gibco-BRL).
3. Ethidium bromide: Stock solution 10 mg/mL of water (*see* **Note 1**).
4. Gel loading solution: TAE buffer containing 50% (v/v) glycerol, 1% bromphenol blue, 1% xylenecyanol FF (Sigma).

3. Methods

3.1. Preparation of Poly A^+-mRNA Using Oligo-dT-Coated Magnetic Beads

1. Wash cells in autoclaved PBS, centrifuge (400*g*, 10 min, 4°C), discard the supernatant, and freeze the dry cell pellet at –70°C prior to use or use immediately.
2. Prepare oligo-dT-coated magnetic beads (90 μL beads/1 × 10^6 cells in 100 μL of 2X binding buffer).
3. Add 100 μL of lysis buffer (containing 500 U/mL RNase inhibitor) to the frozen-cell pellet, vortex vigorously, and incubate for 1 min (*see* **Note 4**).
4. Centrifuge (30 s, 11000*g*, 4°C), transfer the supernatant to the prepared solution of beads, resuspend, and incubate for 4 min on ice (*see* **Note 5**).
5. Put the tubes into the MPC for 20–30 s until all the beads stick to the wall of the tube. Carefully discard the supernatant and add 200 μL of washing buffer I, resuspend carefully, and incubate again in the MPC. Repeat this washing step and add two further washing steps with 200 μL of washing buffer II.
6. Remove washing buffer II, add 10 μL of elution buffer, and incubate for 2 min at 65°C. The beads are immediately incubated for 15 s in the MPC and the mRNA-containing supernatant is quickly transferred to a new tube for RT (*see* **Note 6**).

3.2. Reverse Transcription

1. 1 μL of oligo-dT_{25} solution (0.5 μg/mL) is added to 10 μL of mRNA solution, mixed, spun down, and incubated at 70°C for 10 min to disrupt any secondary structures that may involve the poly-A tail.
2. In the meantime, the RT mix (10 μL/probe) is prepared: 0.4 μL of RNase inhibitor, 4 μL of 5X first-strand buffer, 0.4 μL of 0.1 *M* DTT, 1 μL of 10 m*M* dNTP, 1 μL of Superscript (200 U/mL), and 3.2 μL of H_2O.
3. The RT mix is added to the oligo-dT-mRNA-Mix and incubated for 1 h at 37°C.
4. Incubate at 95°C for 3–5 min. The resultant cDNA (20 μL) is kept at –20°C (*see* **Note 7**).

3.3. PCR

1. Prepare a PCR mix for (n + 2) assays (negative and positive control). The PCR mix for one preparation contains in a volume of 33 μL: 0.7 μL of 10 m*M* dNTP, 3.5 μL of tricine buffer, 26 μL of H_2O, 1.7 μL of 20 μ*M* of sense and antisense primer (*see* **Notes 8–11**), 0.17 μL of *Taq* DNA polymerase (5 U/μL).

2. 1–3 μL of cDNA are added to each probe (*see* **Notes 12** and **13**), mixed, centrifuged down (10 s, 2000*g*, 20°C), and then overlaid with 70 μL of mineral oil.
3. The PCR tubes are closed and put into a preheated thermal cycler (95°C) that is programmed as follows: 1 min at 94°C, 1 min at 58–63°C (dependent on primer used), 1.5 min at 72°C for 35 cycles.
4. After 35 cycles, the machine is kept at 72°C for 15 min before cooling down to 4°C. The PCR products are subsequently analyzed by agarose-gel electrophoresis.

3.4. Touch-Down PCR

cNOS and iNOS mRNA amplification by RT-PCR may sometimes be difficult because of their low abundance. A so-called *Touch-down* protocol ***(4)*** has been very useful in amplifying these mRNAs. The method includes a highly specific preamplification step before the 25- or 30-cycle PCR (*see* **Note 14**). The following cycles are carried out:

(95°C 30 s / 70°C 2 min) × 2
(95°C 30 s / 69°C 1 min 55 s) × 2
(95°C 30 s / 68°C 1 min 50 s) × 2
(95°C 30 s / 67°C 1 min 45 s) × 2
(95°C 30 s / 66°C 1 min 40 s) × 2
(95°C 30 s / 65°C 1 min 35 s) × 2
(95°C 30 s / 64°C 1 min 30 s) × 2
(95°C 30 s / 63°C 1 min 25 s) × 2
(95°C 30 s / 62°C 1 min 20 s) × 2
(95°C 30 s / 61°C 1 min 15 s) × 2
(95°C 30 s / 60°C 1 min 10 s) × 2
(95°C 30 s / 55°C 30s / 72°C 30 s) × 30

3.5. Agarose-Gel Electrophoresis

1. Prepare a 1.5% agarose solution containing 1 μg/mL of ethidium bromide (1:10000 of the stock solution) in TAE buffer, boil twice for 5 s in a microwave. Let it cool down to approx 50°C while slowly stirring, and pour it into the gel cassette containing the comb for the slots.
2. Mix 10 μL of each PCR product with 1 μL of gel-loading solution and fill the slots. Fill one slot with a DNA-standard solution (range: 200–2000 bp) (e.g., Molecular Weight Marker VI, Boehringer Mannheim, Mannheim, Germany).
4. The electrophoresis is performed in TAE buffer at a voltage of 10 V/cm gel width for 1–2 h.
5. Gels are photographed under UV light (302 nm) (Polaroidfilm 667; Polaroid, Cambridge, MA).

4. Notes

1. To prevent RNase contamination, and owing to toxicicty and carcinogenicity of various reagents used in the protocol (e.g., ethidium bromide), it is strongly recommended that one wear gloves at all times.
2. All solutions used for mRNA preparation should be incubated for at least 12 h at 37°C with 0.1% (v/v) DEPC, except those containing Tris. The solution should then be autoclaved to destroy the DEPC. DEPC is highly flammable and must be handled in a fume-hood; it is also thought to be carcinogenic, so it should be handled with care.

3. Lysis buffer, washing buffer I, II, and elution buffer (portions of 50 mL) should be stored at 4°C. 2X Binding buffer should be kept at room temperature to stop the crystallization of SDS.
4. Owing to the instability of mRNA, quick handling is necessary, and if not otherwise stated, all steps are best performed on ice. It is recommended that RT is performed directly after mRNA preparation. Be sure to mix and spin down all small volumes added before each incubation step.
5. Sometimes clotting of the beads during the first washing step is observed. Try to resuspend the beads carefully several times. But don't worry.
6. Collect the magnetic beads after usage. They can be regenerated up to five times by resuspending them in 0.1 *M* NaOH, followed by incubation at 65°C for 2 min. Afterwards, the beads are washed in 2X binding buffer until the pH is below 8.0. Store the recycled beads in 2X binding buffer at 4°C.
7. It is useful to perform RNA preparation, RT, and pipetting of the PCR preparations in one room. The DNA thermal cycler and the analysis of the PCR product should be kept at a different location. You must use a second set of pipets after the PCR-amplification step to minimize DNA contamination.
8. To verify that the mRNA preparation was successful, a PCR of a constitutively expressed house-keeping gene (e.g., β-actin) should be carried out. It is recommended to run only 25 cycles of the β-actin PCR (β-actin sense 5'-AGCGGGAAATCGTGCGTG-3', β-actin antisense 5'-CAGGGTACATGG TGGTGCC-3'; annealing temperature: 55°C). The analysis of the PCR products allows a semiquantitative comparison of the used cDNA amount. Before starting NOS-RT-PCR, the β-actin PCR signal should show a comparable intensity in all preparations that are to be tested.
9. Most of the eNOS and iNOS-specific primers (*see* **Table 1**) are intron-spanning and were selected according to published human eNOS and iNOS mRNA sequences *(5,6)* and the *Oligo* Primer Analysis Software (Vers 4.1; National Biosciences Inc., Plymouth, MN). The specificity of the selected primers was controlled by matching them with the EMBL nucleotide sequence database (European Molecular Biology Laboratory, Heidelberg, Germany) using the *Microgenie* software (Version 7.1, Beckman Instruments, Munich, Germany).
10. For the analysis of NOS expression on a single-cell level, the primer sets IN6sc (*see* **Table 1**) and eNOS2sc can be used. These primers are located in the last 600 bp of the published mRNAs and have been used successfully *(7)*. In contrast to the other primers (*see* **Table 1**), these primer sets are not intron-spanning. As one cannot distinguish between DNA and cDNA based amplifications, this does not necessarily mean that the yield is only owing to the the added cDNA, but may also be a consequence of DNA contamination. To generally exclude this, a parallel-PCR reaction of a probe without RT should be done.
11. Cells expressing high NOS mRNA levels should be used as a positive control (eNOS: human endothelial cells; iNOS: human IL-1-stimulated smooth-muscle cells or the stimulated human adenocarcinoma cell line DLD-1 *(8)*, which can be obtained from the American Type Culture Collection (ATTC, Rockville, MD).

Table 1
Primer Sequences for eNOS and iNOS mRNA Detection by RT-PCR

Primer	Sequence (5'→3')	Position	Exon
iNOS mRNA (6,9)			
IN1 *sense*	TCC GAG GCA AAC AGC ACA TTC A	2263–2284	17
IN1 *anti*	GGG TTG GGG GTG TGG TGA TGT	2704–2724	21
IN2 *sense*	ATG TTT GCG GGG ATG TG	3445–3461	25
IN2 *anti*	AAT CCA GGG TGC TAC TTG TTA	3835–3855	3'-UTR
IN5 *sense*	TCA GCC AGG CCC TCA CCT ACT	2677–2697	20
IN5 *anti*	GGC CAT CTC CAG CAT CTC CTC	3285–3305	24
IN6sc *sense*	AAG TAG CAC CCT GGA TTG A	3837–3857	3'-UTR
IN6sc *anti*	AAG GAA TCA TAC AGG GAA GAC	4087–4107	3'-UTR
ecNOS mRNA (5,10)			
ecNOS *sense*	GTG ATG GCG AAG CGA GTG AAG	1583–1603	12
ecNOS *anti*	CCG AGC CCG AAC ACA CAG AAC	1984–2004	16
ecNOS2sc *sense*	ACG CCT CTT TTC CCT CTC TA	3818–3837	25
ecNOS2sc *anti*	TGG CAC AGT CCC TTA TGG TA	4021–4040	25

Primer combinations	cDNA fragment, bp	Annealing temperature, °C
IN2 *sense* / IN2 *anti*	411	62
IN1 *sense* / IN1 *anti*	462	62
IN5 *sense* / IN5 *anti*	629	61
IN6sc *sense* / IN6sc *anti*	268	60
ecNOS *sense* / ecNOS *anti*	422	61
ecNOS2sc *sense* / ecNOS2sc *anti*	223	57

12. Calculate the amount of added cDNA, so that at all times during the PCR the dNTP-and primer concentration do not become limiting factors.
13. To avoid unspecific amplifications, the PCR tubes should be put into a preheated thermocycler (95°C), or the diluted *Taq* DNA polymerase should be added with a micropipet directly to the preheated solution by passing it through the oil layer.
14. Attempting to amplify low-abundant mRNA copies might lead to unspecific fragments in a 35-cycle PCR and/or a kind of "fog" or "mist" in the agarose-gel electrophoresis. This can be overcome by the use of the described *Touch-down* protocol that starts with a very high-annealing temperature (70°C). After every two cycles, the annealing temperature is lowered by 1°C. This results in a highly specific amplification at the beginning of the PCR. However, product formation is low, because only the full-length primer binds to its specific-binding site. False hybridization is minimized. The lowering of the annealing temperature leads then to a slowly increasing efficiency. A subsequent PCR of 25 or 30 cycles at an annealing temperature of 55°C then leads to a further amplification of these preamplified copies. All this can be done in a "one-pot reaction" without reloading.

References

1. Weinberg, J. B., Misukonis, M. A., Shami, P. J., Mason, S. N., Sauls, D. L., Dittman, W. A., Wood, E. R., Smith, G. K., McDonald, B., and Bachus, K. E. (1995) Human mononuclear phagocyte inducible nitric oxide synthase (iNOS): analysis of iNOS mRNA, iNOS protein, biopterin and nitric oxide production by blood monocytes and peritoneal macrophages. *Blood* **86,** 1184–1195.
2. Reiling, N., Ulmer, A. J., Duchrow, M., Ernst, M., Flad, H. -D., and Hauschildt, S. (1994) Nitric oxide synthase: mRNA expression of different isoforms in human monocytes/macrophages. *Eur. J. Immunol.* **24,** 1941–1944.
3. Ponce, M. R. and Micol, J. L. (1992) PCR amplification of long DNA fragments. *Nucleic Acids Res.* **20,** 623.
4. Charnock-Jones, D. S. (1993) Amplification and direct sequencing of DNA fragments using fluorescently labelled dye-terminators. *Int. Biotechnol. Lab.* **11 (3),** 12.
5. Janssens, S. P., Shimouchi, A., Quertermous, T., Bloch, D. B., and Bloch, K. D. (1992) Cloning and expression of a cDNA encoding human endothelium-derived relaxing factor/nitric oxide synthase. *J. Biol. Chem.* **267,** 14,519–14,522.
6. Geller, D. A., Lowenstein, C. J., Shapiro, R. A., Nussler, A. K., Di Silvio, M., Wang, S. C., Nakayama, D. K., Simmons, R. L., Snyder, S. H., and Billiar, T. R. (1993) Molecular cloning and expression of inducible nitric oxide synthase from human hepatocytes. *Proc. Natl. Acad. Sci. USA* **90,** 3491–3495.
7. Toellner, K. -M., Scheel-Toellner, D., Seitzer, U., Sprenger, R., Trümper, L., Schlüter, C., Flad, H. -D., and Gerdes, J. (1996) The use of reverse transcription polymerase chain reaction to analyse large numbers of mRNA species from a single cell. *J. Immunol. Meth.* **191,** 71–75.
8. Jin, Y., Heck, D. E., De George, G., Tiam, Y., and Laskin, J. D. (1996) 5-Fluorouracil suppresses nitric oxide biosynthesis in colon carcinoma cells. *Cancer Res.* **56,** 1978–1982.
9. Chartrain, N. A., Geller, D. A., Koty, P. P., Sitrin, N. F., Nussler, A. K., Hoffman, E. P., Billiar, T. R., Hutchinson, N. I., and Mudgett, J. S. (1994) Molecular cloning, structure, and chromosomal localization of the human inducible nitric oxide synthase gene. *J. Biol. Chem.* **269,** 6765–6772.
10. Marsden, P. A., Heng, H. H., Scherer, S. W., Steward, R. J., Hall, A. V., Shi, X. M., Tsui, L. C., and Schappert, K. T. (1993) Structure and chromosomal localization of the human constitutive endothelial nitric oxide synthase gene. *J. Biol. Chem.* **268,** 17,478–17,488.

15

Measurement of NOS mRNA by Northern Blotting and the Ribonuclease-Protection Assay

Paloma Martín-Sanz and Lisardo Boscá

1. Introduction

The amount of RNA varies greatly between different cell types and organisms, and the majority of RNA (more than 95%) consists of ribosomal RNA (28S, 18S, and 5S species) and the tRNA. Only a small proportion (usually less than 1% of the total amount) of RNA corresponds to the mRNA species. However, in order to improve mRNA-detection methods, preparations of enriched-mRNA content (poly-A^+) can be obtained after elution from oligo-dT columns of total-RNA samples.

The main problem that may occur when isolating mRNA from various cells types or tissues is the presence of RNAses, which affect the integrity of the RNA. As a result of this degradation, RNA of various sizes is recovered in the Northern blot or by ribonuclease protection assay (RPA).

Northern-blot analysis is a method whereby certain mRNA species can be detected on the basis of size when hybridized with a complementary probe. However, when using this method it is quite difficult to distinguish between mRNA species of similar size. RPA analysis, on the other hand, provides a more quantitative method and allows a precise analysis of the mRNA on the basis of differences in the sequence. The use of alternative-initiation sites for transcription, or differential-splicing products between RNAs can also be detected by RPA.

2. Materials

2.1. Isolation of RNA

1. Diethyl pyrocarbonate (DEPC)-treated water: dissolve 1 mL of DEPC in 1 L of distilled water. Shake, allow to stand for 2 h at room temperature (RT), and autoclave.

From: *Methods in Molecular Biology, Vol. 100. Nitric Oxide Protocols*
Edited by: Michael A. Titheradge © Humana Press Inc., Totowa, NJ

2. Phenol of "molecular biology" quality. Mix 250 g of phenol, 62.5 mL of DEPC-treated water, and 30 mL of 1 m*M* Tris, pH 7.0. Incubate at 50°C to dissolve the phenol and then add 0.3 g of hydroxyquinoline to prevent oxidation. Store at 4°C. Alternatively, quality-tested phenol solutions can be obtained from various commercial suppliers (Biophenol TM, Labgen).
3. 20% Sarcosyl (w/v) in DEPC-treated water. Store at RT. Do not autoclave, otherwise sarcosyl will come out of solution (*see* **Note 1**).
4. β-mercaptoethanol (14 *M* stock solution).
5. 2 *M* Sodium acetate: Add 11.5 mL of glacial acetic acid to 75 mL of distilled water. Adjust pH to 4.0 with 5 *M* NaOH. Make up to 100 mL. Add 1 mL of 10% DEPC in ethanol (*see* **Note 2**). Shake vigorously and autoclave.
6. CHISAM: chloroform:isoamyl alcohol (24:1) (v/v). Store at –20°C.
7. Isopropanol, absolute ethanol, and 75% ethanol (in DEPC-treated water) are stored at 20°C.
8. Solution D: To make up 50 mL, weight 23.63 g of guanidinium thiocyanate in a sterile tube. Add 29 mL of DEPC-treated water, 1.67 mL of sodium citrate (0.75 *M*, pH 7.0), 1.25 mL of 20% sarcosyl, and 0.36 mL of β-mercaptoethanol. Adjust the volume to 50 mL with DEPC-treated water and mix. Store at RT in the dark for 1 mo, and replace β-mercaptoethanol frequently.
9. Homogenization solution: A mixture of one vol of solution D, and 1.1 vol of phenol is prepared before use. Add 0.1 vol 2 *M* sodium acetate. Prepare fresh.
10. Eppendorf tubes: autoclave the tubes with 1 mL/L of volume of 10% DEPC (in ethanol).

2.2. Enrichment of Poly-A^+ RNA

1. Oligo(dT) cellulose. To activate the gel (75 mg), add 2 mL of 0.1 *M* NaOH and mix gently for 2 h at 4°C. Allow the gel to sediment and aspirate the supernatant solution. Fill with an equal volume of 0.1 *M* NaOH.
2. Sterile 1–2 mL syringes.
3. Loading buffer 2X: 1 *M* NaCl, 1 m*M* ethylenediaminetetraacetic acid (EDTA), 50 m*M* Tris-HCl, pH 7.6, 0.2% sodium dodecyl sulfate (SDS) in DEPC-treated water.
4. Washing buffer: 1X Loading buffer.
5. Elution buffer: 1 m*M* EDTA, 10 m*M* Tris-HCl, pH 7.4, 0.05% SDS.
6. Preparation of the column: Load 2 mL of gel suspension (**step 1**). Wash the column with loading buffer until the effluent has a pH < 8.0. The column is ready to be used.

2.3. Probe Labeling for Northern Blot

1. The appropriate endonucleases to cleave the insert from plasmidic DNA are used following the recommendations of the supplier. A 0.2–1 kb probe size is suitable.
2. Isolate the insert of interest after agarose-gel separation (low-melting point) and extraction.
3. Multiprimer labeling kit (Boehringer or Amersham). Plasmidic DNA can be used if the probe is labeled by Nick-translation.

4. 25 ng of DNA probe. Confirm the amount of the DNA by two independent methods (e.g., UV spectrophotometry and quantitation in an agarose gel with standard DNA; *see* **Subheading 3.1.6.**). Mix the probe dissolved in 45 μL of sterile water with Multiprimer DNA labeling system (Amersham) and 5 μL of [α-^{32}P]dCTP (3,000 Ci/mmol). Incubate for 1 h at 37°C and stop the reaction by adding 2 μL of 0.5 *M* EDTA, pH 8.0, distilled water up to a final volume of 100 μL and heating for 10 min at 65°C. Remove unincorporated nucleotides through a Sephadex G-25 pre-packed disposable column (Pharmacia, PD-10). Count the incorporated radioactivity.

2.4. Probe Labeling for RPA

1. Transcription kit containing transcription buffer and RNA polymerases (T7, T3, SP6) (e.g., Riboprobe Core System, Promega, cat. no. P1270).
2. 100 m*M* dithiothreitol (DTT). A fresh solution in DEPC-treated water.
3. RNAsin. Store at 25 units/mL.
4. Nucleotide stock solution: A 2.5 m*M* solution of adenosine triphosphate (ATP), guanosine triphosphate (GTP), and cytidine triphosphate (CTP) in water.
5. 200 μ*M* final concentration of [α-^{32}P]UTP (10 mCi/mL).
6. Linearized-DNA template.
7. Label the probe following the steps of the transcription kit supplier. Usually for 1 h at 37°C
8. Treat with DNAse I following the recommendations of **step 1**. Usually 15 min at 37°C are sufficient to destroy DNA.
9. Extract the labeled probe with 12 μL of phenol and 24 μL of CHISAM. Centrifuge for 2 min in an Eppendorf and repeat the extraction with 24 μL of CHISAM. Take the overlay phase.
10. Precipitate the probe with 50 μL of ethanol and 1.5 μL of 3 *M* sodium acetate. Incubate for 30 min at –20°C and centrifuge. Remove the supernatant and dry the pellet with a nitrogen stream.
11. Resuspend the pellet in 30 μL of DEPC-treated water and count the radioactivity. Prepare a solution at 4–8 × 10^5 cpm/mL. Store the probe at 4°C.

2.5. Northern-Blot Analysis

1. Formamide: Treat 100 mL of formamide with 10 g of BIO-Rad AG501-X8 gel for 30 min at RT and filter through a Whatman 1MM filter.
2. Agarose (medium electroendosmosis grade is sufficient).
3. 10X MOPS: Dissolve 42 g of 4-morpholinepropanesulfonic acid (MOPS) and 4.1 g of anhydrous sodium acetate in 0.8 L of water. Add 10 mL of 0.5 *M* EDTA, adjust the pH to 7.4, and make the volume up to 1 L. Autoclave and store at RT in the dark.
4. Ethidium bromide: Dissolve 20 mg of ethidium bromide in 20 mL of water over a period of 10 h. Filter through Whatman 3MM filter and store at 4°C, light-protected (*see* **Note 3**).
5. Formaldehyde (37%).

6. Glycerol "molecular biology" grade.
7. Sample buffer (SB): Mix 1.25 mL of formamide, 0.25 mL of 10X MOPS, 0.35 mL of formaldehyde (37%), 0.2 mL of glycerol, 0.15 mL of 1% bromophenol blue (w/v), and 0.2 mL of water. Store in 0.5 mL aliquots at –20°C.
8. Nitrocellulose/nylon membranes (0.45 μm mesh).
9. 20X SSC: 3 *M* NaCl, 0.3 *M* sodium citrate, pH 7.4 in distilled water. Autoclave.
10. RNA-transference apparatus.
11. UV-light crosslinker (222 nm).
12. Salmon-sperm stock (10 mg/mL). Dissolve in DEPC-treated water and autoclave.
13. Prehybridization solution: 50% formamide, 0.25 *M* NaCl, 0.5 *M* sodium phosphate, pH 7.2, 7% SDS, 0.1 mg/mL of salmon sperm in DEPC-treated water.
14. Washing solution: 0.1X SSC and 0.1% SDS at 42°C.
15. X-ray film and photographic-development reagents.
16. Dehybridization solution: 0.1X SSC containing 1% SDS.

2.6. RPA Analysis of RNA

1. Acrylamide:bisacrylamide (19:1, w:w) in 100 mL of water.
2. 10X TBE: 108 g Tris-base, 55 g boric acid, 8 g EDTA. Adjust to 1 L.
3. Urea, ammonium persulfate, and TEMED of "molecular-biology" grade.
4. 8% polyacrylamide solution: Weigh 100 g of urea. Add 40 mL of acrylamide: bisacryalmide, 20 mL of 10X TBE and make the volume up to 200 mL with DEPC-treated water. Store protected from light at 4°C.
5. To prepare the polyacrylamide gel, wash all contact surfaces of the electrophoresis plates with DEPC-treated water and dry with absolute ethanol. Prepare the electrophoresis cassette (sequencing-gel apparatus). Fill with running gel by mixing 50 mL of 8% polyacrylamide solution, 0.4 mL of water containing 40 mg of ammonium persulfate, and 15 μL of TEMED. Put the comb on the top of the cassette and allow the gel to polymerize in a near-horizontal position.
6. Running buffer: 0.5X TBE.
7. Sample buffer: 950 μL of formamide, 20 μL of 0.5 *M* EDTA, 30 μL of a solution containing 350 ng/mL of xylene cyanol, and 350 ng/mL of bromophenol blue.
8. RNAse A (store at –20°C), RNAse T1 (store at 4°C).
9. Proteinase K (200 mg/mL). Prepare aliquots of 100 μL and store at –20°C.
10. tRNA (from yeast). Prepare at 500 μg/mL in DEPC-treated water.
11. RNAse digestion buffer: 10 m*M* Tris-HCl, pH 7.6, 5 m*M* EDTA, 300 m*M* NaCl, 40 μg/mL RNAse A, 2 μg/mL RNAse T1.

3. Methods

3.1. Isolation of RNA from Cultured Cells or Tissues

1. A small amount of tissue (approx 50–100 mg) is introduced into a sterile homogenizer containing 4 mL of ice-cold homogenization solution (**Subheading 2.1., step 9**). After homogenization, transfer 1 mL of the suspension to Eppendorf tubes, add 100 μL of CHISAM, and vortex vigorously in a mixer for 30 s. Allow

to stand for 15 min at 4°C and centrifuge in an Eppendorf centrifuge for 15 min at 4°C. When using adherent-cultured cells, wash the plate twice with phosphate-buffered saline (PBS) and add 1 mL of homogenization solution per 6 cm diameter plate. Homogenize the cells by gently passing the suspension through a 1 mL pipet. Transfer 1 mL of the homogenate to an Eppendorf tube and add 100 μL of CHISAM. Proceed as described for the tissue.

2. Take the upper phase of the supernatant, taking care to avoid disturbing the interface, and transfer to a new tube. If the upper layer is not absolutely transparent, a second cycle of RNA extraction with homogenization solution and CHISAM (**steps 1** and **2**) may be used.
3. Fill the tube with at least one vol of isopropanol (at –20°C) and vortex.
4. Store the tubes at –20°C for 2 h and then centrifuge for 15 min at 4°C.
5. Remove the supernatant with a capillary pipet (avoid disrupting the pellet), and add 1 mL of 75% ethanol (at –20°C). Wash with 75% ethanol twice and store at –20°C.
6. The RNA content is evaluated after resuspension of the pellet in 0.1 m*M* EDTA in DEPC-treated water (usually 10–20 μL) and measurement of the absorbance (1–2 μL of RNA sample in 1 mL of DEPC-treated water). The ratio of absorbance at 260 and 280 nm should be around 2.0. Ratios below 1.8 denote the presence of a significant amount of protein and require further purification (*see* **Note 4**). The amount of RNA is calculated according to the formula, RNA(μg/μL) = $O.D_{260} \times 40$.
7. Poly-A^+ RNA can be prepared by passing approx 1 mg of heat-treated total RNA (65°C for 5 min in 1 mL of 1X loading buffer) through the oligo(dT) column (2 mL of gel). The effluent can be applied once again through the column to improve the binding to the oligo(dT)-cellulose.
8. Wash the column with 10 mL of washing buffer (*see* **Subheading 2.2., step 4**).
9. Elute the enriched poly-A^+ RNA with 3 mL of elution buffer (*See* **Subheading 2.2., step 5**). To precipitate the RNA, repeat **step 3** but in the presence of 0.3 *M* sodium acetate (final concentration) to favor the precipitation of the RNA, followed by **steps 4–6**.

3.2. RNA Separation by Electrophoresis, Transference to Membrane, and Hybridization with Labeled Probes

1. For 125 mL of gel, weigh 1.125 g of agarose in a 250-mL sterile Erlenmeyer flask. Add 10.5 mL of 10X MOPS, and make the final volume of up to 125 mL with DEPC-treated water. Heat to ebullition to melt the agarose. With continuous stirring, put the Erlenmeyer in a water bath until the temperature decreases to 50–55°C. Add 6.7 mL of formaldehyde and decant the solution into the laterally sealed cuvet, and put the comb in to form the wells to charge the RNA.
2. When the gel is completely polymerized (1–2 h in the cold room), place the tray into the electrophoresis cuvet and completely cover the gel with 1X MOPS in DEPC-treated water.

3. Resuspend the RNA (1–5 µL; 10–30 µg) in 24 µL of SB and 1 µL of ethidium bromide. Heat at 65°C for 10 min and load into the gel wells.
4. Run the gel at 20–40 mV for 8–12 h.
5. When the blue band corresponding to the bromophenol blue is at 2–5 cm from the bottom of the gel, stop the electrophoresis, and analyze the RNA lanes under UV light. Two main bands corresponding to the ribosomal 28S and 18S RNA should be clearly observed, the first band with double the intensity of the second (*see* **Note 5**). Take a photograph to allow analysis of the lane charge.
6. Cut a nylon/nitrocellulose membrane (Nytran) to the size of the gel. Wet with 10X SSC. Place the membrane in the vacuum device and overlay with the gel. Apply the vacuum as recommended by the supplier.
7. Confirm that the RNA is transferred from the gel to the membrane by direct observation under the UV light. Dry the membrane in an oven at 65°C for one h (alternatively air-dry the membrane for a longer period).
8. Seal the membrane in a bag (alternatively a cylinder roller) with 10–15 mL of prehybridization solution (*see* **Subheading 2.5., step 13**) and incubate at 42°C for 4–6 h.
9. Heat the labeled probe at 95°C for 10 min and immediately place in an ice-water bath. Add the labeled probe and continue the hybridization overnight. Remove the solution containing the probe (it can be re-used during the next two weeks) and wash the membrane once for 10 min at RT, and twice for 20 min at 42°C with 10 mL of washing solution (*see* **Subheading 2.5., step 14**).
10. Cover the membrane with a plastic sheet and check for the radioactivity along the lanes and compare to the background of RNA-untransferred areas. A ratio of 1:5 of intensity between the transferred region and the background is sufficient to proceed for detection of the bands with an X-ray film (Hyperfilm from Amersham or X-OMAT from Kodak). The time of exposure of the membrane to the film may be from 10 h up to more than 1 wk. Different exposures are convenient to ensure that the bands are not overexposed and to allow a quantitation of the band intensities. A more rapid detection can be obtained if a Phosphorimager equipment is available.
11. To normalize for lane charge of the different wells, the membranes are dehybridized with dehybridization solution by boiling the membranes for 15 min. The membrane is rehybridized with a probe that recognizes the mRNA of a gene whose levels remain constant under the experimental conditions investigated. The content of β-actin, ribosomal 18S, albumin (in the case of hepatic samples), or glyceraldehyde phosphate dehydrogenase genes are the most common probes used for normalization.

3.3. Characterization of RNA by RPA

1. The RPA analysis is based on the hybridization in solution of RNA with a cRNA probe that forms a bicatenary molecule resistant to RNAse action.
2. Precipitate the RNA (1–30 µg) overnight at –20°C in ethanol. Centrifuge this solution and remove the ethanol.

3. Prepare immediately 1 mL of a solution containing 250 µL of hybridization buffer (20 m*M* Tris-HCl, pH 7.6, 1 m*M* EDTA, 0.4 *M* NaCl, 0.1% SDS) and 750 µL of formamide. Heat at 45°C. Dissolve the RNA pellet in 29 µL of this solution.
4. Add 1 µL of the labeled riboprobe and incubate at 85°C for 5 min.
5. Leave the tubes at 45°C overnight.
6. RNAse digestion: Add 270 µL of RNAse digestion buffer (*see* **Subheading 2.6, step 11**) to the tubes. A tube without RNAse, but with the other reagents is prepared. Incubate for 1 h at 30°C. During this time, the polyacrylamide gel can be prepared.
7. Add 20 µL of 10% SDS to each tube. After mixing, add 2.5 µL of proteinase K (stock 20 mg/mL) and incubate for 15 min at 37°C with some vortexing.
8. Extract the RNA·RNA hybrids with phenol/CHISAM (150 µL: 300 µL).
9. Precipitate the RNA·RNA hybrids from the upper layer with 700 µL of absolute ethanol, 1.5 µL of 3 *M* sodium acetate, and 4 µL of tRNA. Incubate at –20°C for 30 min. Centrifuge the tubes in a microcentrifuge (Eppendorf) for 30 min in the cold room. Aspirate the supernatant and resuspend the pellet in 5 µL of DEPC-treated water, mix, and add 10 µL of a fresh solution of sample buffer (*see* **Subheading 2.6., step 7**). Centrifuge in an Eppendorf for 5 min.
10. Heat the samples at 95°C for 3 min and rapidly chill on ice for 5 min.
11. A molecular-weight marker (radiolabeled) is suitable.
12. Premigrate the gel at 3,000V (power 40 W) for 1–2 h to reach 50°C. Load the samples and run for 2–3 h. Load the molecular weight markers in a separate lane.
13. Transfer the gel to Whatman 3MM paper and dry under vacuum.
14. Expose the transferred gel to an X-ray film.

4. Notes

1. Do not autoclave detergents.
2. Tris solutions destroy DEPC. DEPC is quite toxic and highly flammable. Handle with care, preferably in a fume-hood.
3. Ethidium bromide is a carcinogenic substance. Liquids containing this dye should be collected and delivered as a hazard solution. Alternatively, solutions can be filtered through an activated charcoal filter that efficiently retains the dye. The filters can be incinerated.
4. When proteins are detected (by the A_{260}/A_{280} ratio), a new cycle of RNA extraction is convenient (*see* **Subheading 3.1., steps 1** and **5**). Proteins might affect the RNA mobility, increase the degradation of RNA, and are also stained by ethidium bromide.
5. The RNA bands corresponding to the 28S and 18S ribosomal species should be very clear, and the density of the upper band double the intensity of the lower. If a smear appears, this could be due to RNA degradation. In this case, clean the electrophoresis cuvet thoroughly with a 10% DEPC solution in ethanol. To avoid these problems, it is convenient to reserve the material intended for RNA isolation and analysis from the bulk of the laboratory equipment (solutions, tubes, glassware, etc.).

References

1. Chirgwin, J. M., Przybyla, A. E., MacDonald, R. J., and Rutter, W. J. (1979) Isolation of biologically active ribonucleic acid from sources enriched in ribonuclease. *Biochemistry* **18,** 5294–5299.
2. Chomczynski, P. and Sacchi, N. (1987) Single-step method of RNA isolation by acid guanidium thiocyante-phenol-chloroform extraction. *Anal. Biochem.* **162,** 156–159.
3. Xie, Q-w., Cho, H. J., Calaycay, J., Mumford, R. A., Swiderek, K. M., Lee, T. D., Ding, A., Troso, T., and Nathan, C. F. (1992) Cloning and characterization of inducible nitric oxide synthase from mouse macrophages. *Science* **256,** 225–228.
4. Gilman, M. (1987) Ribonuclease protection assay, in *Current Protocols in Molecular Biology* (Ausbel, F. M., Brendt, R., Kingston, R. E., Moore, D. D., Seidman, J. G., Smith, J. A., and Struhl, K., eds.), Wiley, New York, pp. 4.7.1–4.7.8.

16

Detection of NOS Isoforms by Western-Blot Analysis

Fiona S. Smith and Michael A. Titheradge

1. Introduction

The amount of nitric oxide synthase (NOS) in tissues and cells can readily be determined by Western blotting. Antibodies are now commercially available for the detection of the inducible (i), neuronal (n), and endothelial (e) NOS. These antibodies are specific for the particular isoforms of NOS and show no cross-reactivity with other isoforms, identifying proteins of apparent Mr 130 kDa, 160 kDa, and 133 kDa respectively *(1)*. This chapter describes a method for the estimation of the amount of iNOS in tissue samples; however, it is easily adapted for the other forms of NOS by the use of the appropriate antibodies. The method details how to extract the protein from the tissue, followed by separation of the proteins by size using sodium dodecyl sulfate (SDS) polyacrylamide-gel electrophoresis (PAGE) and transfer of the proteins to a membrane to allow immunodetection of the iNOS. In order to detect the enzyme on the immunoblot, the blotted-cellular proteins are incubated with a primary antibody specific for the protein to be estimated, e.g., iNOS-polyclonal antiserum, to allow the antibody to bind specifically to its antigen. After a series of washes to remove the excess unbound antibody, the membrane is then exposed to a secondary antibody conjugated to an enzyme [e.g., horse radish peroxidase (HRP), alkaline phosphatase], which specifically recognizes the primary antibody. After washing off excess unbound secondary antibody, the amount of the secondary antibody bound to the membrane is proportional to the amount of the original antigen present in the cell extract and can be determined by the activity of the enzyme to which it is linked, e.g., by enhanced chemiluminescence (ECL). The ECL-detection system is more sensitive than

From: *Methods in Molecular Biology, Vol. 100. Nitric Oxide Protocols*
Edited by: Michael A. Titheradge © Humana Press Inc., Totowa, NJ

other nonradioactive detection systems (the detection limit is 1 pg of antigen) and it produces an excellent signal that can be subsequently removed from the membrane to allow the same samples to be probed for a different protein, e.g., to account for differences in protein loading of the gel or to probe for proteins of a similar molecular weight. However, other visualization systems which rely on HRP, such as the use of 3,3'-diaminobenzidine/nickel chloride as the substrates may be used *(2)*. The principle underlying ECL detection is that luminol is oxidized in an alkaline environment in the presence of chemical enhancers (e.g., phenol), by the HRP linked to the secondary antibody. This enhanced chemiluminescent reaction increases light output 1000-fold and increases the period of light emission. As the maximum-light emission is at a wavelength of 428 nm, it can be detected by a short exposure to blue-light sensitive autoradiography film. The amount of light produced is proportional to the amount of primary antibody bound to the protein of interest *(3)*.

2. Materials

2.1. Sample Preparation

1. Sonication buffer: 0.32 *M* sucrose, 10 m*M* Tris-HCl, pH 7.5, 1 m*M* ethylenediaminetetraacetic acid (EDTA), 1 m*M*-dithiothreitol (DTT). Dissolve 10.95 g sucrose, 0.121 g of Trizma base, 0.037 g EDTA, disosodium salt, and 0.0154 g of DTT in 70 mL of double-distilled water, adjust the pH to 7.5 with HCl and make the volume up to 100 mL. Store at 4°C.
2. Protease inhibitors: Aprotinin, 10 mg dissolved in 1 mL of double-distilled water; leupeptin, 1 mg dissolved in 1 mL of double-distilled water; soybean trypsin inhibitor, 1 mg dissolved in 1 mL of double-distilled water; 200 m*M* phenylmethylsulfonyl fluoride (PMSF), 174 mg dissolved in 5 mL of methanol. Store at –20°C.
3. Extraction buffer: To each mL of sonication buffer, add 1 μL of aprotinin, 10 μL of leupeptin, 10 μL of soybean trypsin inhibitor, and 0.5 μL of PMSF. Owing to the unstable nature of the PMSF, this buffer should be prepared within 30 min of use *(4)*.
4. 1 *M* Tris-HCl, pH 6.8: Dissolve 121.1 g of Trizma base in 800 mL of double-distilled water. Adjust the pH to 6.8 with HCl and make up to 1 L. Store at 4°C.
5. 10% (w/v) SDS: Dissolve 10 g of sodium dodecyl sulfate (SDS) in 100 mL of double-distilled water. Store at room temperature.
6. Loading buffer (x2): Mix 4.0 mL of 10% SDS, 2.0 mL of glycerol, 3.95 mL of water, and 54 μL of 1 *M* Tris-HCl, pH 6.8. Add 308 mg of DTT and 2 mg of bromophenol blue and mix. Store in aliquots at –20°C for up to 6 mo.

2.2. Preparation and Running of SDS-Polyacrylamide Gels (8%)

1. 30% acrylamide: Dissolve 29 g of acrylamide plus 1 g of N,N′-methylene-bis-acrylamide in 100 mL of double-distilled water. Store for up to 3 mo at 4°C in the dark (*see* **Note 1**).

2. 1.5 *M* Tris-HCl, pH 8.7. Dissolve 181.7 g of Trizma base in 800 mL of double-distilled water. Adjust the pH to 6.8 with HCl and make up to 1 L. Store at 4°C.
3. 15% (w/v) ammonium persulphate: Dissolve 75 mg in 500 μL of double-distilled water. Prepare fresh each day.
4. N,N,N′N′-tetramethyl-ethylenediamine (TEMED).
5. Running buffer: Dissolve 29 g of glycine, 1 g of SDS, 6 g of Trizma base in 1 L of double-distilled water. Store at room temperature.
6. Prestained SDS molecular-weight markers (27–180 kDa, Sigma SDS-7B).
7. Bromophenol blue: Saturated solution in Running buffer.
8. A suitable vertical gel-electrophoresis system, e.g., Mini-Protean II from BioRad.

2.3. Electroblotting of Proteins

1. Anode 1 buffer: Dissolve 36.6 g of Trizma base in 700 mL of double-distilled water. Add 200 mL of methanol and make the volume up to 1 L. Store at room temperature.
2. Anode 2 buffer: Dissolve 3.03 g of Trizma base in 700 mL of double-distilled water. Add 200 mL of methanol and make the volume up to 1 L. Store at room temperature.
3. Cathode buffer: Dissolve 36.6 g of Trizma base and 5.25 g 6-amino-n-caproic acid in 700 mL of double-distilled water. Add 200 mL of methanol and make the volume up to 1 L. Store at room temperature (*see* **Note 2**).
4. Whatman 3MM filter paper.
5. Immobilon (PVDF) transfer membrane, pore size 0.45 μ*M* (Millipore, IPVH 00010) (*see* **Note 3**).
6. Methanol (AR grade).
7. Glutaraldehyde.
8. Coomassie-blue stain: Add 1 g of Coomassie Brilliant blue R-250 to 200 mL of methanol. Stir until dissolved. Add 50 mL of glacial acetic acid and make up to 500 mL with double-distilled water. Store at room temperature.
9. Destaining buffer: Mix 200 mL of methanol, 50 mL of glacial acetic acid, and make up to 500 mL with double-distilled water. Store at room temperature.
10. Ponceau S: Dissolve 1 g of Ponceau S in 25 mL of glacial acetic acid. Make up to 500 mL with double-distilled water. Store at room temperature.
11. Tris-buffered saline (TBS)-Tween: Dissolve 5.84 g of sodium chloride in 1 L of 10 m*M* Tris-HCl, pH 7.5 (prepared by dilution of the 1 *M* stock), and add 1 mL of Tween 20. Store at 4°C (*see* **Note 4**).
12 A suitable wet or semi-dry transfer cell, e.g., Trans-Blot SD Semi-Dry Transfer Cell from BioRad.

2.4. Antibody Incubations

1. Blocking buffer: Dissolve 5 g of nonfat, dried-milk powder (Marvel) in 100 mL of TBS-Tween. Prepare fresh (*see* **Note 5**).
2. Anti-iNOS polyclonal antibody: Suitable antibodies for the detection of iNOS and other forms of NOS can be obtained from the Cayman Chemical Co. Ltd or Transduction Laboratories. Store in aliquots at –20°C. Dilute the antibody in TBS-Tween and store at 4°C (*see* **Note 6**).

3. HRP-conjugated anti-rabbit IgG (Sigma A6154) or other appropriate HRP-linked secondary antibody. Store in aliquots at –20°C. Dilute the antibody fresh in TBS-Tween.

2.5. Detection of NOS

1. ECL-detection kit (Amersham, RPN 2106, 2108, 2109, 2209). Store at 4°C.
2. SaranWrap (Dow Chemical Company).
3. Vinyl-covered exposure cassette (Sigma, E8635).
4. Blue-light sensitive X-ray film (available from Fuji, Kodak or Amersham).

3. Methods

3.1. Preparation of Samples

1. Add 1 mL of extraction buffer to approx 100 mg of sample and homogenize or sonicate on ice at low amplitude (three times for 3 s at 14 μm peak to peak) to disrupt the cells and release the enzyme (*see* **Note 7**).
2. Centrifuge the suspension for 30 min at 100,000*g* at 4°C, and transfer the supernatant to a microcentrifuge tube.
3. Reserve an aliquot for the determination of protein and add the remaining supernatant to an equal volume of loading buffer (x2).
4. Boil the samples for 5 min in a water bath to denature the proteins and immediately place on ice for 2 min. Store the samples at –80°C for up to 6 mo.

3.2. Preparation of SDS-Polyacrylamide Gels (8%)

1. Assemble the mini-gel sandwich on the gel casting stand according to the protocol booklet.
2. Check the gel cast for any leaks by the addition of water to the gel assembly. This water must be removed before the acrylamide solution is added to the cast.
3. Place a comb into the assembled-gel sandwich and mark the glass plate about 1 cm below the bottom of the teeth.
4. To prepare an 8% polyacrylamide separating gel, gently mix 4 mL of the 30% acrylamide solution with 4.5 mL of 1.5 *M* Tris-HCl, pH 8.7, 0.2 mL of 10% SDS (w/v), and 6.3 mL of double-distilled water in a 250 mL side-arm vacuum flask. This volume is sufficient to prepare two mini-gels.
5. Stopper the flask and apply a vacuum for several min to degas the solution. This eliminates atmospheric oxygen which would otherwise inhibit polymerization.
6. Add 20 μL of TEMED (a polymerization catalyst), and mix the solution gently, avoiding the generation of air bubbles.
7. Initiate polymerization of the acrylamide gel by adding 100 μL of freshly prepared 15% (w/v) ammonium persulphate.
8 Pipet the solution immediately into each gel sandwich to the level marked previously.
9. Carefully add an even layer of water or water saturated-n-butanol to the surface of the gel solution. This layer eliminates the meniscus and thus prevents the gel

from setting with a curved top. It also excludes oxygen which would otherwise inhibit polymerization at the surface of the gel.

10. After polymerization (approx 15–45 min), pour off the water or water saturated-n-butanol. If alcohol is used, rinse the surface of the gel with distilled water to remove any traces of the alcohol.
11. To prepare the stacking gel (5%), add 1 mL of acrylamide (30%), 2 mL of 1 *M* Tris-HCl, pH 6.8, 83 μL of 10% SDS, 5 mL of water, and 5 μL of bromophenol blue to a 250 mL side-arm vacuum flask (*see* **Note 8**).
12. Degas this solution and add 20 μL of TEMED and 100 μL of 15% ammonium persulfate as in **Steps 6** and **7**.
13. Pipet the acrylamide into the gel sandwich and carefully slot a comb between the two glass plates until the teeth of the comb are completely submerged in the gel. Avoid trapping any air bubbles under the teeth as these will affect the formation of the wells.
14. Remove any stacking gel that spills over the top of the gel sandwich with tissue paper.
15. When the stacking gel has set (approx 15–30 min), carefully remove the combs to avoid distorting the wells, and rinse the wells with double-distilled water.

3.3. Loading the Samples

1. Detach the gel assembly from the casting stand and insert into the chambers of the gel running tank according to the protocol provided with the apparatus.
2. Add running buffer to the lower chamber until at least the bottom 1 cm of the gel sandwich is covered and fill the upper chamber with the same buffer (approx 100 mL if a mini-gel apparatus is used).
3. Clear the wells of any unpolymerized gel by withdrawing and expelling running buffer within the wells with a 1 mL syringe.
4. Add the samples to each well with either a pipet or Hamilton syringe, and run alongside 5 μL of the prestained SDS molecular-weight markers. Any empty wells should be filled with running buffer. The amount of protein added to the wells is typically 10–20 μg; however, this should be optimized for each type of sample (*see* **Note 9**). Protein estimations can be determined from samples without the loading buffer added, using the bicinchoninic acid (BCA) method (Pierce).
5. Apply the power at a constant voltage of 50 V (*see* **Note 10**).
6. Once the dye front has reached the bottom of the gel sandwich, switch off the power supply, and remove the gel clamp assembly from the running tank.

3.4. Transfer of the Proteins to a Membrane

1. Soak four pieces of Whatman 3MM paper in the anode 1 buffer. The Whatman 3MM paper should be slightly larger than the gel (*see* **Note 11**).
2. Pick up the stack of filter papers and carefully squeeze to remove any excess buffer, and place onto the anode of the transfer cell.
3. Soak two more pieces of Whatman 3MM paper in the anode 2 buffer, squeeze out the excess buffer, and place on top of the stack of filter paper on the anode of the transfer cell.

4. Cut a piece of PVDF membrane to slightly larger dimensions than the Whatman 3MM such that it overlaps the filter paper by 5 mm all round. Soak the membrane in methanol, rinse thoroughly in water, and place on top of the stack. If a nitrocellulose membrane is used, it is not necessary to pre-soak it in methanol.
5. Dismantle the gel sandwich and prise the glass plates apart with a spatula.
6. Remove the stacking gel with a razor blade and discard, and carefully transfer the separating gel onto the membrane ensuring that no air bubbles remain under the gel. If PVDF membrane is used, this stage should be carried out as quick as possible to prevent the PVDF from drying out.
7. Using a chinograph pencil mark the position of the top and bottom of the gel on the membrane.
8. Soak four pieces of Whatman 3MM filter paper in the cathode buffer and stack on top of the gel to complete the sandwich.
9. Before engaging the cathode, pick up the sandwich and carefully squeeze the stack to remove any remaining excess buffer and any air bubbles which will prevent complete protein transfer. Return the sandwich to the anode and roll a serological pipet over the surface to squeeze out any excess buffer and dry the edges with tissue.
10. In order to transfer two mini-gels placed side by side, carry out the transfer for $1^1/_4$ h at a constant voltage of 25 V or a current of 0.8 mA/cm^2 (*see* **Notes 12** and **13**).
11. Following transfer, incubate the membrane with gentle agitation in glutaraldehyde (diluted to 1:1000 with water) for 10 min at room temperature. This fixes the proteins to the membrane and thus minimizes the loss of protein during subsequent washes.
12. To determine the efficiency of the transfer, stain the gel with the Coomassie blue stain at room temperature overnight. The detection limit of this stain is 1.5 μg.
13. Discard the excess stain, which can be reused, and replace with the destaining solution. Wash the gel with several changes of destain until the background becomes clear.
14. In addition, to verify transfer efficiency, incubate the fixed membrane in the reversible protein stain, Ponceau S for several min at room temperature. Pour off the excess stain (this can be recycled) and rinse the membrane briefly in TBS-Tween to reveal the transferred proteins. The detection limit of Ponceau S is 1–2 μg of protein.
15. Before continuing to the next stage, the blots should be washed several times in TBS-Tween at room temperature to remove the remaining stain. The membrane can be stored at this stage at 4°C (*see* **Note 14**).

3.5. Immunodetection

1. To minimize any nonspecific binding of the primary or secondary antibody to the membrane, the membrane has to be blocked. Incubate the blot in the blocking buffer with gentle agitation at room temperature for one h or at 4°C overnight and then discard.
2. Remove any remaining blocking buffer by washing the blot briefly in TBS-Tween. Allow the blot to dry or leave in TBS-Tween (*see* **Note 14**).

3. Incubate the blot with gentle agitation for either one h at room temperature or overnight at 4°C, in the anti-iNOS antibody (diluted to 1:1000 in TBS-Tween) (*see* **Note 15**). The conditions for immunodetection have to be optimized for samples from different sources and for each primary antibody used (*see* **Note 9**).
4. To remove any unbound-primary antibody, wash the blot six times with gentle agitation for five min at room temperature with sufficient TBS-Tween to cover the membrane.
5. Incubate the blot with HRP-conjugated anti-rabbit IgG (freshly diluted to 1:1000 in TBS-Tween) with gentle agitation at room temperature for one h (*see* **Note 15**).
6. Discard the secondary antibody and wash the blot as in **Step 4**. The blot should not be allowed to dry out during **Steps 3–6**.

3.6. Western-Blot Detection Using ECL

1. All stages of protein detection should be carried out in a dark room with the aid of a red safety light and gloves should be worn at all times.
2. Mix an equal volume of detection reagent 1 with detection reagent 2 to make sufficient to cover the membrane (0.125 mL/cm^2).
3. Remove the excess TBS-Tween from the washed membrane by holding the membrane vertically with forceps and touching the edge of the membrane against tissue paper.
4. Place the membrane protein side up onto a piece of SaranWrap, and carefully apply the detection reagent so that all the surface of the membrane is completely covered.
5. Incubate the membrane in the reagent for exactly one minute at room temperature and then remove the excess reagent from the membrane as in **Step 3**.
6. Using forceps, transfer the membrane to a new piece of SaranWrap, protein side down, and carefully wrap the SaranWrap around the membrane to form an envelope. Gently smooth out any air pockets and creases.
7. Place the membrane protein side up in a film cassette, and place a sheet of audioradiograph film on top.
8. Close the cassette and after an initial exposure of 15 s, remove the film, and develop immediately.
9. Depending on the appearance of the signal of this first exposure, subject the membrane to a second, longer exposure. **Figures 1** and **2** show scans of typical immunoblots using primary antibodies raised against iNOS and nNOS obtained from the Cayman Chemical Co. Ltd.
10. Following ECL detection, the membrane can be stripped (*see* **Note 16**) and reprobed with another primary antibody, or stored at 4°C.

4. Notes

1. Acrylamide is a neurotoxin and should be weighed out in a fume hood. A mask and eye protection should be worn while weighing out the solid.
2. These buffers should be prepared regularly to prevent the evaporation of the methanol.

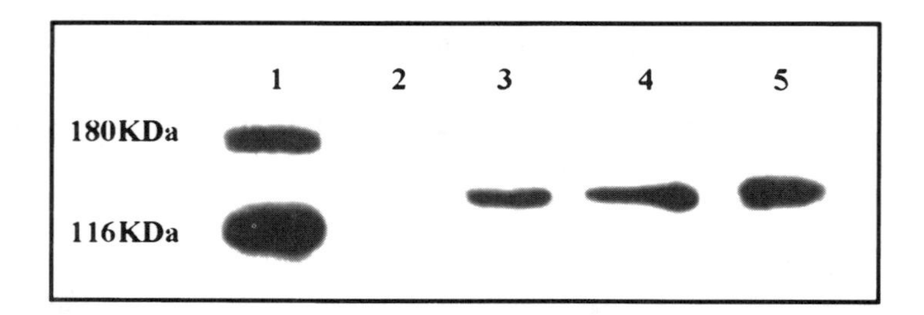

Fig. 1. Detection of iNOS in different tissue samples. The samples were loaded and detected using ECL as follows: Lane 1, molecular weight standards; lane 2, an extract of sonicated cultured hepatocytes; lane 3, an extract of cultured hepatocytes treated with a combination of IFN-γ, IL-1β, TNF-α, and LPS; lane 4; a partially purified extract of hepatocyte iNOS; lane 5, an extract of sonicated J774.2 cells treated with LPS.

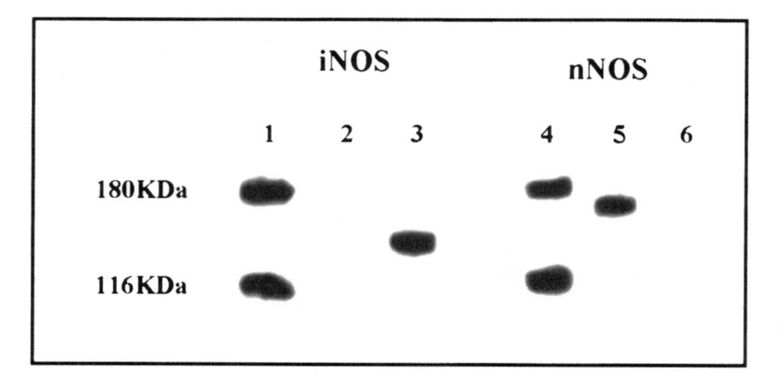

Fig. 2. Discrimination between iNOS and nNOS in cultured hepatocytes and cerebellum. High-speed supernatants were prepared from cerebellum (lanes 2 and 5) and cultured hepatocytes treated with a combination of IFN-γ, IL-1β, TNF-α, and LPS (lanes 3 and 6) and probed with either a primary antibody raised against iNOS or nNOS from the Cayman Chemical Co. Lanes 1 and 4 show the molecular weight markers. No significant cross-reactivity was observed between the nNOS antibody and the iNOS protein in the cultured hepatocytes or the iNOS antibody and the nNOS protein in the cerebellum extract.

3. Zeta-Probe (BioRad) or nitrocellulose membranes may also be used.
4. Numerous alternative formulations for TBS-Tween exist which can be used. The pH of this solution is crucial for the immunodetection steps and it is often the cause of a failure to detect the NOS. Phosphate-buffered saline (PBS) can be used as an alternative.

5. 3% bovine serum albumin (BSA) in TBS-Tween can be used as an alternative blocking reagent.
6. The diluted anti-iNOS is stable for up to 1 yr. The addition of 0.02% (w/v) sodium azide and 1% (w/v) BSA will prevent contamination and increase the stability of the antibody.
7. If cultured cells are to be used, the culture medium should be decanted and the plate rapidly washed in PBS. The extraction buffer can then be added to the plate and the cells scraped off and sonicated. The sonicate can be used directly for the determination of NOS activity using the conversion of radiolabeled-arginine to citrulline (*see* Chapter 7, this volume) or added to an equal volume of loading buffer (x2) for Western-blot analysis. Alternatively, the cells can be solubilized directly in the plate after washing by adding 1 mL per 10 cm diameter dish of hot (70°C) sample buffer that has been diluted 1:1 with 100 m*M* Tris-HCl, pH 6.8.
8. The addition of bromophenol blue to the stacking gel is not necessary for gel polymerization, but it acts to distinguish between the stacking gel and the separating gel and highlights the position of the wells.
9. In order to maximize sensitivity during immunodetection, with the minimal amount of nonspecific binding and background levels, the total protein loaded per sample and the concentration of the primary and secondary antibodies have to be determined for each type of sample and for each primary antibody used. To do this, prepare a gel as described in **Subheading 3.2.**, but carefully destroy all except one of the wells by removing the gel teeth using a syringe needle. Load the area created with a known amount of sample (the equivalent of between 5 μg and 20 μg per original well) and load 5 μL of the prestained molecular-weight marker into the only remaining well. Run and transfer the gel as described in **Subheadings 3.3.** and **3.4.**, respectively, and then fix and stain the blot with Ponceau S. At this stage, cut the blot into five strips of equal size and then block them as described in **Subheading 3.5.** Incubate each strip for one h at room temperature or overnight at 4°C, in primary antibody that has been diluted to either 1:1000, 1:2000, 1:3000, 1:4000, or 1:5000 in TBS-Tween. Following this incubation, wash the strips six times for 5 min at room temperature, before incubating them with HRP-conjugated anti-rabbit IgG diluted to 1:1000 in TBS-Tween for 1 h at room temperature. Finally, wash the strips as above and then develop using the ECL detection kit from Amersham as described in **Subheading 3.6.** The optimal conditions should pick up one clear band corresponding to the NOS protein. If a number of bands are detected in every strip, repeat the procedure but load less protein to the well. If the background levels are too high, repeat the procedure, but dilute the second antibody further. The conditions that produce the maximum signal with the least background, and minimal amount of nonspecific binding to the proteins should be chosen for use with the immunodetection system.
10. Depending on the power supply used, the gels can be run at a range of voltages, typically 50–200 V. We have found that a constant voltage of 50 V ensures the maximum sample resolution and the straightest migration pattern. To reduce the running time, the voltage can be increased up to 200 V.

11. Other buffer systems are available for the transfer and they vary with efficiency depending on the electroblotter used. These methods can be obtained from the relevant manuals that come with the electroblotter.
12. Two mini-gels can be transferred at once if they are of the same thickness. For the most efficient transfer, place the gels side by side on the PVDF membrane, with the minimum amount of membrane showing between them. If only one mini-gel is to be transferred, the voltage can be decreased to between 10–15 V.
13. If PVDF membrane is used, the transfer time may be increased to 2 h. However, if a nitrocellulose membrane is used, excessive transfer may cause the proteins to transfer through the membrane. This is because it has a lower-binding affinity for proteins than PVDF. To check for this, two pieces of nitrocellulose should be used in the stack and both stained for protein using Ponceau S.
14. If PVDF membrane is allowed to dry out after the transfer, it must be soaked in methanol and rinsed thoroughly in double-distilled water followed by TBS-Tween before continuing with the antibody incubations.
15. All antibody incubations and washing steps should be carried out in shallow trays (124 mm long × 80 mm wide × 20 mm deep for mini gels). However, if less than 5 mL of antibody is to be used, the blots can be placed (protein side in) inside a closed 50 mL tube and incubated on a rotatory shaker to ensure complete coverage of the blot.
16. Once the blot has been visualised with the ECL-detection kit, it can be stripped of its primary and secondary antibodies. This enables the same samples to be reprobed for a different protein or to try out different antibody concentrations to obtain a better signal. To strip the blot, immerse it in a buffer containing 62.5 m*M* Tris-HCl, pH 6.8, 2% SDS (w/v) and 100 m*M* 2-mercaptoethanol and incubate at 70–80°C for approx 40 min. After this incubation, wash the blot twice for 10 min in TBS-Tween at room temperature and then add the next primary antibody. Each blot should not be stripped more than four times. If a detection method based on the production of color is used, stripping the membrane will remove the primary and secondary antibodies but not the color from the blot.

References

1. Griffith, O. W. and Stuehr, D. J. (1995) Nitric oxide synthases: Properties and catalytic mechanism. *Ann. Rev. Physiol.* **57,** 707–736.
2. *Protein Electrophoresis Application Guide* (1994) Hoefer Scientific Instruments, San Francisco, CA.
3. *ECL Western Blotting Protocols* (1994) Amersham International PLC, Little Chalfont, Bucks, UK.
4. *Promega Protein Guide: Tips and Techniques* (1993) Promega Corp., Madison, WI.

17

Immunohistochemical Localization of NOS Isoforms

Victoria Cattell and Karen Mosley

1. Introduction

Three distinct isoforms of nitric oxide synthases (NOS) have so far been identified and cloned *(1)*. There are two constitutively expressed NOS—neuronal (n) and endothelial (e), NOS—and one inducible (i) form, iNOS. All three have been demonstrated immunochemically using antibodies raised against purified enzymes or specific-peptide sequences.

Immunochemistry, the visualization of antigen by use of specific antibody, can be applied to cells (immunocytochemistry) or to tissues (immunohistochemistry). The techniques used are slightly different, and for each, there are several different methods that may be successful. We describe here in detail methods that we have used to demonstrate iNOS in a variety of cells and tissues *(2)* and give reference to alternative procedures that have been published and may be equally suitable, depending on the NOS isoform, the nature of the specimens, and their expected level of NOS expression.

There are three main immunochemical techniques:

1. Direct application of a labeled specific (primary) antibody to the preparation.
2. A two-step indirect (sandwich) method where an unlabeled-primary antibody is followed by a labeled-secondary antibody with specificity for immunoglobulin of the species in which the primary antibody was raised.
3. Multistage techniques using enzyme or avidin–biotin complexes. The avidin–biotin system with peroxidase–diaminobenzidene reaction product is the most frequently used. The marker may be a fluorescent dye, an enzyme, or a metal such as colloidal gold.

Technique 1 is relatively insensitive and can only be used where antigen expression is high and has not been affected by fixation or processing, e.g.,

From: *Methods in Molecular Biology, Vol. 100. Nitric Oxide Protocols*
Edited by: Michael A. Titheradge © Humana Press Inc., Totowa, NJ

Table 1
Immunohistochemistry of NOS Isoforms

Isoform	Fixative	Sections		Species	Reference
		Frozen	Paraffin		
nNOS (NOS I)	None	√		R[c]	***6,7,8***
	PFA[a]		√	R	***8***
	Acetone	√		R	***9***
	PLP[b]	√	√	R	***10***
eNOS (NOS III)	None	√		H[d], Ra[f]	***11,12***
	PFA	√		R	***7***
	Acetone	√		R, H	***13,14***
	Ethanol	√	√	H	***15***
iNOS (NOS II)	None	√		R, Ra	***11,16***
	PFA	√		R, M[e]	***17,18,19,20,21,22***
	Acetone	√		R, H	***23,24***
	PLP	√	√	R	***10,25***
	Formalin		√	R	***26,27***
	Bouins		√	R	***28***

[a]PFA = paraformaldehyde, [b]PLP = periodate-lysine-paraformaldehyde. [c]R = rat. [d]H = human. [e]M = mouse. [f] Ra = rabbit.

frozen sections. Although Technique 2 is more sensitive, and has been used for the demonstration of NOS, the most sensitive results can be obtained by Technique 3. Although the methods for Technique 3 are somewhat more laborious, this is usually the method of choice for demonstration of antigens in paraffin-embedded tissue sections. Further techniques are available for enhancing detection of low levels of antigens. (For a more detailed general account and comparison of these methods *see* **refs. *3–5*.**)

Immunoelectronmicroscopy is a highly specialized technique, which will not be dealt with in this chapter.

Carefully controlled immunochemistry can be a powerful tool for identifying the presence and site of antigen expression. The essentials are that the reaction should be consistent, specific, and clearly identified within cells/tissues with good structural preservation. This can only be achieved by careful control of methods of fixation, appropriate controls of specificity, and elimination or reduction in background staining. Problems that may be encountered and suggestions for "troubleshooting" are described in **Subheading 4.** (*See* **Table 1**, which summarizes techniques that have been used to detect NOS isoforms in cells or tissues from several species.)

2. Materials

Phosphate-buffered saline (PBS) or Tris-buffered saline (TBS). PBS is stable and can, therefore, be made up in quantity. TBS is less stable, but is considered by some to be preferable for immunochemistry; it is essential for techniques using alkaline phosphatase labeling.

1. PBS: Dissolve 8 g NaCl, 0.2 g KCl, 0.2 g KH_2PO_4, 1.56 g $Na_2HPO_4 . 2H_2O$ in 1 L of distilled water. Adjust to pH 7.4.
2. TBS: Dissolve 8.7 g NaCl, 1.4 g Tris base, and 6.1 g Tris-HCl in 1 L distilled water. Adjust to pH 7.4.
3. Citrate buffer for microwaving: A x10 stock solution can be made and diluted before use. Dissolve 2.415 g citric acid monohydrate and 26.02 g trisodium citrate in 1 L of distilled water. Adjust to pH 6.0.
4. Hydrogen peroxide blocking reagent: 1% H_2O_2 in methanol. Add 3 mL of stock 30% H_2O_2 (Sigma, UK) to 97 mL of methanol. This solution should be freshly prepared before use.
5. ABC reagent (DAKO, UK): Prepare the avidin–biotin complex according to manufacturer's instructions in PBS with 20% normal swine serum (NSS). Leave for 30 min to allow the complex to form. Prepare immediately before use.
6. Diaminobenzidine tetrahydrochloride (DAB): DAB is possibly carcinogenic and must be handled with caution. Disposable gloves should be used during preparation of the solution. Either:
   a. Keep a stock solution frozen in aliquots at –20°C (150 mg of DAB dissolved in 10 mL of PBS) and add an aliquot to 290 mL of PBS plus 300 µL H_2O_2 immediately before use. This volume allows a rack of slides to be immersed in a glass trough; or
   b. Add 5 mg DAB plus 10 µL of 30% H_2O_2 to 10 mL of PBS immediately before use.

3. Methods

3.1. Immunocytochemistry

Cultured cells or cytospin preparations on glass cover slips can be fixed in acetone for 10 min following air drying and stored at –20°C, wrapped to prevent dehydration. For ease of handling, the cover slips can be adhered to glass slides with UV curing (Loctite, UK) glue. Frozen-tissue sections may also be stained by this method. The immunostaining method is then as in **Subheading 3.2.** omitting **Steps 1** and **2**, and allowing slides to dry completely before immunostaining (*see* **Notes 1** and **2**).

3.2. Immunohistochemistry on Paraffin Sections, Using Avidin–Biotin–Peroxidase Conjugate and Microwaving

1. Dewax and rehydrate: Run the sections sequentially through the following (approx 3 min in each): 100% Xylene x2, 100% ethanol, 70% ethanol, 30% ethanol, and distilled water.

2. Microwave: Use a plastic rack and a container suitable for microwaving with a loose lid. Microwave at 750 W in citrate buffer for 3–5 min, allowing 2 min between microwaves (*see* **Note 2**). Cool the slides to room temperature before proceeding (using a fan, this will take approx 15 min). Wash in water for 5 min, and PBS for 2–5 min. Place the slides flat in a covered moist chamber.
3. Blocking with normal sera: Circle the sections with silicone using a Dakopen (Dako, UK) to prevent the antibodies running off the section. Incubate in 20% normal swine serum in PBS for 30 min at room temperature. Shake off, but do not rinse.
4. Primary antibody: Incubate the slides with primary antibody diluted to an appropriate concentration in 20% NSS in PBS, at 4°C overnight. (*see* **Notes 3**, **6**, and **7**). It is often convenient to incubate overnight, but if necessary the incubation can be shortened to 1–2 h at room temperature. Wash for 3–5 min in PBS.
5. Peroxidase blocking: Add hydrogen peroxide blocking reagent and incubate for 20–30 min at room temperature (*see* **Note 8**). Wash in water. Wash twice in PBS. (If endogenous biotin is present in the tissue under examination, block biotin at this stage—*see* **Note 4**.)
6. Secondary antibody: Add an appropriate biotinylated secondary antibody diluted to an appropriate concentration in serum blocking agent (NSS) + 5% normal serum from the species whose tissue is being immunostained in PBS. Incubate for 60 min at room temperature. (*see* **Note 5** for alkaline phosphatase labeled antibodies). Wash three times in PBS.
7. ABC reagent: Prepare this reagent 30 min before use, according to the manufacturer's instructions. Add to the sections and incubate at room temperature for 30 min. Wash three times in PBS.
8. DAB: Add freshly prepared DAB (or immerse in glass trough). Time the incubation examining a positive control section under the microscope for development of positive staining (the average time is about 3 min). Wash in running tap water for 5 min.
9. Counterstain: Stain in Harris' hematoxylin for 5 s. Wash in running tap water for 5 min.
10. Dehydrate and Mount: Run sections sequentially through (approx 3 min in each): 30% ethanol, 70% ethanol, 100% ethanol, 100% ethanol, 100% xylene, 100% xylene. Mount in DPX mountant. Add 1 drop of DPX to the slide and place the cover slip on top. Press lightly with fiber-free filter paper to remove bubbles and mop-up excess mountant.

4. Notes

1. Cell/tissue storage: Correct conditions of storage are important for the preservation of antigens in material. Cells that are adherent on air-dried cover slips can be stored frozen following appropriate fixation. Drying out can be a problem, if airtight containers or foil are not used. The same applies to tissue and frozen sections. If prolonged storage is necessary, –70°C is preferable to –20°C.

The antigenicity of fixed tissues (e.g., formalin-fixed) is better preserved if the tissue is stored after processing to paraffin blocks.

2. Fixation and processing: Some form of fixation is nearly always essential, even with frozen sections or cell preparations; it acts not only to preserve cell or tissue structure, but also to stabilize antigens, which might otherwise be leached from tissues by the multiple procedures/washings to which they are subjected during immunostaining. There are many types of fixative and trial and error may be required to achieve optimal fixation for a specific antigen. Care should be taken over the length of time and temperature of fixation. Over fixation can destroy antigenic sites. In addition, some fixatives are only suitable for cytochemistry or frozen sections Although some antigens will not survive processing for paraffin sections, this should be the goal, as the tissue preservation is superior to frozen sections. Recently, an increasing number of antigens have been demonstrated in paraffin-embedded tissues, owing to the development of methods for antigen retrieval (see below). **Table 1** lists published methods of fixation that have been used for NOS isoforms.
   a. Antigen retrieval in formalin-fixed paraffin sections: Enzyme digestion (protease or trypsin) can be used to break protein cross links formed by formalin fixation. Optimal digestion times have to be determined for individual laboratories, but approx 30 min is usual. We have not found this method successful for iNOS immunohistochemistry. A more recent technique is microwaving, which is easier to control than enzyme digestion and gives superior tissue preservation. This is described in detail in **Subheading 3.** We have not found that the length of formalin fixation affects the ability of microwaving to reveal iNOS antigen.
   b. Permeabilization: Most fixatives will permeabilize cells allowing entry of antibodies to detect intracellular antigens. For some preparations, additional permeabilization may be necessary and this can be achieved by using detergents, e.g., Triton X-100, Tween-20, or Saponin.
   c. Tissue processing for paraffin embedding: Following formalin fixation, most laboratories employ an automatic tissue-processing schedule of approx 16 h, during which the tissue is dehydrated by graded alcohols, cleared in chloroform, and impregnated in wax. These procedures are not normally the cause of loss of antigenicity, although the routine procedure of drying paraffin sections on a hot plate (60°C) may be injurious, and for some immunohistochemistry it is preferable to dry sections at room temperature overnight. Drying time can be reduced if adhesive-coated slides are used. An adhesive such as poly-L-lysine also prevents tissue detachment during microwaving.
3. Blocking endogenous biotin: Incubate in 0.01% avidin at room temperature for 20 min. Wash three times in PBS. Incubate in 0.01% biotin at room temperature for 20 min. Wash three times in PBS, and then proceed with **Step 6**.
4. Antibody storage: Only lyophilized antibody is stable for prolonged periods irrespective of temperature and freeze/thawing. 1° antibodies should be aliquoted to

avoid repeated freeze thawing and stored frozen at –20°C or –70°C in concentrated form. Dilutions above 1:10 can deteriorate rapidly. 0.1% azide will prevent bacterial contamination without affecting antibody activity. Labeled antibody should be stored at 4°C, again in concentrated form.

5. If alkaline phosphatase (AP) labeled secondary antibody is used, TBS rather than PBS should be used from **Step 6** onwards, substituting AP substrate in **Step 8**, according to manufacturer's instructions, as phosphate will inhibit the enzyme. Blocking endogenous peroxidase is not necessary.
6. Specificity: When the immunochemical specificity of a primary antibody has been well characterized in a laboratory, it is usually sufficient to run the following controls with each batch of immunostaining: (a) A known positive control (cell/tissue), (b) a negative control where the primary antibody is omitted, or preferably, an inappropriate primary antibody with the same species characteristics and concentration as the test antibody is applied.

   If the primary antibody has not been characterized for immunohisto- or cytochemistry, specificity should be established by pre-absorbing it with specific antigen, and demonstrating loss of staining. The simplest method is to add excess antigen to the antibody at an appropriate final dilution and incubate for 30 min and then apply the primary antibody as usual. (For a more detailed account of this procedure, *see* **ref. *3*.**) Note that antibodies that show positive reactions with specific antigens in ELISAs or dot-blots are not necessarily suitable for immunohistochemistry.
7. Background: There are many causes of background staining and some degree of trial and error may be required to eliminate or reduce this. It is quite common with polyclonal antibodies, but usually not a problem when using monoclonal antibodies (MAbs). It may be due to nonspecific binding, Fc receptor binding, or specific cross reactions, as for example in tissues with intrinsic immunoglobulins such as plasma or plaque cells. For a more detailed account *see* **refs. *3*** and ***5*.** Check that the concentration of the 1° antibody is not too high. For polyclonal antibodies, the usual working ranges are between 1/50–1/2000. MAbs may be used at concentrations as high as 1/5. Also check that a suitable blocking serum has been applied. As the background is commonly due to reaction of the secondary antibody, it is standard practice to block with normal serum from the species in which the secondary antibody has been raised. The serum is applied undiluted or up to 1/20 dilution, normally as a pre-incubation before the primary antibody. Cross reaction with host immunoglobulin may be reduced by adding serum of the host species to the secondary antibody. In a particular system, some sera seem to have better blocking abilities than others, and the rule of using species appropriate sera does not always apply.

   Note that poorly dried-cell preparations, or tissues with fixation artefact will also give false positive localization. Care should be taken that tissue sections do not dry out at any stage of the immunostaining protocol.
8. Endogenous peroxidase: Endogenous peroxidase can be high in some normal tissues, (principally in red cells and leucocytes) and cause problems when using peroxidase labeled antibodies. This activity can be inhibited by adding excess

hydrogen peroxide. Hydrogen peroxide is usually applied before the primary antibody, but if this destroys the reaction, it can be used after the incubation with the primary antibody, as described in **Subheading 3.2.** If inhibition is incomplete, check that the hydrogen peroxide used has not decayed (with typical use it has a shelf life of 6–9 mo) and has been applied at the correct concentration. (*See* also in the Methods Section the note on endogenous biotin, which may give false positive results when using biotinylated antibodies.)

References

1. Nathan, C. (1992) Nitric oxide as a secretory product of mammalian cells. *FASEB J.* **6,** 3051–3064.
2. Cattell, V. and Jansen, A. (1995) Inducible nitric oxide synthase in inflammation. *Histochemical Journal* **27,** 777–784.
3. van Noorden, S. (1993) Problems and Solutions, in *Immunocytochemistry: A Practical Approach* (Beesley, J. E., ed.), Oxford University Press, New York, pp. 207–240.
4. Jackson, P. and Blythe, D. (1993) Immunolabelling techniques for light microscopy, in *Immunocytochemistry: A Practical Approach* (Beesley, J. E., ed.), Oxford University Press, New York, pp. 15–42.
5. Bancroft, J. D. and Stevens, A. (1996) *Theory and Practice of Histological Techniques.* Churchill Livingstone, London.
6. Bredt, D. S., Hwang, P. M., and Snyder, S. H. (1990) Localization of nitric oxide synthase indicating a neural role for nitric oxide. *Nature* **347,** 768–770.
7. Bachmann, S., Bosse, H. M., and Mundel, P. (1995) Topography of nitric oxide synthesis by localizing constitutive NO synthases in mammalian kidney. *Am. J. Physiol.* **268,** F885–F898.
8. Fischer, E., Schnermann, J., Briggs, J. P., Kriz, W., Ronco, P. M., and Bachmann, S. (1995) Ontogeny of NO synthase and renin in juxtaglomerular apparatus of rat kidneys. *Am. J. Physiol.* **268,** F1164–F1176.
9. Wilcox, C. S., Welch, W. J., Murad, F., Gross, S. S., Taylor, G., Levi, R., and Schmidt, H. H. H. W. (1992) Nitric oxide synthase in macula densa regulates glomerular capillary pressure. *Proc. Natl. Acad. Sci. USA.* **89,** 11,993–11,997.
10. Tojo, A., Gross, S. S., Zhang, L., Tisher, C. C., Schmidt, H. H. H. W., Wilcox, C. S., and Madsen, K. M. (1994) Immunocytochemical localization of distinct isoforms of nitric oxide synthase in the juxtaglomerular apparatus of normal rat kidney. *J. Am. Soc. Nephrol.* **4,** 1438–1447.
11. Wildhirt, S. M., Dudek, R. R., Suzuki, H., Pinto, V., Narayan, K. S., and Bing, R. J. (1995) Immunohistochemistry in the identification of nitric oxide isoenzymes in myocardial infarction. *Cardiovascular Res.* **29,** 526–531.
12. Myatt, L., Brockman, D. E., Eis, A. L. W., and Pollock, J. S. (1993) Immunohistochemical localization of nitric oxide synthase in the human placenta. *Placenta* **14,** 487–495.
13. Conger, J., Robinette, J., Villar, A., Iaaij, L., and Shultz, P. (1995) Increased nitric oxide synthase activity despite lack of response to endothelium-dependent vasodilators in postischemic acute renal failure. *J. Clin. Invest.* **96,** 631–638.

14. Thomsen, L. L., Miles, D. W., Happerfield, L., Bobrow, L. G., Knowles, R. G., and Moncada, S. (1995) Nitric oxide synthase activity in human breast cancer. *Br. J. Cancer* **72,** 41–44.
15. Shaul, P. W., North, A. J., Wu, L. C., Wells, L. B., Brannon, T. S., Lau, K. S., Michel, T., Margraf, L. R., and Star, R. A. (1994) Endothelial nitric oxide synthase is expressed in cultured human bronchiolar epithelium. *J. Clin. Invest.* **94,** 2231–2236.
16. Kobzik, L., Bredt, D. S., Lowenstein, C. J., Drazen, J., Gaston, B., Sugerbaker, D., and Stamler, J. S. (1993) Nitric oxide synthase in human and rat lung: immunocytochemical and histochemical localization. *Am. J. Respir. Cell Mol. Biol.* **9,** 371–377.
17. Buttery, L. D. K., Evans, T. J., Springall, D. R., Carpenter, A., Cohen, J., and Polak, J. M. (1994) Immunochemical localization of inducible nitric oxide synthase in endotoxin-treated rats. *Lab Invest.* **71,** 755–764.
18. Buttery, L. D. K., Springall, D. R., Andrade, S. P., Riveros-Moreno, V., Hart, L., Piper, P. J., and Polak, J. M. (1993) Induction of nitric oxide synthase in the neovasculature of experimental tumours in mice. *J. Pathol.* **171,** 311–319.
19. Connor, J. R., Manning, P. T., Settle, S. L., Moore, W. M., Jerome, G. M., Webber, R. K., Tjoeng, F. S., and Currie, M. G. (1995) Suppression of adjuvant-induced arthritis by selective inhibition of inducible nitric oxide synthase. *Eur. J. Pharm.* **273,** 15–24.
20. Van Dam, A.-M., Bauer, J., Man-a-Hing, W. K. H., Marquette, C., Tilders, F. J. H., and Berkenbosch, F. (1995) Appearance of inducible nitric oxide synthase in the rat central nervous system after rabies virus infection and during experimental allergic encephalomyelitis but not after peripheral administration of endotoxin. *J. Neuro. Res.* **40,** 251–260.
21. Corbett, J. A. and MacDaniel, M. L. (1995) Intraislet release of interleukin 1 inhibits cell function by inducing cell expression of inducible nitric oxide synthase. *J. Exp. Med.* **181,** 559–568.
22. Campbell, I. L., Samimi, A., and Chiang, C.-S. (1994) Expression of the inducible nitric oxide synthase. *J. Immunol.* **153,** 3622–3629.
23. Nicholson, S., Bonecini-Almeida, Lapa e Silva, J. R., Nathan, C., Xie, Q., Mumford, R., Weidner, J. R., Calaycay, J., Geng, J., Boechat, N., Linhares, C., Rom, W., and Ho, J. L. (1996) Inducible nitric oxide synthase in pulmonary alveolar macrophages from patients with tuberculosis. *J. Exp. Med.* **183,** 2293–2302.
24. Kleeman, R., Rother, H., Kolb-Bachofen, V., Xie, Q., Nathan, C., Martin, S., and Kol, H. (1993) Transcription and translation of inducible nitric oxide synthase in the pancreas of prediabetic BB rats. *FEBS.* **328,** 1,2. 9–12.
25. Sato, K., Miyakawa, K., Takeya, M., Hattori, R., Yui, Y., Sunamoto, M., Ichimori, Y., Ushio, Y., and Takahashi, K. (1995) Immunohistochemical expression of inducible nitric oxide synthase (iNOS) in reversible endotoxic shock studied by a novel monoclonal antibody against rat iNOS. *J. Leukoc. Biol.* **57,** 36–44.
26. Jansen, A., Cook, T., Michael-Taylor, G., Largen, P., Riveros-Moreno, V., Moncada, S., and Cattell, V. (1994) Induction of nitric oxide synthase in rat immune complex glomerulonephritis. *Kid. Int.* **45,** 1215–1219.

27. Cook, H. T., Bune, A. J., Jansen, A. S., Taylor, G. M., Loi, R. K., and Cattell, V. (1994) Cellular localisation of inducible nitric oxide synthase in experimental shock in the rat. *Clin. Science.* **87,** 179–186.
28. Cattell, V., Smith, J., Jansen, A., Riveros-Moreno, V., and Moncada, S. (1994) Localisation of inducible nitric oxide synthase in acute renal allograft rejection in the rat. *Transplantation.* **58,** 1399–1402.

18

A Practical Protocol for the Demonstration of NOS Using *In-Situ* Hybridization

G. Michael Taylor and H. Terence Cook

1. Introduction

In-situ hybridization (ISH) is a technique that allows the demonstration of nucleic-acid sequences of interest within tissues and their localization to subpopulations of cells within those tissues.

The aim of this chapter is to describe an ISH protocol suitable for the demonstration of nitric oxide synthase (NOS) in animal and human tissues that we have used for inducible NOS (iNOS).

Although the specific probe sequences were designed to identify only iNOS, the minutiae apply equally to all target sequences and can easily be adapted to other methods.

To detect the expression of mRNA for iNOS, which like other acute-phase proteins, may be transient or follow a set time-course, requires a method of medium to high sensitivity. This can best be achieved with a radioisotopic method. Owing to the improved resolution offered by nonisotopic methods these are being increasingly used in the expression of medium-high abundance proteins and in viral replication and for identifying the cellular compartment of the target sequences. Our own, limited experience with nonisotopic methods has shown that these are not sufficiently sensitive for low-copy mRNA's and do not perform well on routine archival tissue. In any event, we suggest the first approach should be with a radioisotopic probe to identify a positive control tissue expressing the gene of interest, in this case iNOS.

When this method was first developed, antibodies to iNOS protein were not generally available, and so ISH using probe sequences deduced from the published nucleotide sequence was given added impetus in the study of its expres-

From: *Methods in Molecular Biology, Vol. 100. Nitric Oxide Protocols*
Edited by: Michael A. Titheradge © Humana Press Inc., Totowa, NJ

sion in septic shock *(1)*. If an antibody is available this may first be used to demonstrate the protein immunohistochemically. ISH is a time-intensive procedure and should not be entered into lightly. It demands some technical expertise in the handling of tissue specimens. Isotopic methods can be expensive and there are safety considerations in the handling and disposal of the radionuclides. There must be periodic access to a darkroom, molecular biology-grade autoclaved solutions dedicated to ISH and ideally, a small laboratory area set aside to provide an RNase-free environment.

Although these considerations are important, the method described below using synthetic DNA probes is robust and does not require more than routine care to guard against RNase contamination. A successful batch of ISH with 12–20 slides can provide much information on the tissue expression, time course, and the identity of cells expressing the mRNA. In addition, it can confirm local synthesis as opposed to uptake of a protein and in combination with immunohistochemistry, can compare their concordance of expression.

The method we advocate is the use of synthetic oligoncleotide-DNA probes designed to hybridize to the messenger-sense polarity RNA. We have used primers 30 bases in length with a G + C content from 45% to 55%. This assures adequate specificity, allows hybridization to be performed at ambient temperature and avoids loss of sections and damage due to high-wash temperatures. The probes are 3' end-labeled with a polymeric tail of ^{35}S-deoxyadenosine triphosphate (^{35}SdATP). The enzyme terminal deoxynucleotidyl transferase (tdt) is used to add up to 15 labeled bases onto the 3' end of the primers. When several labeled probes are used as a "cocktail" to hybridize to different regions of the mRNA, the result is a very sensitive method. The use of ^{35}SdATP is a compromise between radionuclides of high (^{32}P) and low energy of emission (^{3}H), combining relatively short exposure times with medium resolution. Once a method has been established, the greater resolution possible with tritiated probes (0.1–2 µm cf. 5–15 µm with ^{35}S) can be used to more confidently identify cell types if required.

DNA probes are readily available to order from commercial sources and do not require cloning and other molecular-biology experience. We have found that they behave predictably according to the equation $t_{melt} = 16.6 \log [M] + 0.41[P_{gc}] + 81.5 - P_m - B/L - 0.65 [P_f]$, where

1. *M* is the molar concentration of sodium to a maximum of 0.5;
2. P_{gc} is the percent of G + C bases in the probe;
3. P_m is the percent of mismatched bases;
4. P_f is the percent of formamide in the buffer;
5. B is 675 (for probes up to 100 bases in length); and
6. L is the probe length in bases *(2)*.

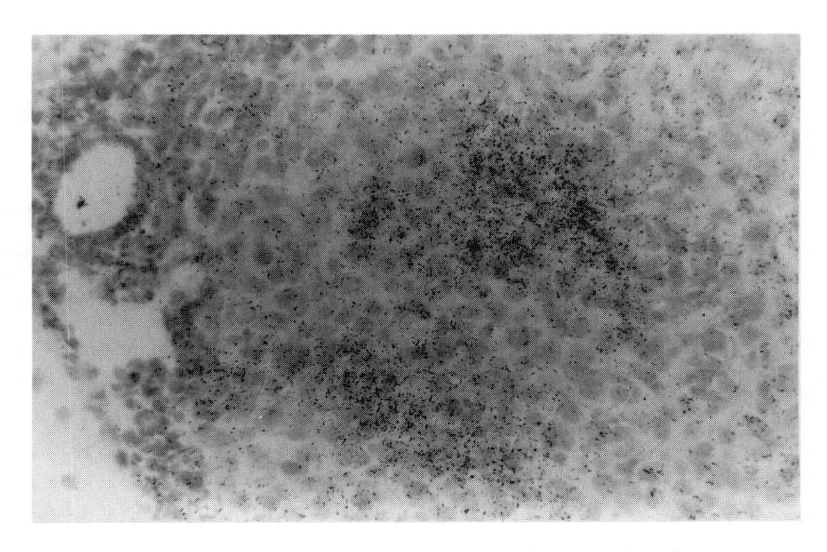

Fig. 1. Localization of iNOS mRNA in a frozen liver section from a mouse with endotoxic shock induced with iv *C. parvum* and 7 d later with intraperitoneal LPS. Hybridization signal is seen in a cytoplasmic distribution overlying two granulomata comprised predominantly of macrophages. Hematoxylin counterstain, magnification ×235.

They also have a number of other advantages that will only be mentioned briefly here. First, probes of this length have good penetration into tissue sections. This makes them ideal candidates for eventual use on paraffin-embedded sections which offers better morphology and interpretation of cell types. In addition, this may be a convenient source of rarer material, not easily gathered prospectively.

Nearly all oligonucleotide probes (oligos) we have tried will work on cryostat sections, provided samples have been collected with reasonable precautions to minimize RNA degradation. However, not all oligos that work on frozen sections perform equally well when the protocol is modified for use on paraffin-embedded material. This observation applies to a variety of ISH methods. We believe this is owing to some aspects of fixation, possibly cross-linking of regions of RNA to proteins and incorporation of the additive fixative formalin, into the tissue matrix. Of five DNA probes evaluated, the three to mouse/rat iNOS described below work well on protease-treated paraffin sections as well as on frozen sections. **Figure 1** shows the localization of iNOS mRNA in frozen-liver sections of a rat treated with *C. parvum* and lipopolysaccharide (LPS). The human probes are designed to homologous regions of the hepatocyte iNOS gene. Using a BLAST search available on the internet (*see* http://www.ncbi.nlm.nih.gov:/) the five probes have been found to have significant homology only to iNOS and under the stringent conditions used here would not hybridize to other sequences.

Second, the use of oligonucleotide probes facilitates inclusion of control experiments, particularly important when attempting a new ISH method. Controls we used initially in validation experiments included the use of a large excess (50-fold) of unlabeled antisense (AS) probes in competition with the radiolabeled moieties to abolish hybridization signal. This contrasts with competition with a similar excess of "irrelevant" oligonucleotide or randomer, which should have no effect on a specific signal. Probes of sense-strand polarity may be purchased and end-labeled to confirm the specificity as indeed may end-labeling of the irrelevant oligo, both of which should highlight any non-specific binding. A probe of reverse AS polarity (same sequence as AS but synthesized in the 5'-3' direction) makes an ideal irrelevant oligo in this context, having the same nucleotide base sequence. The iNOS method described below has been fully validated, but to assist in interpretation of detected signal it is important to include some controls with every experiment. RNAse-treated slides and pre-absorption experiments with AS and "irrelevant" oligonucleotides as described are usually satisfactory for this purpose.

2. Materials

2.1. Solutions

Prepare in advance the following diethyl pyrocarbonate (DEPC)-treated solutions (*see* **Subheading 3.1.**).

1. Water, 2 L.
2. Phosphate-buffered saline (PBS) pH 7.4, 5 L.
3. Triethanolamine, 0.1 *M*, pH 8.0, containing 9 g sodium chloride per L.
4. 20X SSC: sodium chloride/sodium citrate, pH 7.0, containing 175.3 g NaCl and 88.2 g Na_3 citrate·$2H_2O$ per L.

2.2. Section Cutting and Preparation

1. Acetic anhydride (Sigma A-1606).
2. Acetone (Hipersolv™, BDH/Merck 15296).
3. Glass troughs 10 × 250 mL (BDH/Merck).
4. Methanol (Analar®, BDH/Merck 10158). Prepare an alcohol series (30, 50, 70, 90, and 100%) with DEPC-treated water.
5. Microscope slides (pre-washed, single or twin-frosted) and cover slips.
6. Paraformaldehyde (Sigma, P-6148).
7. Plastic slide storage boxes (Philip Harris Scientific, London, UK).

8. PBS, DEPC treated, pH 7.4, 5 L.
9. Proteinase K (Boehringer Mannheim, Lewes, E. Sussex, UK).
10. RNases A and T1 (Sigma R-8251 and R-5250).
11. Slide staining racks.
12. Tissue-Tek® O.C.T. compound (Raymond Lamb, London, UK).
13. Triethanolamine, 0.1 *M*, pH 8.0, containing 9 g sodium chloride per L.
14. Vectabond® (Vector Laboratories, Peterborough, UK).
15. Xylene (low in sulphur, BDH/Merck 3607).

2.3. Prehybridization

1. Denhardt's solution (Sigma D-2532).
2. Dextran sulfate (Sigma D-8906).
3. Dithiothreitol (DTT) (Sigma D-9779).
4 Mixed-bed resin AG 501-X (D) to deionize formamide (BioRad).
4. Ethylenediaminetetraacetic acid (EDTA), 100 m*M* stock, stored at –20°C in 1 mL aliquots.
5. Formamide (Fluka Chemie AG, Glossop, UK).
6. Polyadenylic acid, 12.5 mg/mL stock in DEPC water, stored at –20°C.
8. Salmon-sperm DNA, 10 mg/mL stock, sheared by passage through a 23-gage needle. Make 10 mL and store in 1 mL aliquots at –20°C.
9. 20X SSC (1 L):sodium chloride/sodium citrate, pH 7.0, containing 175.3 g NaCl and 88.2 g Na_3 citrate·$2H_2O$.
10. 100 m*M* Tris-HCl buffer stock, pH 7.4.
11. Yeast transfer (t)-RNA (Sigma R-9001).
12. Moist chamber.
13. Hybridization buffer: 10 m*M* Tris-HCl buffer pH 7.4, with 50% deionized formamide, 1X Denhardt's solution, 1 m*M* EDTA, 100 µg/mL denatured salmon-sperm DNA, 10% dextran sulfate, 250 µg/mL yeast transfer RNA, 6X SSC.

In a well-ventilated fume hood, mix 100 mL of formamide (toxic) in a beaker with 10 g of mixed-bed resin for 30 min at RT. Filter the mixture through Whatman paper to remove the resin. To 50 mL of the deionized formamide, add 10 mL of the stock 100 m*M* Tris-HCl and 30 mL of 20X SSC. Add 10 g of dextran sulfate, 25 mg of yeast (t)-RNA, 2 mL of 50X Denhardt's, 1 mL of denatured salmon-sperm DNA, and 1 mL of stock EDTA. Make the solution up to a final volume of 100 mL with DEPC-treated water and store 5 mL aliquots at –20°C (does not freeze). Stable for up to 1 yr. Before use, heat a vented aliquot in a boiling bath for 3 min, quench on ice for 10 min, and add polyadenylic acid (Poly [A]) and DTT to produce final concentrations of 250 µg/mL and 10 m*M*, respectively. Alternatively, several hybridization fluids are available commercially (e.g., Sigma) to which Poly [A] and DTT may be added. Hybridization buffer is in contact with sections for prolonged times and must be RNAse free.

2.4. Oligonucleotide Probe Labeling and Purification

1. DNA-purification cartridges (NEN Du Pont, NLP 022X).
2. ^{35}S-deoxyadenosine 5' (α-thio) triphosphate (^{35}S-dATP, Amersham code SJ 1334).
3. Ethanol.
4. Methanol (Hipersolv™, BDH/Merck 15250).
5. Nensorb™ A DNA-column-wash buffer: 0.1 *M* Tris, pH 7.4 containing 10 m*M* triethylamine, and 1 m*M* EDTA. Store at 4°C. Stable up to 1 mo.
6. Nensorb™ B (20% ethanol).
7. Plastic syringe, 10 mL.
8. Terminal deoxynucleotidyl transferase (TdT) and end-labeling buffer (Promega, M-1871).
9. Water [high-pressure liquid chromatography (HPLC) grade, BDH/Merck 15273].
10. Probes: We have used three antisense 30-base synthetic-oligonucleotide probes corresponding to the cDNA sequences of both the inducible mouse-macrophage gene and the rat-inducible vascular smooth-muscle gene from regions where these sequences are identical ***(3,4)***. Probes NOS-1, NOS-2, and NOS-3 were synthesized by Oswel DNA, Edinburgh, Scotland, and probes hNOS-1 and hNOS-2 by Genosys, Cambridge, UK. The sequences of these probes are as follows:

    NOS-1 (amino acids 1–10 inclusive):

    5'-TTT-GAA-GAG-AAA-CTT-CCA-GGG-GCA-AGC-CAT-3'.

    NOS-2 (aa's 17–26):

    5'-GTT-GTT-AAT-GTC-CTT-TTC-CTC-TTT-CAG-GTC-3';

    NOS-3 (aa's 152–161):

    5'-GGC-CAG-ATG-TTC-CTC-TAT-TTT-TGC-CTC-TTT-3'.

    For ISH on human tissue, we have designed two further probes hNOS-1 and hNOS-2, which hybridize to equivalent regions of NOS-1 and NOS-2 ***(5)***. These were synthesized by Genosys, Cambridge, UK. The sequences of these are:

    hNOS-1:

    5'-CTT-GAA-CAG-AAA-TTT-CCA-AGG-ACA-GGC-CAT-3'

    and hNOS-2:

    5'-GTT-GTT-GAT-GTC-TTT-TTC-CCC-ATT-CAT-TGC-3'.

NOS-3 contains no mismatches with the human-hepatocyte iNOS sequence *(5)* and is used with hNOS-1 and hNOS-2.

2.5. Hybridization and Stringency Washes

1. Hybridization buffer as for prehybridization step.
2. Labeled and purified AS DNA probes.
3. Unlabeled AS and control probes as required.
4. 1X SSC diluted from stock solution of 20X SSC.

2.6. Autoradiography

1. Amber plastic concertina bottles (BDH/Merck).
2. Autoradiography boxes (BDH/Merck).
3. Brown glass bottle, 50 mL.
4. Chromium (III) potassium sulfate (chrome alum, BDH/Merck 27758).
5. DEPC (Sigma, D5758).
6. DPX mountant (BDH/Merck 36125).
7. Gelatin.
8. Glacial acetic acid.
9. Glycerol (Sigma G-5516).
10. Mayer's hemalum.
11. Nuclear track emulsion (K5, Ilford Ltd, Knutsford, Cheshire, UK).
12. Phenisol® developer (Ilford Ltd).
13. Silica desiccant sachets.
14. Slide mailer, plastic 20 mL.
15. Sodium citrate.
16. Sodium thiosulfate.
17. UV-curing adhesive (RS-505-189, RS Components, Corby, Northants, UK or Loctite®357, Loctite Ltd. Watchmead, Welwyn, Herts, UK).

3. Methods

3.1. Solutions

A number of solutions are common to more than one step of the procedure and may conveniently be prepared in advance (*see* **Subheading 2.1.**). To denature any contaminating ribonucleases (RNAses), these should be treated with 200 μL of diethyl pyrocarbonate (DEPC)/L of liquid for 24 h and autoclaved. Store at room temperature. A residual sweet odor is normal. Note that DEPC is toxic. Follow handling precautions. Wear gloves when handling stock solution and dispense in a fume hood.

3.2. Cutting and Preparation of Sections

This method applies to prospectively collected tissues, isolated as rapidly as possible, and frozen in liquid nitrogen within embedding compound to prevent drying. Store in suitable screw-cap test tubes at –70°C (*see* **Note 1**).

3.2.1. Cryostat Sections

1. Cryostat sections should be cut at a thickness of 6–10 µm on Vectabond® coated slides (*see* **Note 2**). If block size permits, it is useful to place 2–3 sections on the lower half of each slide to facilitate later dipping in photographic emulsion.
2. Fix slides in freshly prepared 4% paraformaldehyde, pH 7.4 for 5 min (*see* **Note 3**).
3. Wash slides in PBS for 5 min to remove fixative.
4. Treat RNase controls at this point in PBS containing 20 µg/mL of ribonuclease A and 80 units/mL of ribonuclease T1 for 30 min at 37°C (*see* **Note 4**).
5. Wash thoroughly in PBS.
6. Treat all sections for 20 min in 250 mL of 0.1 *M* triethanolamine, pH 8.0, containing 500 µL of acetic anhydride (*see* **Note 5**).
7. Wash in PBS and dehydrate through the alcohol series, 2 min/trough.
8. Allow to air dry and store in plastic slide boxes at –20°C (*see* **Note 6**).

3.2.2. Paraffin Sections

1. Cut at a thickness of 2–4 µm from prospectively gathered material, which has been fixed in neutral-buffered formalin for 12–24 h (*see* **Note 7**). Archival material may be suitable for some mRNA's expressed in medium to high abundance.
2. Dewax in xylene for 10 min.
3. Dry around the sections to remove excess xylene and rehydrate through the alcohol series. Wash in PBS, 2–5 min.
4. Treat with proteinase K, 50 µg/mL for 10 min at room temperature (RT) (*see* **Note 8**).
5. Wash in PBS, 2–5 min.
6. Follow **Steps 4–7** as for frozen sections.

3.3. Prehybridization

1. Remove an aliquot of hybridization buffer from the freezer, heat in a boiling water bath for 3 min to denature the DNA, quench on ice, and add poly [A] and DTT as described in **Subheading 2.3.**
2. Mix well and overlay sections with sufficient buffer to cover the tissue totally. Leave sections for 3 h in moist chamber at RT (*see* **Note 9**).

3.4. Probe Labeling and Purification

The DNA oligonucleotide probes are 3'-end labeled using tdt to add a polymeric tail of radioligand, in this case ^{35}S-dATP, to their 3' end. Nonincorporated ligand is removed using a Nensorb™ DNA-purification column.

3.4.1. Labeling Procedure

1. Add 10 pmol of each probe in a volume of 5 µL of HPLC-grade water to a 0.5 mL sterile Eppendorf tube.
2. Add 8 µL of 5X reaction buffer (*see* **Note 10**).

3. Add 4 μL of ^{35}S-dATP/probe to be labeled (*see* **Note 11**).
4. Add 35 units of tdt (Promega).
5. Add HPLC water to a final volume of 40 μL (*see* **Note 12**).
6. Incubate at 37°C for 1 h in water bath or PCR-machine block on manual control.
7. Stop the reaction with the addition of 400 μL of Nensorb™ A column-wash buffer.

*3.4.2. Purification Procedure (*see ***Note 13***).*

1. Clamp a Nensorb™ purification push-column in a retort stand and wash through with 2 mL of HPLC grade methanol, collecting the waste into a universal tube (do not let the column run completely dry).
2. Follow this with 2 mL of Nensorb™ reagent A.
3. Run the labeling reaction contents slowly onto the column and follow it with an additional 2 mL of Nensorb™ A. The eluate should contain unincorporated ^{35}S-dATP.
4. Wash the column through with 2 mL of HPLC grade water (optional).
5. Elute the probe(s) with 1 mL of 20% ethanol (Nensorb™ B), collecting the first 10 drops into a sterile Eppendorf tube (*see* **Note 14**).
6. Check the tube for efficient labeling and collection of probe using a hand-held radiation monitor. Similarly, monitor the working area for accidental contamination and dispose of all tips, gloves, reaction tubes, and spent column in appropriate bins for incineration. Unincorporated ^{35}S-dATP should be disposed of via a sink suitable for this purpose.
7. Count duplicate 5 μL aliquots of the purified probe in a beta-scintillation counter to determine labeling efficiency (*see* **Note 15**).

3.5. Hybridization

1. Dilute the labeled probes to give between 2.0 and 3.0 × 10^6 cpm/100 μL of hybridization fluid.
2. Meanwhile, pour off prehybridization buffer from the slides, and dry around the tissue sections.
3. Add sufficient hybridization fluid to cover each section with a "puddle" (*see* **Note 16**).
4. Incubate overnight at RT in a slide moist chamber. Ambient temperature is near the optimum-incubation temperature for the probes, t_i (found from $t_i = t_{melt}$) −15°C, where t_{melt} is the calculated melting temperature of duplex DNA at which 50% of the strands are dissociated *(2)*.

3.5.1. Controls

The use of synthetic-DNA oligonucleotide probes allows a number of controls to be easily performed at this stage. These include:

1. Ribonuclease treated sections (described earlier), probed with antisense-polarity oligonucleotides.

2. Preabsorption of hybridization signal with an excess (50–100 fold) of unlabeled antisense probe in competition with labeled probe(s).
3. Competition with a similar excess of "irrelevant" or randomer nucleotide of similar length and G + C content to ensure that signal is not abolished.
4. 3' end-labeling of the "irrelevant" or randomer probe, which should not result in hybridization.
5. 3' end-labeling of messenger sense polarity probe(s), which should not result in hybridization.
6. Tissue unlikely to be expressing the mRNA of interest.

For the iNOS ISH method described here, controls 1–4 and 6 have been used to validate the method.

3.5.2. Stringency Washes

1. After overnight hybridization, wash the sections in 1X SSC for 30 min at RT followed by four more washes in 1X SSC at 55°C (*see* **Note 17**).
2. Allow the slides to cool thoroughly and rinse in deionized water to remove traces of PBS.
3. Immerse in chrome alum/gelatin, stand upright to drain, and allow to air dry (*see* **Note 18**).

3.5.3. Autoradiography-Section Dipping

1. Under darkroom conditions (*see* **Note 19**), dispense about 10 mL of Ilford K5 nuclear track emulsion into a clean, dark-brown glass bottle, seal the cap, and wrap in aluminium foil to exclude all light. Pipet 10 mL of deionized water and 300 µL of glycerol into a plastic slide mailer and wrap in foil.
2. Place both containers into a water bath set at 43°C and leave for 30 min (*see* **Note 20**).
3. Pour the melted emulsion into the slide mailer, snap the lid shut, and mix gently to avoid frothing. Leave for 30 min.
4. Dip the slides singly, to a depth which just covers the sections. This will minimize emulsion overrun. Leave immersed for 5 s and remove slowly from the emulsion, keeping the slide upright (10 s). Check that the slide has coated evenly (*see* **Note 21**). Wipe the back of the slide and leave to drain in an autoradiography-box rack. When all the slides have been dipped, refit the box lid, and leave to dry slowly for one h at RT.
5. Again under safelight conditions, remove the lid and insert two desiccant sachets behind the lid-drainage tray which forms part of the Merck autoradiography box. Before leaving the darkroom, wrap the box in aluminium foil or black plastic to ensure it is lightproof.
6. Store at 4°C away from radiochemicals, preferably in perspex boxes to minimize accidental exposure to beta emitters. Exposure times will vary, but are typically 6–12 d for cryostat sections and 17–25 d for formalin-fixed and archival material

3.5.4. Development

1. Allow the slides to reach ambient temperature slowly (2 h).
2. Prepare four glass troughs to contain
   a. Phenisol® developer diluted 1:5;
   b. acid-stop solution consisting of 1% glacial acetic acid and 1% glycerol;
   c. 0.3 *M* sodium thiosulfate clearing and fixing reagent, and
   d. water.

   Use deionized water in the preparation of the developing solutions. Solutions b–d may be prepared in the laboratory, but diluted Phenisol® must be prepared under safelight conditions from a stock solution previously equilibrated to ambient temperature (*see* **Note 22**).
3. In the darkroom, check all solutions are within 1°C of each other.
4. Transfer the slides to a staining rack and immerse in diluted Phenisol® (2 min). Agitate once after 1 min. Transfer sequentially to the acid-stop solution (1 min), sodium thiosulfate (4 min), and water (2 min).
5. Lights may be turned on at this stage. Wash sections thoroughly in several changes of water over 20 min, finishing in deionized water.
6. Counterstain in Mayer's hematoxylin, "blue" in tap water, rinse in deionized water, and air dry.
7. Mount in DPX or equivalent (*see* **Note 23**).

4. Notes

1. Fresh tissue may conveniently be frozen by embedding in Tissue Tek® or equivalent reagent. Cut small cubes of tissue (5 mm^3), e.g., kidney or liver, and place into aluminium foil molds (previously prepared by wrapping foil around the end of a circular glass rod or pencil) containing one or two drops of Tissue Tek®. Add a further two drops of embedding compound and, with long forceps, touch the bottom of the mold onto the surface of isopentane cooled in liquid nitrogen in a small dewar-flask. When completely frozen (opaque), transfer the sample to a screw-cap vial, and store at –70°C. Alternatively, tissue and embedding compound may be frozen on cork boards (22 mm diameter, Speci-microsystems Ltd., Carshalton, Surrey, UK); the boards can be easily attached to the cryostat chuck.
2. To improve adhesion of sections, we would strongly recommend that the slides be coated with Vectabond®, which comes with full instructions (Vector Laboratories, Peterborough, UK). The Vectabond® reagent (which is light sensitive) will be sufficient to prepare between 300–500 slides. In our experience, slides prepared thus will be stable for up to a year.
3. Paraformaldehyde is difficult to dissolve at neutral pH. In a fume hood, add 4 g paraformaldehyde to 50 mL of water and add a few drops of 1 *M* NaOH while stirring. When dissolved, slowly adjust the pH back to 7.4 with dilute HCl and make up to 100 ml with PBS. Check the pH before use.
4. Dedicate a Coplin jar or plastic equivalent solely for use with RNases to avoid contaminating other glassware. Treat several slides for use as controls in later

batches and store at –20°C. Do not recombine treated slides with the remainder until thoroughly washed in several changes of PBS. RNase treatment tends to increase the background signal, particularly on cryostat sections. This will be evenly distributed.

5. Acetic anhydride is hazardous. Wear gloves and follow safety instructions. Open the bottle and dispense in a fume cupboard. Add to triethanolamine while stirring. Allow a few min for the acetic anhydride to disperse before adding section rack.
6. If ISH is to be delayed for some days or weeks, store at –70°C. Plastic boxes will prevent slide racks becoming heavily frosted.
7. For reasons that are unclear, tissues fixed in paraformaldehyde and routinely processed and embedded in paraffin give poor or absent signal. Therefore, we use neutral-buffered formalin to fix paraffin sections. There is no need to float paraffin sections on DEPC-treated water for mounting on slides, as in our experience RNases do not gain access to cells prior to de-waxing and limited proteolysis. However, as they must resist protease and high-stringency washes, it is most important that paraffin sections are cut onto coated slides, and we have found that Vectabond® is the best reagent for this. If planning a number of controls, consider cutting numbered-serial sections to compare the same regions on adjacent slides.
8. This concentration is a guideline only for formalin-fixed kidney and liver tissue. Protease treatment should be optimized for the tissue of interest, e.g, lung requires minimal unmasking. Material that has received prolonged fixation (d) may require longer unmasking. A useful range of proteinase to try is 50–200 µg/mL for 10 min at RT. Higher concentrations begin to result in loss of specific ISH signal and increased background.
9. Humid chambers may be purchased commercially (Sigma H-6644) or improvized from Tupperware containers with glass rods to support slides above dampened filter paper.
10 5X Reaction buffer is supplied with the Promega tdt enzyme.
11. The Amersham product SJ-1334 is aqueous based and compatible with most 3' end-labeling systems. It gives the best 3' end-labeling in our experience. Other good performers are the NEN Du Pont 3' end-labeling kit used with their own label (NEP-100 and NEG-034H, NEN Du Pont, Stevenage, UK). (Products from other suppliers may cause precipitation of cobalt salts from the Promega reaction buffer.)
12. We have occasionally had difficulties with "in-house" water-purification systems and DEPC-treated water, which, if not completely removed, may inhibit the tdt enzyme.
13. More detailed instructions are given in the column kit. Nensorb™ A reagent should be stable for up to 1 mo. Nensorb™ B (20% ETOH) should be made fresh for each experiment.
14. Provided the column has not been allowed to run dry, the first two drops of the 20% ETOH wash will reflect the column-void volume and should not contain labeled DNA probes. They may be discarded by the practised operator to give slightly more concentrated eluate.

15. Typical 5 μL counts for successful labeling experiments are in the region 6×10^5–2×10^6 cpm for a β liquid-scintillation counter operating at about 70% efficiency. Labeling efficiency should be sufficient to permit probe dilution by at least 1:6 to dilute the ethanol in the column eluate and still retain $20–30 \times 10^6$ cpm/mL of hybridization buffer.
16. Hybridization fluid is viscous and should not "creep" off the section. If trouble is experienced, cover sections with a Parafilm® coverslip cut to required size. These readily float off during the first SSC wash without causing damage.
17. The temperature of the solution around the slides determines the wash stringency, not the setting on the water bath, which may have to be increased by a few degrees to bring the SSC up to the required value. The washes will contain small amounts of ^{35}S-dATP and should be disposed of down sinks designated for radioactive use.
18. Dissolve the gelatin in 250 mL of deionized water using a stirring hot plate. Add the chrome alum and allow to cool to RT. The surface of the solution should be free of air bubbles and dust, both of which may cause uneven coating of emulsion. dip blank slides to test.
19. Ilford dark-brown filter 904 with a 15-W bulb. A light-brown filter 902 will suffice for indirect illumination.
20. Avoid water baths with exposed elements that produce a red glow and mask any power and heater element indicator lights. Dipping vessels are available commercially, but are wasteful of emulsion.
21. These may be rescued if redipped before the emulsion cools, but mark the slide to assist interpretation.
22. The state of the developer is the single most important determinant of background signal. To minimize oxidation, we aliquot new Phenisol® in a darkroom into 5 × 1 L amber plastic concertina bottles from which the air may be excluded as the solution is used (BDH). These are labeled and stored at 4°C. If kept in this way, Phenisol® is stable for several mo. Before developing a batch of slides, check the color of the solution. Phenisol® oxidizes to a mahogany-brown colored solution if incorrectly stored. Use only developer that is off-white or pale yellow.
23. A good alternative is UV-curing adhesive (RS or Loctite®) that has a refractive index similar to glass. Note that it is much more difficult to remove the coverslips should the need arise.

References

1. Cook, H. T., Bune, A. J., Jansen, A. S., Taylor, G. M., Loi, R. K., and Cattell, V. (1994) Cellular localization of inducible nitric oxide synthase in experimental endotoxic shock in the rat. *Clin.Sci.* **87,** 179–186.
2. Davis, L. G., Dibner, M. D., and Battey, J. F., eds. (1986) Hybridisation with synthetic end-labelled probes, in, *Basic Methods in Molecular Biology*, Elsevier, New York, pp 75–78.
3. Lyons, C. R., Orloff, G. J., and Cunningham, J. M. (1993) Molecular cloning and functional expression of an inducible nitric oxide synthase from a murine macrophage cell line. *J. Biol. Chem.* **267,** 6370–6374.

4. Nunokawa, Y., Ishida, N., and Tanaka, S. (1993) Cloning of inducible nitric oxide synthase in rat vascular smooth muscle cells. *Biochem. Biophys. Res. Commun.* **191,** 89–94.
5. Geller, D. A., Lowenstein, C. J., Shapiro, R. A., Nussler, A. K., Disilvio, M., Wang, S. C., Nakayama, D. K., Simmons, R. L., Snyder, S. H., and Billiar, T. R. (1993) Molecular cloning and expression of inducible nitric oxide synthase from human hepatocytes. *Proc. Natl. Acad. Sci. USA* **90,** 3491–3495.

19

Use of NO Donors in Biological Systems

Jayne M. Tullett and Daryl D. Rees

1. Introduction

The endogenous production of nitric oxide (NO) plays a key role in many bioregulatory systems, including the control of vascular tone, inhibition of platelet aggregation, neurotransmission, and macrophage toxicity. Many of these pathways are activated by the stimulation of soluble guanylyl cyclase *(1)*. Owing to the inconvenient handling of solutions of NO, there is an increased interest in compounds capable of generating NO *in situ*. These compounds can be divided into different groups that include organic nitrates [e.g., glyceryl trinitrate (GTN)], organic nitrites (e.g., iso amyl nitrite), inorganic nitroso compounds [e.g., sodium nitroprusside (SNP)], sydnonimines [e.g., molsidomine (SIN-1)], and S-nitrosothiols (RSNO) [e.g., S-nitrosoglutathione, (GSNO)]. All these compounds differ in their need for specific cofactors required to release NO *(2)*.

Organic nitrates are metabolized by enzymatic and nonenzymatic pathways. The activity of glutathione-S-transferase and cytochrome P450-related enzymes are thought to be involved. Thiols present in the cytosol are likely to account for the nonenzymatic pathway, and in both cases an unstable thionitrate may be the common intermediate in the production of NO. The main drawback of using these compounds clinically is the occurrence of tolerance to the drug. This is thought to be owing to a down-regulation of the enzymes involved or a depletion in the amount of available thiol groups *(3,4)*. Organic nitrates have been shown to be less potent in the inhibition of platelet aggregation in vitro than the relaxation of vascular smooth muscle, presumably owing to the absence of the metabolizing enzymes in platelets *(5)*. Inorganic nitrites are thought to react with available thiol groups, forming RSNOs that decompose to give NO. The extent of NO generation will depend on the rate of formation

From: *Methods in Molecular Biology, Vol. 100. Nitric Oxide Protocols*
Edited by: Michael A. Titheradge © Humana Press Inc., Totowa, NJ

of the RSNO and its rate of decomposition, which will vary with temperature and pH. SNP spontaneously liberates NO by an as-yet-unknown mechanism. This NO donor is very potent and has a very short biological half-life (approx 2 min in humans). The main drawback of SNP treatment is the concomitant production of cyanide, which appears to be obligatory for the release of NO *(6)*. The sydnonimine SIN-1 requires metabolism by liver esterases to give the active metabolite SIN-1. This then undergoes hydrolysis to yield the open-ring form SIN-1A, which releases NO via a radical process involving molecular oxygen. In addition to the release of NO, there is a stoichiometric formation of superoxide anions that may give rise to oxidative side reactions, owing to hydrogen peroxide and hydroxyl radical generation. The sydnonimines are highly susceptible to light and oxygen and should therefore be prepared just before use and kept protected from light and heat. RSNOs liberate NO spontaneously by homolytic fission of the S–NO bond *(4,6)* (*see* **Table 1**).

The biological effects of the majority of these NO donors has already been extensively covered, except for RSNOs, which are still very much an experimental class of compound, mainly owing to their attributed instability. This varies depending on the pH, oxygen tension, redox state, transition-metal contamination, and the presence of thiol groups. RSNOs decompose both thermally and photochemically, yielding NO and the corresponding disulphide. The decomposition of certain RSNOs has been shown to be catalyzed by trace amounts of copper *(7)*. RSNOs are usually red or green in color, thus their stability can be assessed by monitoring the UV absorbance (320–360 nm) of the compound at one of the $\lambda_{max.}$, normally approx 335 nm with a detection limit of approx 0.1–10 m*M*.

RSNOs can be synthesized by acid-catalyzed nitrosation of the corresponding thiol with nitrite in mineral acid *(8)*. With certain RSNOs, it is often difficult to obtain the compound in a pure form owing to it's instability, and is often easier to use the original reaction solution. For biological assays, using an RSNO-reaction solution generated from mineral acid is generally not suitable, owing to the resulting low pH. An alternative method has been outlined in **Subheading 3.1.**

Two of the major properties of NO, and hence NO donors, is their ability to inhibit platelet aggregation and relax vascular smooth muscle both in vitro and in vivo. This chapter outlines the methods used in assessing these properties in vitro.

NO has often been cited as being a toxic molecule *(9)*, thus NO donors may also fall into this category. When investigating a new NO donor, it may be desirable to assess its toxicological potential. A method has been outlined in **Subheading 3.4.** It should be pointed out that this assay does not distinguish between cytotoxicity or cytostasis *(10)*.

2. Materials

2.1. Preparation of NO Donors

1. Citrate buffer pH 2: The concentration of citrate buffer used should be the same as the concentration of RSNO being synthesized, in order to ensure sufficient buffer capacity. Generally, it is prepared by dissolving citric acid (monohydrate) in deionized water and adjusting to pH 2.0 with conc. HCl.
2. 0.9% w/v Saline solution: Dissolve 9 g of NaCl in 1 L of deionized water.
3. The more stable NO donors, i.e., GTN, are available commercially and should be made up according to the manufacturer's recommendation.

2.2. Inhibition of Platelet Aggregation

1. 3.15% w/v Trisodium citrate: Filter sterilize and store at 4°C.
2. At least 400 mL of blood collected from healthy volunteers who have not taken any drugs (in particular aspirin-based products, which prevent aggregation) in the previous 10 d.
3. 1 *M* Tris buffer: Dissolve 52.8 g Trizma HCl and 80.6 g Trizma base in 1 L of water.
4. 1 mg/mL Prostacyclin (PGI_2) in 1 *M* Tris buffer: Solutions stored at –80°C.
5. Tyrodes solution containing 0.1% w/v sodium bicarbonate: Dissolve 9.5 g of Tyrodes salt (Gibco) and 1 g of sodium bicarbonate in 1 L of water. Filter sterilize and store at 4°C.
6. 1 mg/mL Collagen diluted to 100 and 10 μg/mL in buffer specified by manufacturer. Store solutions at 4°C.

2.3. Relaxation of Vascular Smooth Muscle

1. Krebs Ringer solution (10X) diluted 1 in 10 with deionized water containing the following:
   a. 10 μ*M* Cycloheximide, to inhibit the induction of inducible nitric oxide synthase (iNOS);
   b. 10 μ*M* Indomethacin, to inhibit prostanoids, e.g., PGI_2; and
   c. 2.5 μ*M* Calcium chloride.
2. 25 m*M* Phenylephrine in saline.
3. 25 m*M* Acetylcholine in saline.
4. 0.9% w/v Saline solution (*see* **Subheading 2.1., Item 2**).

2.4. MTT-Toxicity Assay

1. Glycine buffer: Dissolve 375.4 mg glycine and 292.2 mg of NaCl in 50 mL of water. Adjust pH to 10.5 with 10 *M* NaOH. Store at 4°C.
2. Dimethyl sulphoxide (DMSO).
3. DMSO/Glycine buffer: 3 mL Glycine buffer, pH 10.5, added to 24 mL DMSO.
4. Phosphate-buffered saline (PBS).
5. Methylthiouracil (MTT) solution: Dissolve 100 mg of MTT in 20 mL of media.
6. Hams F12 media supplemented with 10% fetal calf serum (FCS), penicillin 100 iu/mL, streptomycin 100 μg/mL, and glutamine (2 m*M*).

Table 1
Properties of Common NO Donors

Class	Compound	Stability of Solid	Stability of Solution	Notes
Organic Nitrates	Glyceryl trinitrate (GTN)	NA	Stable stock solutions prepared in ethanol or DMSO. Protect. from light.	Metabolized by enzymatic and nonenzymatic pathways. Thionitrate common intermediate in the release of NO concomitantly with nitrite. Less potent inhibitor of platelet aggregation.
	Isosorbide dinitrate (ISDN)	NA	As above	As above
Iron Nitrosyls	Sodium nitroprusside	Store desiccated in dark at RT.	Prepare aqueous solutions fresh and protect from light.	Enzymatic and photochemical release of NO. Produces concomitant amounts of cyanide.
	Roussin's black salt (heptanitrosyltri-μ3-thioxotetraferrate)	Store at –20°C under argon.	Water soluble. Prepare in assay buffer immediately use. Protect from light.	Highly photosensitive. NO formation depends on intensity of llumination.
Organic Nitrites	Amyl nitrite	NA	Store in sealed glass containers at 4°C.	Release of NO requires presence of thiol groups. RSNO are active intermediates, and rates of NO release are a function of rate of formation and metabolism of RSNO involved.
Sydnonimines	3-Morpholinosydnonimine (SIN-1)	HCl salts stable as solids. Store desiccated at 4°C. Protect from light.	Water soluble. Stable in acidic solution, pH 5.0. Keep cool and protected from light.	SIN-1 undergoes hydrolysis to yield the open-ring form, SIN-1A. NO released from

				SIN-1A via a radical process. Forms stoichiometric amount of superoxide ions. NO release enhanced by superoxide dismutase. Peroxynitrite and hydroxyl radicals may be formed.
	Molsidomine	Stable solid. Store at RT and protected from light.	Prepare stock solutions in DMSO.	Inactive in vitro. Converted to active metabolite SIN-1 by liver esterases.
S-Nitrosothiols	S-Nitrosoglutathione (GSNO)	Store desiccated at –20°C. Protect from light.	Prepare fresh using citrate/HCl buffer, pH 2.0 (*see* **Subheading 2.1.**), or 0.5-1M HCl. Store at 4°C and protected from light.	Rapid decomposition to yield the disulphide and NO. Thiol radicals may form. Decomposition catalysed by trace amounts of Cu^{2+}.
	S-Nitroso-N-acetyl-D,L-penicillamine (SNAP)	Store desiccated at RT. Protect from light.	Prepare fresh. Store at 4°C.	As above.
NONOates	Diethylamine/NO (DEA/NO)	Store under argon or nitrogen at –80°C.	Water soluble. Prepare fresh in dilute NaOH. Store on ice and under argon.	Generate NO spontaneously, independent of tissue. High concentrations of thiols decrease release. Release pH-dependent, stable at alkaline pH, rapid decomposition at pH 5.0. Predictable NO release (*see* Chapter 27).
	Spermine/NO (SPER/NO)	As above.	As above.	Produces spermine which may have biological activity.

Data obtained from **ref.** ***3***.

3. Methods

3.1. Preparation of NO Donors

The following method prepares a 1 m*M* solution of RSNO, all concentrations will have to be altered according to the concentration of RSNO required (*see* **Notes 2–5**).

1. Prepare a solution of thiol (20 m*M*) in 1 m*M* citrate buffer pH 2.0 (Solution A).
2. Prepare a solution of $NaNO_2$ (20 m*M*) in saline (Solution B).
3. Take 250 μL of solution A and B and add to 4.5 mL of citrate buffer pH 2.0. This should yield a colored solution of the appropriate RSNO.
4. Assess the concentration of RSNO by obtaining a UV spectrum of the compound. It is usual to dilute the RSNO in deionized water and scan between 200 and 600 nm. Determine the absorbance at the λ_{max} at approx 330 nm and calculate the concentration of RSNO using the Beer-Lambert law and the appropriate ε value (e.g., ε values for the following RSNOs in water: GSNO 0.79 $mM \cdot cm^{-1}$, S-nitroso-L-cysteine 0.67 $mM \cdot cm^{-1}$ and S-nitroso-N-acetyl-L-cysteine 0.87 $mM \cdot cm^{-1}$: see **ref. *11*** for further values).

3.2. Inhibition of Platelet Aggregation

3.2.1. Preparation of Human Washed Platelets (HWPs)

1. Collect blood using 3.15% w/v trisodium citrate (as the anticoagulant), in the ratio 1 part citrate to 9 parts blood.
2. Centrifuge the blood at 220*g* for 20 min at room temperature. The supernatant is termed as platelet rich plasma (PRP).
3. Transfer the PRP into clean centrifuge tubes and add PGI_2 to a final concentration of 0.3 μg/mL. Care is needed so as not to cause foaming of the PRP or to disturb the collected cells from the pellet. PRP should be dispensed into the tube by running the PRP down the wall of the tube, this avoids turbulence which could activate the cells
4. Centrifuge the mixture at 700*g* for 10 min at room temperature, no brake must be used to ensure that the platelet pellet is not disrupted.
5. The resultant pellet contains platelets and some erythrocytes. The supernatant, platelet poor plasma (PPP), should be removed using a 5000 μL pipet and finished off with a 1000 μL pipet. Take care so as not to disturb the pellet.
6. Prepare 20 mL of Tyrode solution and add PGI_2 to a final concentration of 0.3 μg/mL. Carefully pour 5 mL of this solution onto the pellet.
7. Re-suspend the pellet using a 1000 μL pipet; do this gently and try to leave the red cells at the bottom of the tube.
8. Transfer the re-suspended platelet solution into the remaining 15 mL of Tyrode solution and swirl gently.
9. Centrifuge at 680*g* for 10 min at room temperature.

10. Remove the supernatant carefully. Wash the surface of the platelets with 3 × 1 mL of Tyrode solution without PGI_2. Take care as not to disturb the pellet during the washes. Remove any traces of PGI_2 from the walls of the tube with tissue paper.
11. Add 5 mL of Tyrode solution and re-suspend the pellet gently again, discarding any region that contains erythrocytes.
12. Obtain a platelet count and adjust the final volume with Tyrode solution (PGI_2 free) in order to have a final platelet count of 200–250 × 10^{-9} /L (*see* **Note 6**).

3.2.2. Platelet Aggregation Assay

Platelet aggregation is measured in an aggregometer by the method of Born and Cross ***(12)***.

3.2.2.1. Determination of Collagen EC_{90}

1. Place 0.5 mL of the platelet suspension into an aggregometer at 37°C and incubate for 1.5 min. The platelet suspension should be stirred continuously at a rate of 900 rpm.
2. Start recording aggregation once the platelet suspension has reached an equilibrium that is normally after 1.5 min.
3. After 2 min, add collagen (1–10 μg/mL) and monitor aggregation for 4 min.
4. From the dose-response curve obtained for collagen, the effect of the NO donors can then be assessed using a submaximal concentration of collagen. This is normally an EC_{90} and is described as the dose that gives a 90% response of the maximum aggregation induced by collagen (*see* **Note 8**).

3.2.2.2. Determination of Potency (IC_{50}) of Individual NO Donors

1. Prepare NO-donor stock solution as detailed in **Subheading 2.1.** Dilute down accordingly in Tyrodes solution prior to use. Store all solutions on ice. With the RSNO solutions, ensure the compound is also protected from light.
2. Repeat **Steps 1** and **2** from **Subheading 3.2.2.1.**
3. At 2 min, add the NO donor (0.003–30 μ*M*) and incubate for 1 min. Care should be taken so as not to suck any platelets back up into the pipet.
4. Add collagen (EC_{90} previously determined) at 3 min and monitor aggregation for a further 3 min.
5. Inhibition of aggregation is expressed as a percentage of the maximal aggregation induced by collagen. Determine potency (IC_{50}) values for each compound (*see* **Notes 9–11**).

3.3. Relaxation of Vascular Smooth Muscle

The following method is a general principle used for smooth muscle from Male Wistar rats, but can be adapted for other vessels.

1. Male Wistar rats (250–300 g) are anesthetised briefly with isofluorane and killed by exsanguination.

2. Remove the thoracic aorta and carefully trim off all adhering fat and connective tissue and cut into 4 mm rings. Prepare rings of aorta both with the endothelium intact and denuded, because the endothelium generates NO that can interfere with exogenous-NO donors. Endothelial cells are removed by gently rubbing the internal surface with a cut down pipe-cleaner. Care must be taken to avoid removing the endothelial cells from the intact tissues.
3. Mount rings under 1 g of resting tension (other vessels will require the optimal tension for that tissue), on stainless-steel hooks, in organ baths filled with Krebs buffer containing indomethacin, cycloheximide, and gassed with 95% O_2/5%CO_2 at 37°C (*See* **Notes 12 and 13**).
4. Allow the tissues to equilibrate for approx one h, which will allow the tissues to reach a stable baseline (this will differ depending on the tissue used). During this time, the Krebs buffer should be changed at regular intervals in order to remove vasoactive mediators released from the tissue after removal from the animal.
5. Carry out a dose response curve to phenylephrine (1–10000 n*M*), or other contractile agents, and determine the EC_{90} dose to be used for the relaxation studies. The EC_{90} will be less for endothelium-denuded tissues than intact tissues owing to the removal of the continuous dilatory actions of NO.
6. Pre-contract tissues by addition of the EC_{90} dose of phenylephrine. Induce relaxation of the tissues by addition of acetylcholine or other endothelium-dependent relaxant (0.1–10 μ*M*). Failure of the tissue to relax is taken as an indication of endothelium removal. Rings showing less than 80% relaxation at this stage should be discarded.
7. Flush the organ bath with Krebs solution at regular intervals to remove acetylcholine and phenylephrine and allow to equilibrate for a further 60 min.
8. Prepare NO-donor stock as detailed in **Subheading 2.1.** Dilute down accordingly in Krebs buffer prior to use. Store all solutions on ice.
9. Again pre-contract tissues by addition of phenylephrine EC_{90} and obtain cumulative relaxation curves for the NO donors in question (0.1–30,000 n*M*).
10. Determine EC_{50} values for each compound. The EC_{50} is described as the concentration of NO donor required to relax the tissue to 50% of the maximum contraction induced by phenylephrine.

3.4. MTT Assay

The following assay described below uses the human-lung carcinoma-cell line A549, but can be adapted for other cells.

1. Seed A549 cells at 1.1×10^3 cells/well in a 96-well plate. Incubate for 24 h.
2. Remove the medium, being careful not to dislodge any cells (*see* **Note 15**). Replace with medium (200–250 μL) containing the NO donor (0.05–0.5 m*M*). (For preparation, *see* **Subheading 2.1.**) Incubate for 72 h.
3. Remove medium after 72 h and wash cells with 200 μL PBS.
4. Add 200 μL of the MTT solution to each well and incubate for a further 4 h at 37°C.
5. Remove the MTT solution and wash the cells twice with PBS.

6. Add 200 μL of the DMSO/Glycine buffer to dissolve the formazan product.
7. Shake the plate for 20 min to ensure complete dispersal of the blue product.
8. Read the absorbance of each well at 540 nm.
9. The absorbance obtained is proportional to the number of viable cells, because it is only living cells that can convert the yellow tetrazolium salt to the blue formazan product. Results are expressed as the percent of inhibition of growth as compared to control cells.

4. Notes

1. Unless otherwise stated, water used in all of these experiments should be of the ultra-pure grade.
2. RSNO solutions should be protected from light and heat by wrapping the containing vessel in aluminium foil and storing on ice (4°C). Solutions should always be made up fresh on the day of the study, just prior to use.
3. The lifetime of RSNO can be prolonged by adding a copper chelator such as DTPA (approx 100 μ*M*) to the reaction solution.
4. In the method outlined in **Subheading 2.1.** for the preparation of RSNO, a dilution of 1 in 20 is made of both the thiol and $NaNO_2$ in the citrate buffer. This can be adjusted accordingly, but it is always advisable to keep the dilution of $NaNO_2$ to a maximum. This will keep the amount of saline added to a minimum in order to ensure that the pH remains at 2.0.
5. It is advisable to make the RSNO stock at least five times more concentrated than the upper dose required. This will allow for a dilution to be made in the appropriate buffer, which will increase the pH before addition to the tissue/platelets.
6. When preparing HWP, always leave platelet suspensions for approx 2 h before using in order for the anti-aggregatory effects of PGI_2 to wear off. Although PGI_2 is unstable, its anti-aggregatory effects, once initiated, last longer.
7. Platelets and aortic rings must be used on day of preparation.
8. Care must be taken when choosing the EC_{90} of collagen in the platelet-aggregation assay and phenylepherine with the aortic rings to ensure that the concentration is submaximal. If the EC_{90} used is too high, this will result in maximum aggregation and contraction, which makes it difficult to observe any anti-aggregatory or relaxation effects.
9. Tyrodes solution should be used as a blank in the aggregometer with HWP.
10. PRP can be used instead of HWP, but PPP must be used as a blank.
11. With the platelet-aggregation assay, run a control at the beginning and the end of the dose response curve.
12. Aortic rings must be kept in oxygenated Krebs buffer at all times.
13. When mounting the aortic rings, care must be taken not to over-stretch the vessels as this can result in damage.
14. Unless otherwise stated, all solutions used in the MTT assay must be pre-warmed to 37°C before use. All solutions must also be sterilized prior to use.
15. Remove media, PBS etc., from a 96-well plate by aspiration, using a hypodermic needle attached to a vacuum line.

References

1. Moncada, S, Palmer, R. M. J., and Higgs, E. A. (1991), Nitric oxide: Physiology, pathophysiology and pharmacology. *Pharmacol. Rev.* **43,** 108–142.
2. Feelisch, M. (1991) The biochemical pathways of nitric oxide formation of nitrovasodilators: Appropriate choice of exogenous NO donors and aspects of preparation and handling of aqueous NO solutions. *J. Cardiovascular Pharmacol.* **17(Suppl. 3**), S25–S33.
3. Feelisch, M. and Stamler, J. S. (1996) Donors of nitrogen oxides, in *Methods in Nitric Oxide Research* (Feelisch, M. and Stamler, J. S., eds.), Wiley, New York, pp. 71–115.
4. Feelisch, M. (1993) Biotransformation to nitric oxide of organic nitrates in comparison to other nitrovasodilators. *Eur. Heart J.* **14(Suppl. 1**), 123–132.
5. Aissa, J. and Feelisch, M. (1992) Generation of nitric oxide accounts for the antiplatelet effects of organic nitrates-the role of plasma components and vascular cells, in *The Biology of Nitric Oxide, I: Physiological and Clinical Aspects*, (Moncada, S., Marletta, M. A., Hibbs Jr., J. B., and Higgs, E. A.) Portland Press, London, pp. 142–144.
6. Bauer, J. A. and Fung, H. L. (1995) Non-nitrate nitric oxide donors. *J. Myocard. Ischem.* **7(1),** 17–22.
7. Dicks, A. P., Swift, H. R., Williams, D. L. H., Butler, A. R., Al-Sa'doni, H. H., and Cox, B. G. (1996) Identification of Cu^+ as the effective reagent in nitric oxide formation from S-nitrosothiols (RSNO). *J. Chem. Soc. Perkins Trans.* **2,** 481–487.
8. Hart, T. W. (1985) Some observations concerning the S-nitroso and S-phenylsulphonyl derivatives of L-cysteine and glutathione. *Tet. Lett.* **26(16),** 2013–2016.
9. Liu, R. H. and Hotchkiss, J. H. (1995) Potential genotoxicity of chronically elevated nitric oxide: A review. *Mutation Res.* **339,** 73–79.
10. Mosmann, T. (1983), Rapid colorimetric assay for cellular growth and survival: Application to proliferation and cytotoxicity assays. *J. Immunol. Methods* **65,** 55–63.
11. Oae, S., Kim, Y. H., Fukuhima, D., and Shinhama, K. (1978) New syntheses of thionitrites and their chemical reactivities. *J. Chem. Soc. Perkin I,* 913–917.
12. Born, G. V. R. and Cross, M. J. (1963) The aggregation of blood platelets. *J. Physiol.* **168,** 178–195.

Barry W. Allen

20

Making and Working with Peroxynitrite

Roger White, John Crow, Nathan Spear, Steven Thomas, Rakesh Patel, Irene Green, Joseph Beckman, and Victor Darley-Usmar

1. Introduction

Peroxynitrite ($ONOO^-$) is formed by the reaction of nitric oxide (NO) with superoxide (O_2^-) *(1)*. Under normal circumstances, these two free radicals are not formed at similar rates in the same cellular or extracellular compartment; however, under some pathological conditions, such as inflammation, this does appear to occur *(2–3)*. Peroxynitrite is a reactive compound that has the potential to modify biomolecules through several different mechanisms, and is a good candidate as a mediator of NO-dependent inhibition of key physiological or biochemical processes *(4)*. There are, of course, several steps in testing the role of $ONOO^-$ in an experimental model, including assessing whether the process is both sensitive to NOS inhibitors and scavengers of O_2^-, as well as examining the sample for specific evidence of nitration reactions. One example, the nitration of tyrosine, may be used as a marker, and this will be covered in Chapter 28, this volume. Here, we focus on the practical aspects of determining the effects of $ONOO^-$ addition on biochemical, cell culture, or tissue systems. Testing the biochemical basis of the hypothesis that $ONOO^-$ plays an important role in the pathophysiology of human disease must be approached with a keen awareness of its physicochemical properties.

1.1. Properties of Peroxynitrite

Peroxynitrite rapidly decays under physiological conditions via its protonated form (pKa 6.8) peroxynitrous acid (HOONO), with a half-life of approx 1 s *(5)*. Peroxynitrite anion ($ONOO^-$) is relatively stable under strongly alkaline conditions, but is itself a reactive species, e.g., able to oxidize sulfhydryl

From: *Methods in Molecular Biology, Vol. 100. Nitric Oxide Protocols*
Edited by: Michael A. Titheradge © Humana Press Inc., Totowa, NJ

groups 1000 times faster than does hydrogen peroxide *(6)*. It may modify biological substrates through mechanisms that involve either a one- or two-electron oxidation, via intermediates with the characteristics of hydroxyl radical and nitrogen dioxide. It is also a potent and efficient nitrating agent. Evidence to support the biological relevance of all these processes have been reported, including the oxidation of lipids, protein thiol groups, and iron–sulfur/zinc–thiolate centers, and the nitration of proteins, carbohydrates, and nucleic acids *(2–3,5–19)*.

1.1.1. Synthesis of Peroxynitrite

A commonly used method of synthesis is to use the reaction between acidified nitrite (nitrous acid) and hydrogen peroxide:

$$HONO + H_2O_2 \rightarrow HOONO + H_2O. \quad (1)$$

Owing to the instability of HOONO, the reaction has to be rapidly quenched with sodium hydroxide *(20)*. Prepared in this way, solutions range in concentration from 150–300 m*M*, which is an advantage when making dilutions for addition to cell or tissue preparations as it minimizes artifacts owing to changes in pH or salt concentration. Particularly important is the fact that the molarity of NaOH (0.3–0.78 *M*) in the final stock solution of $ONOO^-$ may exceed the buffering capacity of cell culture or perfusion media, and the pH of the solution after addition must be routinely checked.

As an alternative, $ONOO^-$ can be prepared by reacting ozone with sodium azide, resulting in preparations that are essentially free of hydrogen peroxide and contain only traces of azide *(21)*. These preparations are of low ionic strength, have a pH of approx 12, and are easily buffered at neutral pH values.

After decomposition, solutions of $ONOO^-$ are contaminated with a mixture of nitrite, nitrate, and hydrogen peroxide. These compounds, although generally innocuous, are present in high concentrations, and their presence needs to be controlled. This is achieved by the so-called "reverse-order addition" protocol in which $ONOO^-$ is added to the buffer and allowed to decompose before addition to the system in which it is being studied. In this short overview, we will illustrate the application of these principles to several model systems.

1.1.2. Peroxynitrite Generating Systems

The rate of the reaction between NO and O_2^- is close to the diffusion limit (6.7×10^9 $M^{-1}s^{-1}$) *(1)* and in principle, any combination of systems that produce both NO and O_2^- simultaneously and in equimolar amounts will produce $ONOO^-$. Examples include the enzyme xanthine oxidase and substrates such as acetaldehyde or pterin to generate O_2^- when incubated together with NO-releasing compounds such as S-nitroso glutathione (GSNO), S-nitroso-Nacetylpenicillamine (SNAP), or the NONOates. The NONOate compounds have the advantage that rates of decomposition are less dependent on the precise conditions of the assay (e.g., contamination with transition metals) than the S-nitrosothiols. The compound 3-morpholinosydnonimine-HCl (SIN-1) is particularly useful because it decomposes to form O_2^- and NO at equal rates ***(11–12,22)***.

Compounds that require metabolic activation, such as the organic nitrates or nitrites, are not recommended for this application. The mechanism of NO release from these compounds is different and depends both on the particular NO donor and the reaction conditions such as pH, light, and temperature. For example, all nitrosothiols and SNP release NO and thiyl radicals on exposure to light, a fact that can be used to control the rate of NO release. Alternatively, metals such as copper can promote release from SNAP, although the addition of transition metals to lipid systems may also promote lipid peroxidation independently of NO-dependent reactions. With NO release mechanisms that depend on the precise reaction mixture, the rate of NO (and O_2^-) release should be measured for the experimental conditions used. Controls for the decomposition products of the NO donors must be performed to check that they do not have an effect on the system under investigation.

1.1.3. Detection of Peroxynitrite In Vitro *Through Oxidation of Dichlorodihydroflourescein and Dihydrorhodamine*

Dichlorodihydrofluorescein (DCDHF) and dihydrorhodamine (DHR) 123 are often used to detect the production of oxidants in cells via oxidation to their respective fluorescent products. Exposure of DCDHF and DHR to a number of biological oxidants in vitro revealed that efficient oxidation is accomplished only by $ONOO^-$, H_2O_2 plus horseradish peroxidase, and, to varying degrees, by hypochlorous acid. Thus, in systems where hypochlorous-acid production can be ruled out, oxidation of either DCDHF or DHR, which is inhibitable by superoxide dismutase (SOD) or NOS inhibitors, is highly suggestive of $ONOO^-$ formation and can be exploited for this purpose. DCDHF and DHR also are useful for quantifying low levels of $ONOO^-$ formed in vitro by either fluorescence or absorbance (500 nm) measurements ***(23)***.

2. Materials

2.1. Synthesis of Peroxynitrite

2.1.1. Method 1

These solutions are sufficient to prepare approx 8 batches of $ONOO^-$. It is suggested that reagents of the highest purity available are used to avoid trace-metal contamination.

1. 2 *M* $NaNO_2$: 6.92 g of $NaNO_2$ dissolved in 50 mL of double-distilled water.
2. 2.11 *M* H_2O_2 in 1.85 *M* HNO_3: Add 12.9 mL of an 8.2 *M* stock solution of H_2O_2 to 25 mL of double-distilled water. Add 8.33 mL of concentrated nitric acid and dilute to 50 mL with double-distilled water.
3. 4.2 *M* NaOH: 8.33 g of NaOH dissolved in 50 mL of double-distilled water.
4. Two 10-mL glass syringes taped together with a connecting "T" piece. Flexible tygon tubing (1-cm lengths) can be used to connect the "T piece" that forms the reaction chamber for the reagents. The tubing can be clamped shut with crocodile clips.
5. A small column of granular manganese dioxide is made by placing a plug of glass wool in the bottom of a 5-mL syringe or pipet tip and pouring on top approx 2–3 cm of manganese dioxide.

2.1.2. Method 2

Ozone is an extremely strong oxidizing agent. Reactions must be carried out behind a safety shield in a fumehood and exposure eliminated. A solution of 10% (w/v) potassium iodide in 0.07 *M* phosphate buffer will absorb and destroy any excess unreacted ozone. Sodium azide is toxic if swallowed, and skin and eye contact should be avoided.

1. Ozonator, such as the Wallace and Tierman ozonator (model BA 023).
2. 0.2 *M* Sodium azide solution in distilled water (100 mL). Adjust the pH of the solution to 12.0 with 1 *M* sodium hydroxide. Chill solution to 0–4°C in an ice-water mixture immediately before and during the preparation procedure.

2.2. Simultaneous Generation of NO and O_2^-

1. Solutions of SIN-1 hydrochloride are prepared immediately before use as a 10–100 m*M* stock solution in water, which should give a pH of ~4.5 and be kept in the dark on ice.
2. The buffer to be used for the experiment should contain diethylenetriamine pentaacetic acid (DTPA) (100 m*M*) if nonspecific metal-dependent oxidation reactions are to be avoided. It may then be added to the oxidation system at the desired concentration (typically 0.1–1 m*M*).

2.3. Measurement of the Oxidation of DHR and DCDHF by Peroxynitrite

1. Stock solutions of DCDHF (10–20 m*M*) are prepared by dissolving DCDHF diacetate in 20 m*M* NaOH and mixing vigorously. Total dissolution of the compound indicates that diacetate groups have been hydrolyzed (this requires 5–10 min at room temperature).
2. A 14.5 m*M* stock of DHR is prepared by dissolving the compound in acetonitrile (10 mg DHR per 2 mL ACN) or virtually any water-miscible organic solvent can be used (e.g., ethanol, methanol, acetone, DMSO, DMF). DCDHF and DHR are both light sensitive and should be protected from light and kept on ice.

2.4. Buffers Used to Study the Effects of Peroxynitrite on Blood-Vessel Relaxation

1. Prepare a Krebs-Henseleit (KH) solution of the following composition: 118 m*M* NaCl, 4.6 m*M* KCl, 27.2 m*M* $NaHCO_3$, 1.2 m*M* KH_2PO_4, 1.2 m*M* $MgSO_4$, 1.75 m*M* $CaCl_2$, 0.03 m*M* Na_2EDTA, and 11.1 m*M* glucose (pH 7.4). Continuously aerate the solution with a 95% O_2/5% CO_2 gas mixture and maintain at 37°C in a water-jacketed tissue bath.
2. Because bicarbonate may accelerate the decomposition of $ONOO^-$ *(24)*, additional relaxation studies may be performed using a HEPES-buffered KH solution: 145 m*M* NaCl, 4.8 m*M* KCl, 10 m*M* HEPES, 1.0 m*M* KH_2PO_4, 1.75 m*M* $CaCl_2$, 1.2 m*M* $MgSO_4$, $7H_2O$, 5.5 m*M* glucose, 0.03 m*M* Na_2Ca EDTA (pH 7.4).

2.5. Preparation of a Saturated Solution of Nitric-Oxide Gas

1. Fill a gas-sampling tube with double-distilled H_2O. Bubble the H_2O solution with argon gas for 30 min, allowing the gas to vent through the top stopcock. This step purges oxygen from the solution and the gaseous phase.
2. Connect a gas line from a NO cylinder to the sampling tube and bubble for approx 5 min. Once the solution becomes saturated with dissolved NO gas, a brown vapor can be detected exiting the tube through the stopcock. Turn off the NO gas cylinder and immediately seal the stopcocks on the sampling tube to prevent oxygen from entering the gas phase. The concentration of NO in the saturated solution will be approx 1.8–2.0 m*M*.
3. The NO solution can then be sampled by piercing a self-sealing cap on the side arm of the tube with a needle or syringe.

2.6. Buffers Used to Study the Effect Peroxynitrite on Fura 2-Sensitive Calcium Mobilization in PC12 Cells

1. Cells are detached from culture flasks and loaded with fura 2 in suspension.
2. Calcium-sensitive emission profiles are obtained from cells suspended in a phosphate buffer of the following composition: 50 m*M* Na_2HPO_4, 90 m*M* NaCl, 5.0 m*M* KCl, 1.0 m*M* $CaCl_2$, 0.8 m*M* $MgCl_2$, 5 m*M* glucose, pH 7.3. Addition of $ONOO^-$ adjusts the pH of the solution to 7.4. Cells are maintained at 37°C throughout the experiment in a water-jacketed cuvet.

3. Methods

3.1. The Synthesis of Peroxynitrite

3.1.1. Method 1

This is a potentially hazardous procedure. It is recommended that a lab coat, gloves, and safety glasses are worn, and that the reaction is performed in the fume hood.

1. Cool the solutions on ice for at least 30 min before commencing the reaction.
2. Clamping one arm of the T piece, place 6.6 mL of $NaNO_2$ in one syringe and then, after clamping the arm of the filled syringe, place 6.6 mL of acidified H_2O_2 in the other syringe. Take care not to mix the samples.
3. Immerse the syringes in ice for at least 30 min while cooling a beaker containing 6 mL of NaOH with a magnetic stirrer.
4. Once cool, hold the syringes with the "T" piece just above the NaOH solution (which should be stirred and kept in the ice bath) and press both plungers down with equal force using one hand (*see* **Note 1**).
5. To remove unreacted hydrogen peroxide, pass the yellow $ONOO^-$ solution through a small column (1 cm × 3 cm) of granular manganese dioxide at 4°C (*see* **Note 2**).
6. Collect the sample of $ONOO^-$, dilute 400 times in 0.1 *M* NaOH and measure the UV/visible absorption spectrum in a quartz cuvet (previously blanked with the NaOH solution alone) between 245 and 400 nm. An absorption band at 302 nm should be evident. Decompose the $ONOO^-$ solution by the addition of 10 μL of conc HCl to both the reference and sample cuvets and measure the absorption spectra between 245 and 400 nm, recording any residual absorbance at 302 nm. The clear absorption band for peroxynitrite at 302 nm should no longer be apparent. Substract the absorbance of the acidified solution at 302 nm from that of the sample prior to the addition of acid.
7. Calculate the concentration of $ONOO^-$ by dividing this absorbance by 1670, which will give the molar concentration of $ONOO^-$ in the cuvet (ε_{302} $ONOO^-$ = 1670 $M^{-1}cm^{-1}$). Under these conditions, a 300 m*M* solution of $ONOO^-$ will give an absorbance of 1.25 at this wavelength (*see* **Notes 3** and **4**).
8. Divide the samples into 1-mL aliquots in glass-reagent tubes and store at –80°C. They may be thawed and used under these conditions over a 2–3 wk period (*see* **Notes 5** and **6**).

3.1.2. Method 2

1. To generate ozone, pass dry oxygen (using a flow rate of 50 mL/min) through the ozonator undergoing a silent electric discharge (such as 180 V) (*see* **Note 7**).
2. Pass this ozone-in-oxygen mixture through a glass frit into 100 mL of 0.2 *M* sodium azide in water (the pH of the water being previously adjusted to 12 with 1 *M* NaOH) chilled to 0–4°C in an ice-water mixture.
3. Withdraw aliquots of ozonized azide solution into glass tubes at intervals of 5 or 10 min.
4. Estimate $ONOO^-$ concentration spectrophotometrically (*see* **Step 6** in the Method section, **item 1**).
5. Continue ozonation even after a maximal yield of $ONOO^-$ is obtained to ensure the total oxidation of azide. Further decomposition of $ONOO^-$ does occur but is not significant. Using this system, maximum formation of $ONOO^-$ is typically obtained after a reaction time of 70 min and is typically 50–60 m*M* (*see* **Note 8**).

6. Divide the sample into usable aliquots in glass-reagent tubes and store at –80°C. They may be thawed out and used under these conditions over a 2–3 wk period.

3.2. Calculating the Exposure of a System to Preformed Peroxynitrite

The concentrations of $ONOO^-$ used to inactivate enzymes or kill cells may be as high as 1 m*M*, which seems at first glance to be rather extreme. However, $ONOO^-$ decomposes within seconds at pH 7.4, so the actual exposure is quite brief. To take account of the decomposition of a short-lived oxidant in exposure experiments, it is helpful to measure concentration as a function of time, rather than concentration alone. It is relatively easy to calculate the exposure to $ONOO^-$, if one knows how fast $ONOO^-$ decomposes in the particular buffer system of interest. The net exposure is simply the initial concentration of $ONOO^-$ divided by the rate of decomposition. At 37°C and at pH 7.4 in 50 m*M* potassium-phosphate buffer, $ONOO^-$ decomposition is first order, occurring at a rate of 0.64 s^{-1}. If the initial concentration of $ONOO^-$ was 1 m*M*, the net exposure becomes 1560 n*M*/s or changing the time units, 26 μ*M*/min or 0.43 μ*M*/h. Consequently, exposure to a steady-state concentration of 0.43 μ*M* $ONOO^-$ for 1 h is, in principle, equivalent to exposure to a bolus addition of 1 m*M* $ONOO^-$. The pKa of $ONOO^-$ in 50 m*M* phosphate buffer is 6.8 and is largely temperature-independent. The rate of $ONOO^-$ decomposition can be readily calculated at any pH in the range of 4–9 by the simple equation

$$k_1 = k_{HA} * H^+/(H^+ + K_a)$$

where H^+ is simply $10^{(-pH)}$ and K_a is $10^{(-pKa)}$. At 25°C, k_{HA} is 1.3 s^{-1} and at 37°C, k_{HA} is 4.5 s^{-1} *(25)*. The net exposure to $ONOO^-$ can be calculated at any pH by combining and rearranging the above equation as follows:

$$\text{Exposure} = [ONOO^-]_o\,(1 + 10^{(pH - pKa)})/\,k_{HA}.$$

At acidic pH well below the pKa of $ONOO^-$, the rate of decomposition is independent of pH and the equation reduces to $[ONOO^-]_o/k_{HA}$. At the pKa of $ONOO^-$, the exposure is twice as great as at acidic pH. At alkaline pH, the exposure becomes much greater because $ONOO^-$ is longer lived (*see* **Note 9**).

3.3. Peroxynitrite Generating Systems

1. SIN-1 is prepared immediately before use as a 10 m*M* stock solution in the buffer to be used for the experiment, which should contain DTPA (100 μ*M*) if nonspecific, metal-dependent oxidation reactions are to be avoided. Keep the solution on ice and in the dark. It may then be added to the oxidation system at the desired concentration (typically 0.1–1 m*M*) (*see* **Notes 10–14**).

2. As an alternative $ONOO^-$ generating system, xanthine oxidase (0.01–0.05 U/mL) + acetaldehyde (1 m*M*) or pterin (0.2 m*M*) can be combined with SNAP (0.1–1.0 m*M*). In this case, the rates of O_2^- and NO are not necessarily constant over prolonged periods of time (greater than 20 min), and this should be monitored as part of the protocol (*see* **Notes 15** and **16**).

3.3.1. Measuring Peroxynitrite Formation by Reaction with DHR or DCDHF

1. For real-time spectral measurement of $ONOO^-$ formation in vitro, add 50 μ*M* of either DCDHF or DHR to a stirred, 3-mL cuvet containing 0.1 m*M* DTPA in 100 m*M* KPi, pH 7.4. The limit of DHR solubility is ~100 μ*M*; DCDHF is considerably more water soluble.
2. Monitor formation of dichlorofluorescein (DCF) or rhodamine at 500 nm (ε = 59,500 and 78,780 $M^{-1}cm^{-1}$, respectively). The relative oxidative efficiencies for bolus additions of $ONOO^-$ are 38% of added $ONOO^-$ for DCDHF and 44% for DHR (*see* **Note 17**).
3. Alternatively, measure the formation of DHR or DCF oxidation products using a spectrofluorometer (excitation = 500 nm, emission = 536 nm) (*see* **Note 18**).

An example of the formation of $ONOO^-$ from the decomposition of SIN-1 measured by the oxidation of rhodamine is shown in **Fig. 1**.

3.4. Initiation of Lipid Peroxidation by Peroxynitrite

1. Take the lipid system to be investigated (typically 0.2–1 mg protein/mL for a lipoprotein preparation or 0.2 mg lipid/mL for liposomes) and dilute in a buffer containing 100 μ*M* DTPA and at least 50 m*M* phosphate at the desired pH (*see* **Note 19**).
2. Pre-incubate the solution to the desired temperature and initiate the reaction by the addition of $ONOO^-$ solution to the side of the reaction tube. Mix rapidly. If working with lipoproteins, avoid violent vortexing or stirring, which denatures the protein. In determining the concentration dependence of oxidation reactions with $ONOO^-$, adjust the stock solutions by dilution into water immediately prior to addition, such that identical volumes are added (typically a 20 μL addition to a 1-mL sample).
3. Incubate for the required time period and stop further oxidation before analysis by the addition of butylated hydroxytoluene (1 m*M*) in ethanol (final ethanol concentration 1% v/v). With preformed $ONOO^-$, the reaction is usually complete within a few s of addition, but the secondary propagation of lipid peroxidation may persist for another 30–60 min. A time course is recommended.
4. To initiate lipid peroxidation with the $ONOO^-$ generating systems, these are added at **Step 2**, and the reaction monitored for longer periods of time.
5. Experimental design should include controls for oxidation promoted by the decomposition products of $ONOO^-$. One way to achieve this is to add the $ONOO^-$ to the buffer before the lipid sample, wait 5 min, and then add the lipoprotein. In

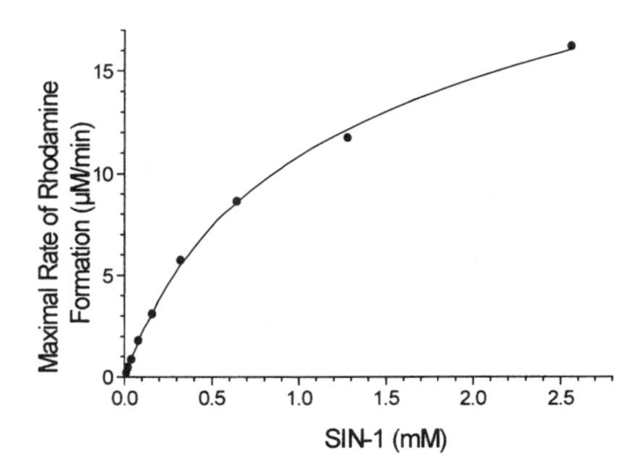

Fig. 1. SIN-1 hydrohydrochloride was dissolved in 100 m*M* potassium phosphate, pH 5.0 to a stock concentration of 100 m*M* and added at the indicated final concentrations to stirred cuvets containing 100 µ*M* dihydrorhodamine 123 (DHR) and 0.1 m*M* DTPA (redox inactive-metal chelator) in 50 m*M* potassium phosphate, pH 7.4 at 37°C. Following a lag of approx 5 min, DHR oxidation (500 nm) becomes linear. Rates equivalent to the maximal rate of oxidation were determined for the 6–10 min interval and plotted as a function of SIN-1 concentration. The rates of $ONOO^-$ formation were calculated from the rates of DHR oxidation following correction for the 44% oxidative efficiency of $ONOO^-$ determined previously following bolus additions of synthetic $ONOO^-$ to DHR solutions.

the case of the simultaneous generation of NO and O_2^-, each generating system should be added alone and the effects of varying the ratio of NO:O_2^- determined (*see* **Notes 16**, **20**, and **21**).

6. Indices of oxidation may then be measured on the sample. These may include lipid oxidation products and, in the case of lipoproteins, relative-electrophoretic mobility *(11)*.

3.5. Measurement of Peroxynitrite-Mediated Vessel Relaxation

1. Mount vascular rings in an isolated tissue bath using standard protocols for the measurement of isometric tension (*see* Chapter 19, this volume).
2. Contract vessels with phenylephrine (~3×10^{-8}–10^{-7} *M*) to generate active tension in vascular-ring segments.
3. Prepare serial dilutions of $ONOO^-$ stock using 0.1 *M* NaOH. Stock dilutions between 10^{-6} and 10^{-10} *M* are recommended.
4. Add $ONOO^-$ to the tissue bath cumulatively after the PE-induced contraction reaches a stable plateau. Add appropriate volumes of each stock solution to achieve final tissue-bath concentrations of $ONOO^-$ between 10^{-7} and 10^{-3} *M*. Full relaxation of PE-contracted vessels will occur over this concentration range (*see* **Note 22**).

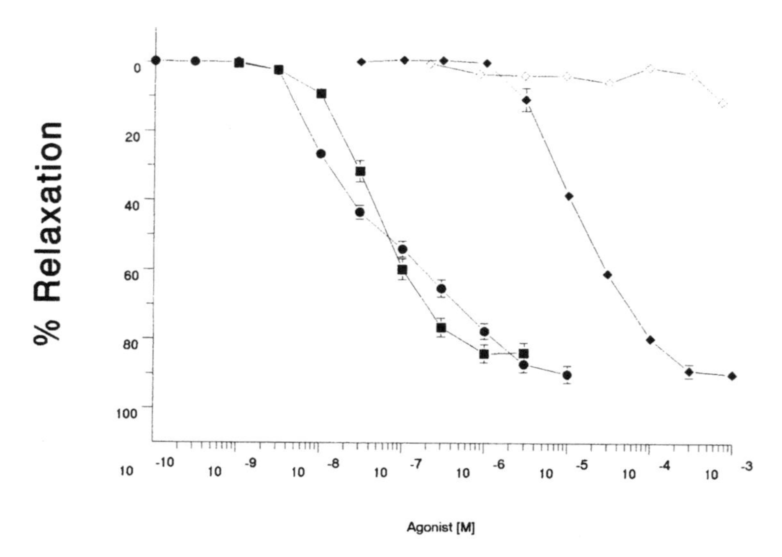

Fig. 2. Effects of $ONOO^-$, NO, and acetylcholine on the in vitro relaxation of rabbit aortic-ring segments. Ring segments were contracted with phenylephrine followed by cumulative administration of $ONOO^-$ (♦) or decomposed $ONOO^-$ (◊). As seen, $ONOO^-$ induced a dose-dependent relaxation of aortic-ring segments, whereas decomposed $ONOO^-$ had no effect on vessel tone. Control experiments showing the dose-dependency of acetylcholine (■), an endothelium-dependent vasodilator, and a saturated solution of NO gas (•) on vessel relaxation are included for comparison of the effects of these agents to responses elicited by $ONOO^-$. Data are mean ± SEM.

5. Appropriate control experiments may be performed using decomposed $ONOO^-$ or a saturated solution of nitric oxide (**Fig. 2**).
6. Experiments may be performed in either HEPES- or bicarbonate-buffered saline solutions to assess the reactivity of $ONOO^-$ in different buffers (*see* **Note 19**).

3.6. Using Peroxynitrite to Study Calcium Mobilization Responses in Cultured PC12 Cells

1. Culture PC12 cells in either T-75 flasks or 100 mm dishes to approx 90% confluence. Wash cells three times with serum-free RPMI media to remove debris and serum. Cells may be readily detached by treatment with calcium-free, Hank's balanced salt solution containing 5 m*M* EDTA. Centrifuge and resuspend cells in serum-free RPMI. Add fura 2 to a final concentration of 2 μ*M*. Incubate for 30 min, followed by two washes with serum-free RPMI containing 1 mg/mL bovine serum albumin (BSA). Aliquot into individual sample tubes containing approx 5×10^5 cells.
2. Wash cells after $ONOO^-$ addition and resuspend in phosphate-buffered saline (PBS). Using a fluorescence-recording device, monitor changes in cellular

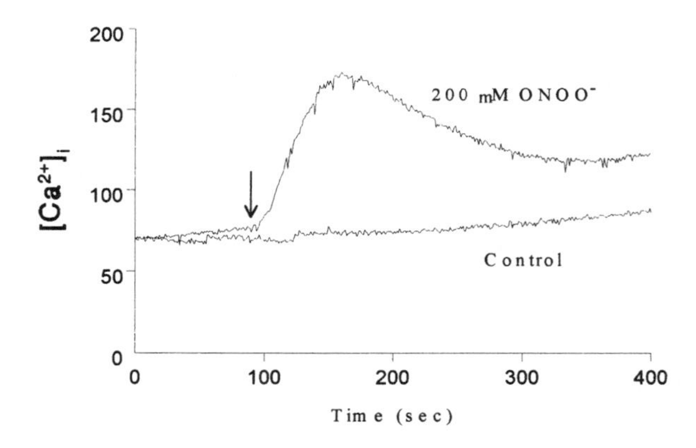

Fig. 3. Intracellular-calcium transients in PC12 cells. Fura-2-loaded-cell suspensions were treated with bolus addition of $ONOO^-$ at the time point indicated by the arrow. Addition of 200 μ*M* $ONOO^-$ induced a significant increase in $[Ca^{2+}]_i$. Control experiments, performed to determine the effect of NaOH solvent on intracellular Ca^{2+} showed no effect.

calcium by alternately exciting samples at 340 and 380 nm. Emission spectra are monitored at 510 nm.

3. Add $ONOO^-$ (10–200 μ*M*) as a bolus to the cell suspension. **Figure 3** shows the results of addition of 200 μ*M* peroxynitrite on intracellular calcium concentration in PC12 cells. Appropriate control experiments using decomposed peroxynitrite are recommended (*see* **Note 23**).

4. Notes

1. In making $ONOO^-$, it is important that the contents of the syringes mix thoroughly and that the resultant HOONO is quickly stabilized in the NaOH solution. As the reaction is highly exothermic and releases a brown gas (probably nitrogen dioxide), it is important that the reaction is performed in the fume hood.
2. Typically, the concentration of $ONOO^-$ may decrease 20% on passage through the manganese-dioxide column. In some cases, this step may be omitted, but the necessity of performing control experiments with the decomposition mixture remains crucial.
3. Slight modifications of the original synthesis of $ONOO^-$ reported by Hughes and Nicklin ***(20)*** routinely produce yields of 150–160 m*M* $ONOO^-$ with nitrite and H_2O_2 contaminants in the range of 40–50 m*M*, 230 m*M* NaCl, and a final NaOH concentration of 0.3 *M* ***(5)***. This synthesis maximizes the yield of $ONOO^-$ while maintaining the unavoidable nitrite and H_2O_2 contaminants at concentrations considerably below that of $ONOO^-$; H_2O_2 can be removed by manganese-dioxide treatment. This method has been employed by many investigators and may be preferred provided the stock concentration of $ONOO^-$ is adequate.

4. The maximum theoretical yield of $ONOO^-$ that can be prepared under these conditions is approx 690 m*M* and is typically 250–350 m*M* prior to addition to the manganese-dioxide column.
5. The concentration of $ONOO^-$ slowly decreases on storage, so it is important to measure the precise concentration prior to use in each experiment. It is not advisable to use solutions at a dilution of lower than 1/200, because artifacts associated with pH changes become more likely.
6. If the $ONOO^-$ is left in the freezer, it concentrates towards the top of the glass-reagent tube with concentrations as high as 1 *M* achieved through gentle thawing and extracting the solution. This does not increase the "window" for biological experiments in which the decomposed $ONOO^-$ may be having artifactual effects, because this also increases the salt and base concentrations. Do not store $ONOO^-$ or transport it on dry ice, as it reacts with carbon dioxide even at low temperatures.
7. The optimal-flow rate and current will vary depending on the type and model of ozonator used. Refer to the user manual to optimize these parameters.
8. The $ONOO^-$ concentrations that can be obtained depend on the starting concentration of azide. Concentrations up to 70 m*M* can be achieved using 0.2 *M* sodium azide. Residual azide is to be avoided because it is a ligand for heme proteins and can inhibit a number of metabolic pathways.
9. There are several limitations in the use of the equations to calculate exposure to $ONOO^-$. First, the rate of $ONOO^-$ decomposition is significantly influenced by buffer composition and concentration. This can change the rate of $ONOO^-$ decomposition by orders of magnitude, particularly if a bicarbonate buffer is used. Peroxynitrite reacts directly with carbon dioxide to form a potent nitrating agent *(24)*. Another consideration is the rate of reaction with the target species, which can substantially accelerate the decomposition of $ONOO^-$. A final point is that the actual concentration of the target molecule should be compared to that of the added $ONOO^-$. One often reports the amounts of proteins in terms of mg, without considering the M concentration. If the concentration of an enzyme is 10 μ*M*, then it would not be too surprising to find that greater than 10 μ*M* peroxynitrite was required for inactivation.
10. Solid SIN-1 is stable at room temperature when protected from light. Its molecular weight is 206.5 and should be made up as a stock solution of 10–100 m*M*. Solid SIN-1 exists as the hydrochloride salt so that aqueous solutions will be somewhat acidic (~pH 5.0).
11. The rate of $ONOO^-$ formation is a function of the SIN-1 concentration such that, at pH 7.4 and 37°C, approx 1% of the SIN-1 concentration per min breaks down to form the oxidant, e.g., 1 m*M* SIN-1 produces ~10 μ*M* $ONOO^-$ per min and decomposition is essentially complete after 3–4 h (*see* **Fig. 1**).
12. Ideally, solutions should be made fresh just prior to use; however, we have found that solutions protected from light and kept on ice are stable for many hours and can be frozen and reused. The primary decomposition product is SIN-1C, which appears to be relatively inert in systems we have examined thus far. Typically,

SIN-1 is added in large excess so that old solutions that are partially (1–2%) decomposed are often suitable for $ONOO^-$ generation.

13. SIN-1 produces $ONOO^-$ via an initial base-catalyzed oxidation to generate a radical species that liberates NO. In simple buffered systems, molecular oxygen carries out the initial oxidation thereby generating O_2^-. In more complex systems where other (better) oxidants such as H_2O_2 are present, the generation of O_2^- by SIN-1 can be significantly decreased, with a resulting decrease in the production of $ONOO^-$.
14. Initial oxidation of SIN-1 (upon addition to slightly alkaline solutions) produces O_2^- and thus consumes oxygen, so that concentrations of SIN-1 and actual reactions conditions (e.g., open/closed vessel, stirred/unstirred, etc.) must be taken into consideration to prevent oxygen depletion.
15. Avoid the use of hypoxanthine or xanthine as a substrate for xanthine oxidase because the urate that is produced is known to have antioxidant properties and is a particularly effective peroxyl-radical scavenger.
16. The concentrations of NO donors and O_2^- generators and the experimental conditions used will determine the rate of generation of NO or O_2^-, and the period of $ONOO^-$ exposure. The rate of NO production can be measured either by a chemiluminesence technique or by an NO electrode. The true rate of NO production under these conditions can only be measured in the presence of 4000–5000 U/mL of SOD. The commonly used method of monitoring oxidation of oxyhemoglobin by NO cannot be used owing to interference by $ONOO^-$ in this assay ***(10)***.
17. Because addition of NO to DHR inhibits oxidation, regardless of whether DHR is oxidized by UV light, HOCl, HRP/hydrogen peroxide or $ONOO^-$, the addition of NOS inhibitors could result in an apparent increase in DHR oxidation.
18. Measurement of the fluorescent products DCF and rhodamine using fluorescent spectroscopy allows detection of as little as 1–10 n*M* $ONOO^-$. However, visible spectroscopy at 500 nm (ε = 78,700 $M^{-1}cm^{-1}$) is easier to employ and is only about 10-fold less sensitive than fluorescence measurement. Synthetic $ONOO^-$ readily oxidizes both compounds with an efficiency equal to about 40% of added $ONOO^-$. SIN-1-mediated oxidation of both compounds is inhibitable by SOD, consistent with its ability to form $ONOO^-$.
19. The buffers selected for experiments with $ONOO^-$ need to be chosen with care because they can alter the course of the reaction.
20. Both O_2^- and H_2O_2 can cause lipid oxidation, although this reaction is likely to be dependent on trace-metal ion contamination. This can, however, be suppressed with the inclusion of the metal-chelating agent DTPA; we routinely use a concentration of 100 μ*M* (made from a stock solution of 1 m*M* in water). Other metal chelators such as desferrioxamine or ethylenediamine tetraacetic acid are not used because they either react with $ONOO^-$ or promote metal-dependent oxidation reactions.
21. The rate constant for SOD catalyzed O_2^- dismutation is 2×10^9 $M^{-1}s^{-1}$, about three times lower than the reaction between NO and O_2^-. Hence, relatively high

levels of SOD (2000–5000 U/mL) are used to inhibit $ONOO^-$ formation. Denatured SOD (prepared by boiling) should also be used to ensure the random interception of lipid radicals by the protein is not the main inhibitory pathway. SOD is not an easy enzyme to inactivate, and it is therefore recommended that the activity of the nominally denatured form is checked before addition to the assay.

22. Peroxynitrite and NO should be added to isolated vascular-ring segments using a gas-tight Hamilton syringe to minimize decomposition and exposure to air. Additionally, application of $ONOO^-$ or NO in close proximity to the ring segment is suggested to maximize delivery and prevent degradation prior to reaction with the vessel. This technique may be used in ring segments with either intact or denuded endothelium.
23. Caution should be exercised to add the $ONOO^-$ bolus in the cell suspension as rapidly as possible.

Acknowledgments

The authors are grateful for support from the National Institutes of Health (HL 48676 and HL54815), the American Heart Association (CRW), Muscular Dystrophy Association (JC), United Soybean Board (VDU), and Amyotrophic Lateral Sclerosis Association (JC).

References

1. Huie, R. E. and Padmaja, S. (1993) The reaction of NO with superoxide. *Free Rad. Res. Comm.* **18,** 195–199.
2. Beckman, J. S., Ye, Y. Z., Anderson, P. G., Chen, J., Accavitti, M. A., Tarpey, M. M., and White, C. R. (1994) Extensive nitration of protein tyrosines observed in human atherosclerosis detected by immunohistochemistry. *Biol. Chem. Hoppe-Seyler.* **375,** 81–88.
3. Haddad, I. Y., Pataki, G., Hu, P., Galliani, C., Beckman, J. S., and Matalon, S. (1994) Quantitation of nitrotyrosine levels in lung sections of patients and animals with acute lung injury. *J. Clin. Invest.* **94,** 2407–2413.
4. Brunelli, L., Crow, J. P., and Beckman, J. S. (1995) The comparative toxicity of nitric oxide and peroxynitrite to Escherichia coli. *Arch. Biochem. Biophys.* **316,** 327–334.
5. Beckman, J. S., Chen, J., Ischiropoulos, H., and Crow, J. P. (1994) Oxidative chemistry of peroxynitrite. *Methods Enzymol.* **233,** 229–240.
6. Radi, R., Beckman, J. S., Bush, K. M., and Freeman, B. A. (1991) Peroxynitrite-mediated sulfhydryl oxidation: the cytotoxic potential of superoxide and nitric oxide. *J. Biol. Chem.* **266,** 4244–4250.
7. Crow, J. P., Beckman, J. S., and McCord, J. M. (1995) Sensitivity of the essential zinc-thiolate moiety of yeast alcohol dehydrogenase to hypochlorite and peroxynitrite. *Biochemistry* **34,** 3544–3552.
8. Beckman, J. S., Beckman, T. W., Chen, J., Marshall, P. M., and Freeman, B. A. (1990) Apparent hydroxyl radical production by peroxynitrite: implications for

endothelial injury from nitric oxide and superoxide. *Proc. Natl. Acad. Sci. USA* **87,** 1620–1624.

9. Hogg, N., Joseph, J., and Kalyanaraman, B. (1994) The oxidation of alpha-tocopherol and trolox by peroxynitrite. *Arch. Biochem. Biophys.* **314,** 153–158.
10. Schmidt, K., Klatt, P., and Mayer, B. (1994) Reaction of peroxynitrite with oxyhaemoglobin: interference with photometrical determination of nitric oxide. *Biochem. J.* **301,** 645–647.
11. Darley-Usmar, V. M., Hogg, N., O'Leary, V. J., Wilson, M. T., and Moncada, S. (1992) The simultaneous generation of superoxide and nitric oxide can initiate lipid peroxidation in human low density lipoprotein. *Free Rad. Res.Comms.* **17,** 9–20.
12. Hogg, N., Darley-Usmar, V. M., Wilson, M. T., and Moncada, S. (1992) Production of hydroxyl radicals from the simultaneous generation of superoxide and nitric oxide. *Biochem. J.* **281,** 419–424.
13. Beckman, J. S., Ischiropoulos, H., Zhu, L., Wored, M., Smith, C., Chen, J., Harrison, J., Martin, J. C., and Tsai, M. (1992) Kinetics of superoxide dismutase- and iron-catalyzed nitration of phenolics by peroxynitrite. *Arch. Biochem. Biophys.* **298,** 438–445.
14. Graham, A., Hogg, N., Kalyanaraman, B., O'Leary, V. J., Darley-Usmar, V. M., and Moncada, S. (1993) Peroxynitrite modification of low density lipoprotein leads to recognition by the macrophage scavenger receptor. *FEBS Lett.* **330,** 181–185.
15. Van der Vliet, A., Smith, D., O'Neill, C. A., Kaur, H., Darley-Usmar, V. M., Cross, C. E., and Halliwell, B. (1994) Interactions of peroxynitrite with human plasma and its constituents. Oxidative damage and antioxidant depletion. *Biochem. J.* **303,** 295–301.
16. Moro, M. A., Darley-Usmar, V. M., Goodwin, D. A., Read, N. G., Zamora-Pino, R., Feelisch, M., Radomski, M. W., and Moncada, S. M. (1994) Paradoxical fate and biological action of peroxynitrite on human platelets. *Proc. Natl. Acad. Sci. USA* **91,** 6702–6706.
17. Villa, L. M., Salas, E., Darley-Usmar, V. M., Radomski, M., and Moncada, S. (1994) Peroxynitrite induces both vasodilatation and impaired vascular relaxation in the isolated perfused heart. *Proc. Natl. Acad. Sci. USA* **91,** 12,383–12,387.
18. Moore, K., Darley-Usmar, V. M., Morrow, J., and Roberts, L. J. (1995) Formation of F_2 isoprostanes during the oxidation of human low density lipoprotein and plasma by peroxynitrite. *Circ. Res.* **77,** 335–341.
19. Moro, M. A., Darley-Usmar, V. M., Lizasoain, I., Su, Y., Knowles, R. G., Radomski, M. W., and Moncada, S. M. (1995) The formation of nitric oxide donors by peroxynitrite. *Brit. J. Pharmacol.* **116,** 1999–2004.
20. Hughes, M. N. and Nicklin, H. G. (1968) The chemistry of pernitrites. Part I. Kinetics of decomposition of pernitrous acid. *J. Chem. Soc.* 450–452.
21. Pryor, W. A., Cueto, R., Jin, X., Ngu-Schwemlein, M., Squadrito, G..L, Uppu, P. L., and Uppu, R. M. (1995) A practical method for preparing peroxynitrite solutions of low ionic strength and free of hydrogen peroxide. *Free Rad. Biol. Med.* **18,** 75–83.

22. Feelisch, M. (1991) The biochemical pathways of nitric oxide formation from nitrovasodilators: appropriate choice of exogenous NO donors and aspects of preparation and handling of aqueous NO solutions. *J. Cardiovasc. Pharmacol.* **17,** S25–S33.
23. Crow, J. (1997) Dichlorodihydrofluorescein and dihydrorhodamine 123 are sensitive indicators of peroxynitrite *in vitro*: implications for intracellular measurement of reactive oxygen and nitrogen species. *Nitric Oxide: Biol. Chem.* **1,** 145–147.
24. Lymar, S. V. and Hurst, J. K. (1997) Rapid reaction between peroxynitrite ion and carbon dioxide: implications for biological activity. *J. Am. Chem. Soc.* **111,** 8867–8868.
25. Koppenol, W. H., Moreno, J. J., Pryor, W. A., Ischiropoulos, H., and Beckman, J. S. (1992) Peroxynitrite, a cloaked oxidant formed by nitric oxide and superoxide. *Chem. Res. Toxicol.* **5(6),** 834–842.

21

The Use of NO Gas in Biological Systems

Neil Hogg and B. Kalyanaraman

1. Introduction

The methodology for nitric oxide[1] (NO) delivery into biological systems has both a biochemical and a pharmacological\toxicological perspective. From the biochemical point of view, the NO delivery system is required to mimic physiological NO production. From the pharmacological vantage point, no such restriction exists. The rate of NO production, and the consequent steady-state concentration, strongly dictates the mechanism of NO decomposition in biological systems *(1–3)*. This is primarily owing to the kinetics of the reaction between NO and oxygen, as the rate of this reaction is proportional to the squared power of the concentration of NO *(4,5)*. Consequently, if NO is delivered to a system at a faster rate or at a higher concentration than would ever be achieved physiologically, the importance of the reaction of NO with oxygen will be overemphasized. This is of major importance, as NO is relatively unreactive to most biological molecules, whereas oxidation products of NO may have complex biological reactivity. For example, dinitrogen trioxide (N_2O_3) is a potent nitrosating agent that reacts with amines and thiols to generate N-nitroso and S-nitroso derivatives, respectively *(6–8)*. This problem has been addressed by the development of slow-releasing NO donor compounds that attempt to imitate the biological release of NO *(9,10)*. All such compounds have problems and in some cases, such as with the S-nitrosothiols, it is not always clear that NO is the active agent *(11,12)*. The use of aqueous solutions of NO gas is unambiguous in terms of the chemical species present; however, sample handling is more complex and it is difficult to avoid high, non-

[1]Nitrogen monoxide in IUPAC nomenclature.

From: *Methods in Molecular Biology, Vol. 100. Nitric Oxide Protocols*
Edited by: Michael A. Titheradge © Humana Press Inc., Totowa, NJ

physiological concentrations without severely limiting the experimental design. This chapter presents a method for the production of aqueous solutions of NO for use in biological systems.

2. Materials

1. NO gas cylinder.
2. Argon gas cylinder.
3. Metmyoglobin (usually, though not necessarily, from horse heart).
4. 100 m*M* phosphate buffer, pH 7.4.
5. 1 *M* sodium hydroxide.
6. Sodium dithionite.
7. Sephadex G25.

3. Methods

3.1. Preparation of Aqueous NO Solutions

The following procedure will generate an aqueous solution of NO with a concentration of between 1.7 and 2 m*M* depending on the water/buffer used.

1. Half fill a 50 mL round-bottomed flask with either distilled-deionized water or buffer and bubble with argon for approx 1 h (*see* **Note 1**).
2. Seal the flask with an appropriately sized rubber seal (such as a "Suba-seal" from Aldrich). Remove any residual oxygen by 2–3 cycles of vacuum degassing followed by refilling with argon. This can be achieved using small-bore (~16 gage) needles, attached to the vacuum line and the argon supply, to deliver both the vacuum and the argon through the rubber seal.
3. Vacuum degas a final time and refill the vessel with NO gas. When using NO gas it is important to:
   a. Use stainless-steel tubing to deliver NO to the system.
   b. Place a trap containing 10–20 mL of 1 *M* sodium hydroxide in the gas stream to remove trace quantities of N_2O_3 and N_2O_4 that are always present in commercially available ·NO. These contaminants, if not removed, will hydrolyze in aqueous solution to nitric and nitrous acids, lowering the pH of the solution and contaminating the final preparation with nitrite and nitrate ions.
   c. Make sure all oxygen is blown out of the NO gas line before administration to the flask by leaving the gas flow running for several seconds. The presence of a brown gas in the line indicates oxygen is still present.
   d. Perform all operations in a fume hood as NO and its gaseous oxidation products are highly toxic (*see* **Note 2**).
4. Thoroughly shake the solution to dissolve NO in the aqueous phase. The gas will be under positive pressure inside the flask and this should be restored to atmospheric pressure by piercing the seal with a small-bore needle for several seconds (*see* **Note 3**). After allowing re-equilibration between the aqueous and gaseous phases, the concentration of NO can be measured.

3.2. Measurement of NO Concentration

The concentration of NO in aqueous solution can most easily be measured by monitoring the conversion of either oxyhemoglobin or oxymyoglobin to their ferric derivatives (**Eq. 1**).

$$Hb^{2+}O_2 + \cdot NO \rightarrow Hb^{3+} + NO_3^- \quad (1)$$

Such oxidation is accompanied by a well-characterized change in the visible absorption spectrum of the hemeprotein, which can be used to quantify NO concentration. The following method allows the accurate quantitation of NO solutions using oxymyoglobin.

1. Myoglobin is usually obtained in the ferric or met form. Reduction to the ferrous derivative can be accomplished by adding excess sodium dithionite (also called sodium hydrosulfite). Add 2–4 mg of sodium dithionite to about 1 mL of myoglobin (1–2 m*M*). This results in an immediate color change as the myoglobin is reduced from ferric to the ferrous form. Sodium dithionite will also consume oxygen in solution and the myoglobin will be in the deoxy-form. Separate the protein from excess sodium dithionite by passage down a sephadex G-25 size exclusion column (1 × 25 cm) that has been pre-equilibrated with 100 m*M* phosphate buffer, pH 7.4. Once separated from the dithionite, ferrous myoglobin rapidly picks up oxygen from solution to form the oxy derivative (*see* **Note 4**). Determine the concentration of oxymyoglobin by spectrophotometry at 580 nm (ε = 14,400 $M^{-1}cm^{-1}$).
2. Dilute the oxymyogobin to about 70 μ*M* (giving an absorbance of about 1 at 580 nm) in a sealable 3 mL cuvet. Seal the cuvet with a rubber seal, apply a vacuum through the seal to degass the solution, and replace the gas with argon through the seal. This process lowers the oxygen concentration in solution, thus preventing side reactions of NO (*see* **Note 5**). Take the visible-absorption spectrum between 450 and 650 nm and record the absorbance at 580 nm.
3. Titrate the oxymyoglobin solution with 10 μL aliquots of NO solution and take the visible spectrum between 450 and 650 nm after each addition, making sure to record the value at 580 nm. NO gas should be added to the cuvet using a gas-tight Hamilton syringe (*see* **Note 6**). Titrate the oxymyoglobin until all the oxymyoglobin has been converted to metmyoglobin (*see* **Note 7**).
4. Convert the spectral changes observed during titration to concentration of nitric oxide added using $\Delta\varepsilon$ = 10,300 $M^{-1}cm^{-1}$ for oxymyoglobin–metmyoglobin at 580 nm (*see* **Note 8**). Assuming a 1:1 stoichiometry for the oxidation of oxymyoglobin by ·NO, a simple back calculation will then give the concentration of stock ·NO.

3.3. The Use of NO Gas Solutions

Aqueous solutions of NO are best delivered to the system under study using gas-tight Hamilton syringe. NO solution should be withdrawn through the eal and administered as soon as possible. This methodology is ideal for measuring the kinetics of reactions of NO with other biological molecules. Donor molecules cannot be used for this purpose, as the rate of release of NO from the compound will complicate the kinetics of the reaction. When performing such an experiment, it is important to adhere to the following guidelines:

1. Make sure all samples are thoroughly deoxygenated before the addition of ·NO.
2. Minimize the headspace volume of the reaction vessel and continuously agitate the reaction vessel.
3. When analyzing products, ensure that the reaction vessel is degassed before exposure to oxygen. This will remove any unreacted NO and prevent oxygen-dependent reactions.
4. Avoid low pH (<5) in product analysis. Any nitrite contamination, or nitrite formed during the reaction, will decompose to yield NO at low pH. If this occurs in the presence of oxygen, oxidation products of NO may confound the analysis. Moreover nitrous acid can act as an NO^+ donor. One example of this problem is the formation of S-nitrosothiols from a mixture of thiol and nitrite, which can occur during high-pressure liquid chromatography (HPLC) separation using trifluoroacetic acid as the mobile phase.

3.4. Improper Use of NO Gas

Improper use of NO gas has led to erroneous conclusions concerning the reactivity of ·NO. It is essential to differentiate between the chemistry of NO and that of N_2O_3 in order to understand the biological reactivity of ·NO. The addition of NO at high concentration in the presence of oxygen cannot be regarded as being biologically relevant. This is primarily owing to the fact that higher oxides of nitrogen can be formed in the system under study. Moreover, the practice of bubbling NO gas through the system is a totally unacceptable method of NO delivery and any conclusions drawn from such studies cannot be extrapolated to the biological systems.

4. Notes

1. Solutions of NO can be prepared in any-sized glass vessel as long as it can be sealed.
2. When handling NO gas, it is strongly advised that gloves be worn, as oxidation products of NO can form nitrous and nitric acids in skin moisture.
3. The concentration of NO in the gaseous phase is about 20 times higher than that in the aqueous phase. One result of this is that the unavoidable contamination with small quantities of oxygen, through either slow leakage through the seal or from repeated withdrawals of solution, will only slightly affect the NO concentration in the aqueous phase. The major consequence of oxygen leakage will be an increase in nitrate and nitrite in the NO solution.
4. The change from deoxy- to oxymyoglobin that occurs as the protein is separated from dithionite can often be seen as a color change as the myoglobin passes down the Sephadex column.

5. The reaction of NO with oxymyoglobin is fast and low levels of oxygen will not interfere with the assay. One advantage of using myoglobin over hemoglobin is that its higher affinity for oxygen allows dissolved oxygen in myoglobin solutions to be lowered significantly without dissociating the bound oxygen from the heme protein.
6. It is important to refill the syringe for each titration point and add the NO solution to the cuvet as rapidly as possible, as NO will degrade in the barrel of the Hamilton syringe.
7. The visible-absorbtion spectrum of metmyogobin is strongly pH dependent. As NO solutions have a tendency to be acidic, it is important that the myoglobin solution is well buffered, e.g., 100 m*M* phosphate, pH 7.4.
8. The reaction of NO with oxymyoglobin also results in changes in absorbance at lower wavelengths (~400–430 nm). Changes in the region are almost an order of magnitude more sensitive than those at 580 nm. However, for the sake of determining the concentrations of NO solutions (of approx 2 m*M*), sensitivity is not an issue. Use of a more sensitive wavelength is disadvantageous as the number of titration points that cover a particular absorption change is reduced and most spectrophotometers have a working range of about two absorbance units.

References

1. Rubbo, H., Radi, R., Trujillo, M., Telleri, R., Kalyanaraman, B., Barnes, S., Kirk, M., and Freeman, B. (1994) Nitric oxide regulation of superoxide and peroxynitrite-dependent lipid peroxidation. *J. Biol. Chem.* **42,** 26,066–26,075.
2. Goss, S. P. A., Hogg, N., and Kalyanaraman, B. (1995) The antioxidant effect of spermine NONOate in human low-density lipoprotein. *Chem. Res. Toxicol.* **8,** 800–806.
3. Miles, A. M., Bohle, S., Glassbrenner, P. A., Hansert, B., Wink, D. A., and Grisham, M. B. (1996) Modulation of superoxide-dependent oxidation and hydroxylation reactions by nitric oxide. *J. Biol. Chem.* **271,** 40–47.
4. Wink, D. A., Darbyshire, J. F., Mims, R. W., Saavedra, J. E., and Ford, P. C. (1993) Reaction of the bioregulatory agent nitric oxide in oxygenated aqueous media: determination of the kinetics for oxidation and nitrosation by intermediates generated in the $\cdot NO/O_2$ reaction. *Chem. Res. Toxicol.* **6,** 23–27.
5. Kharitonov, V. G., Sundquist, A. R., and Sharma, V. S. (1994) Kinetics of nitric oxide autoxidation in aqueous solution. *J. Biol. Chem.* **8,** 5881–5883.
6. Wink, D. A., Nims, R. W., Darbyshire, J. F., Christodoulou, D., Hanbauer, I., Cox, G. W., Laval, F., Laval, J., Cook, J. A., Krishna, M. C., LeGraff, W. G., and Mitchell, J. B. (1994) Reaction kinetics for nitrosation of cysteine and glutathione in aerobic nitric oxide solutions at neutral pH. Insights into the fate and physiological effects of intermediates generated in the NO/O_2 reactions. *Chem. Res. Toxicol.* **7,** 519–525.
7. Goldstein, S. and Czapski, G. (1996) Mechanism of nitrosation of thiols and amines by oxygenated ·NO solutions-the nature of the nitrosating intermediates. *J. Am. Chem. Soc.* **118,** 3419–3425.

8. Miwa, M., Stuehr, D. J., Marletta, M. A., Wishnok, J. S., and Tannenbaum, S. R. (1987) Nitrosation of amines by stimulated macrophages. *Cacinogenesis* **8,** 955–958.
9. Feelisch, M. (1991) The biochemical pathways of nitric oxide formation from nitrovasodilators: appropriate choice of exogenous NO donors and aspects of preparation and handling of aqueous NO solutions. *J. Cardiovasc. Pharm.* **17(suppl. 3),** S25–S33.
10. Maragos, C. M., Morley, D., Wink, D. A., Dunams, T. M., Saavedra, J. E., Hoffman, A., Bove, A. A., Isaac, L., Hrabie, J. A., and Keefer, L. K. (1991) Complexes of NO with nucleophiles as agents for the controlled biological release of nitric oxide. Vasorelaxant effects. *J. Med. Chem.* **34,** 3242–3247.
11. Park, J.-W., Billman, G. E., and Means, G. E. (1993) Transnitrosation as a predominant mechanism in the hypotensive effect of S-nitrosoglutathione. *Biochem. Mol. Biol. Int.* **30,** 885–891.
12. Hogg, N., Singh, R. J., and Kalyanaraman, B. (1996) The role of glutathione in the transport and catabolism of nitric oxide. *FEBS Lett.* **382,** 223–228.

22

A Microtiter-Plate Assay of Human NOS Isoforms

John Dawson and Richard G. Knowles

1. Introduction

Nitric oxide synthase (NOS) catalyzes the conversion of L-arginine, molecular oxygen, and nicotinamide adenine dinucleotide phosphate (NADPH) to NO, citrulline, and $NADP^+$ (reviewed in **ref. *1***). The neuronal (n) and endothelial (e) NOS isozymes are highly regulated by Ca^{2+} and calmodulin (CaM), whereas the iNOS has CaM tightly bound. NOS are heme proteins that also contain tightly bound flavin adenine dinucleotide (FAD) and flavin mononucleotide (FMN) and require tetrahydrobiopterin (BH_4) for activity, whereas NADPH is used as a substrate.

This chapter describes a microtiter-plate assay for measuring NOS activity developed from the oxyhemoglobin assay described by Salter and Knowles (*see* Chapter 6, this volume). The assay was developed to enable the study of inhibitors of the three human-NOS isoforms under the same conditions. With the appropriate microtiter-plate reader, this permits 96 simultaneous kinetic measurements. This assay is run at 37°C and is ideal for the study of NOS inhibitors and their time dependence, which can also be temperature-dependent. Using a similar method, it is possible to measure enzyme rates under different conditions simultaneously. This is useful, for example, for the study of the cofactor and substrate dependence of NOS preparations.

Other microtiter-plate assays of NOS activity have been described. The oxymyoglobin assay *(2,3)* should on theoretical grounds be very similar to that described here, kinetically measuring NO formation. Assays of NOS by measuring NADPH oxidation *(4)* are much less sensitive and may not accurately reflect NO synthesis, since NADPH consumption is under some circumstances uncoupled from NO synthesis. Assays of NOS using the Griess reaction to

From: *Methods in Molecular Biology, Vol. 100. Nitric Oxide Protocols*
Edited by: Michael A. Titheradge © Humana Press Inc., Totowa, NJ

measure nitrite formed *(5)* are end-point rather than kinetic, are somewhat less sensitive, and are subject to the concern that only one of the two major products of NO breakdown (nitrate and nitrite) is being measured: the ratio of nitrite to nitrate formed can vary depending on what is present in the assay. Microtiterplate assays based on oxyhemoglobin or oxymyoglobin are therefore the most sensitive and reliable assays for NOS activity.

This version of the assay is somewhat less sensitive than the optimized spectrophotometric (*see* Chapter 6, this volume) or radiometric (*see* Chapter 7, this volume) assays, requiring approx 10 pmol/min of NOS activity to achieve a 1 mODU/min rate. This sensitivity is quite adequate for many studies, e.g., assays of brain-cytosol NOS, of recombinant-expressed NOS or of purified NOS.

2. Materials

1. Extraction buffer (EB): The extraction buffer should be prepared in advance. The basic buffer is first prepared by dissolving sucrose (250 m*M*), Tris base (50 m*M*), and ethylenediaminetetraacetic acid (EDTA) (1 m*M*) in double-distilled or milliQ-grade water and bringing its pH to 7.4 at room temperature by the addition of HCl. The following constituents are then added to the final concentrations indicated: 0.1 m*M* D/L-dithiothreitol (DTT), 0.5 μ*M* leupeptin, 0.5 μ*M* pepstatin A, and 10 μ*M* Antipain; the buffer is then made up to its final volume with water. This EB is then distributed into aliquots (typically 50 mL per aliquot) and stored at –20°C until required.
2. 10 mg/mL Phenylmethylsulphonyl fluoride (PMSF): Because it is unstable in aqueous solution, PMSF is not included in the buffer at this stage, but prepared as a solution in absolute ethanol, stored at –20°C, and added to the EB during the extraction procedure (*see* below). The composition of this EB is designed to permit extraction of NOS from cells or tissues without breaking intracellular organelles and minimizing proteolysis by chelating divalent cations with EDTA and the inclusion of protease inhibitors.
3. Assay buffer: 100 m*M* HEPES is dissolved in double-distilled or milliQ-grade water and brought to pH 7.4 by the addition of NaOH. This can be stored for several weeks at 4°C. Prior to the assay of NOS, DTT is added to the buffer to give 100 μ*M*. $CaCl_2$, $MgCl_2$, and hemoglobin are added as required and the buffer mixture warmed to the required temperature. The cofactor and substrate stocks are kept in the dark on ice until usage (*see* **Notes 1–3**).
4. 1 *M* stock solutions of $CaCl_2$ and $MgCl_2$ can be stored for several weeks at 4°C.
5. Cofactor stocks: 7.5 m*M* L-arginine (HCl), 250 m*M* NADPH, 250 μ*M* FMN, 250 μ*M* FAD, and 25 μ*M* CaM (Sigma P2277, from Bovine Brain, assumed M.W. 17,000) are dissolved in HEPES buffer and stock solutions kept at –20°C. A "cocktail" of these can be made and stored in the same way to reduce the number of additions to the assay.
6. (6R)-5,6,7,8-Tetrahydro-L-biopterin hydrochloride (BH_4): On the day of assay, DTT is added to 10 m*M* HCl (stock stored at room temperature) to give 500 μ*M*.

BH_4 is dissolved in this to give 500 μ*M* and the solution kept on ice in the dark until usage. It can be used for approx 4 h.

7. NOS inhibitors: Inhibitors are dissolved in 25% (v/v) DMSO/100 m*M* HEPES buffer pH 7.4 at 12.5 times the final concentration required. A stock solution of 12.5 m*M* N^G-methyl-L-arginine (L-NMMA) for blanks, dissolved in this way, can be stored at –20°C. Up to 2% DMSO can be tolerated in the assay. For IC_{50} determination, we routinely use 10 concentrations with twofold dilution.
8. Ion-exchange resin: Dowex 50W Na^+ form, 200–400 mesh, 8% cross-linked (prepared as described in Chapter 7, this volume).
9. 3 m*M* oxyhemoglobin (monomer). (*See* Chapter 8, this volume.)
10. Enzyme: Baculovirus-expressed human-recombinant iNOS, nNOS, and eNOS *(6)*. 1U = 1 nmol/min per plate at 37°C under the conditions of the assay.
11. Microtiter plates: Costar 3598 flat bottom.
12. Microtiter-plate reader: A dual-wavelength microplate reader with 405 and 420 nm filters, temperature control, and kinetic software is required. We have used the Molecular Devices THERMOmax, Dynatech MR7000, and Dynatech DIAS readers with SOFTmax, Biolinx, and Revelation software, respectively.
13. A microtiter-plate incubator and incubated orbital-plate shaker, both set at 37°C, are also required.

3. Methods

3.1. Enzyme Preparation

We routinely use crude extracts of the human recombinant NOS isoforms expressed in *Spodoptera frugiperda* Clone 21 (*Sf*21) cells using the Baculovirus expression system *(6)*.

1. Scrape *Sf*21 cells grown as monolayers into a small volume of EB using a rubber policeman or a purpose-made cell scraper with the plate or flask standing on ice. Centrifuge cells grown in suspension to collect them as a cell pellet and resuspend the cell pellet in a small volume of ice cold EB.
2. Disrupt the cell extracts by sonication. As with the procedures with tissue extraction, it is important to keep the temperature at 0–4°C. Most probe sonicators should be suitable for this purpose; we have routinely used a MSE Soniprep 150 with a pre-cooled 5 mm tip probe, sonicating three times for 10 s at 10 μm amplitude with 30 s cooling in ice water in between. PMSF is added after the first sonication to give 100 μg/mL.
3. Centrifuge the cell extract at 100,000*g* for 30 min at 4°C to remove insoluble material and store at –70°C.

3.2. Removal of Endogenous Arginine

The arginine concentration in the standard assay is 30 μ*M*. However, high concentrations of arginine can be present in the cell extracts. We routinely remove endogenous arginine by a pre-treatment with Dowex 50W (Na^+ form) ion-exchange resin, prepared as described, and then washed in EB.

1. Add 2 vol of enzyme extract to 1 vol of ice-cold, packed resin in microfuge tubes and mix;
2. Briefly centrifuge the mixture (e.g. 10,000*g* for 1 min at 0–4°C) and collect the supernatant for assay. It is crucial that no resin is carried forward into the assay: it will adsorb the substrate!

3.3. The Assay Procedure

1. Add 20 μL of NOS inhibitor, 12.5 m*M* L-NMMA (for blanks, to produce 100% inhibition) or 25% (v/v) DMSO /100 m*M* HEPES buffer pH 7.4 (controls) to the 96-well plate.
2. Warm the plate in the 37°C incubator for ~30 min (*see* **Note 4**).
3. Dilute the enzyme stock to the required working strength, 12.5 times the final concentration, in HEPES buffer (2.5 mL of working strength per plate, 0.5 U per mL) (*see* **Note 5**).
4. Add 0.6 mL of cofactor mix and 0.6 mL of BH_4 to 23.3 mL of the pre-warmed oxyhemoglobin, $CaCl_2$, $MgCl_2$, and buffer mixture. Final concentrations of the effectors used with the human recombinant enzyme are 5 μ*M* oxyhemoglobin; 1 m*M* $MgCl_2$; 30 μ*M* L-arginine (or as required); 100 μ*M* NADPH; 1 μ*M* FMN; 1 μ*M* FAD; 100 nM CaM; and 10 μ*M* BH_4 (*see* **Note 6**).
5. Add 210 μL of this complete mixture to each well of the plate, keeping it at 37°C.
6. Rapidly add 20 μL of working strength enzyme per well and mix the plate with an incubated (37°C) orbital shaker for 15 s or a multipipet, taking care not to cross contaminate wells.
7. Place the plate in the reader (pre-warmed to 37°C) and start reading.
8. Follow the reaction by reading the difference in absorbance between 420 and 405 nm, using a dual-wavelength kinetic-plate reader at 37°C. Rates are determined as mOD/min by linear regression. [*See* **Fig. 1** for representative results with L-NMMA and N^{ω}-nitro-L-arginine (NitroArg).]

3.4. Calculation of Results

The L-NMMA-blank rate is subtracted from all the control and test rates and the percentage inhibition calculated if required. Absolute NOS activity (enzyme units per weight of tissue or protein) can be calculated using the apparent-extinction coefficient (*see* **Note 7**). The path length will depend on the assay volume and type of plate.

4. Notes

1. Phosphate buffer can also be used for the assay (*see* Chapter 6, this volume).
2. We have found that concentrations of DTT above 500 μ*M* inhibit the enzyme, when using the crude human-recombinant isozymes (unpublished data). This is in contradiction with the literature reporting stabilization of purified NOS by higher concentrations of DTT *(7,8)*.

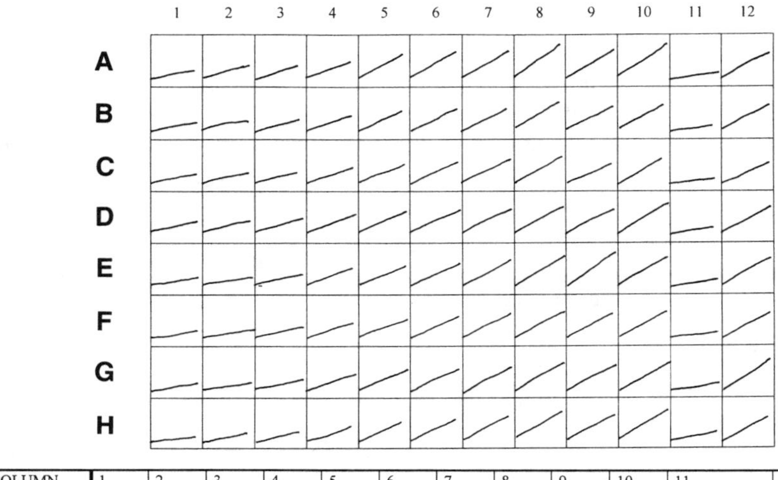

COLUMN	1	2	3	4	5	6	7	8	9	10	11	12
ROWS A - D L-NMMA	50	25	12.5	6.25	3.13	1.56	0.78	0.39	0.20	0.10	L-NMMA BLANKS	CONTROLS
ROWS E - H NitroArg	10	5	2.5	1.25	0.625	0.313	0.160	0.080	0.040	0.020	L-NMMA BLANK	CONTROLS

Fig. 1. (Top) Time-courses of NO synthesis by nNOS and its inhibition by L-NMMA and NitroArg. (Bottom) 96-well plate layout. Concentration of inhibitor (μ*M*), Human Recombinant nNOS, 37°C, 30 min. Scales: x axis time 0–30 min, y axis absorbance 0–100 mOD.

3. It is not necessary to add calcium under these conditions, since the calcium concentration present in the buffers as a contaminant (probably low micromolar) is sufficient for expression of full activity. Addition of high concentrations (e.g., 200 μ*M*) can be detrimental, decreasing the linearity of the enzyme when using unpurified human-recombinant preparations.
4. Plates at room temperature cool rapidly. The temperature settings of incubators and plate readers should be checked. We routinely monitor the well temperature with a miniature thermocouple (Physitemp BAT-12 with IT-23 thermocouple microprobe).
5. It is crucial to ensure that under the particular conditions of the assay, temperature and enzyme source, rates are linear both with time and concentration of enzyme extract added.
6. The "complete-assay mix" can only be used for ~1 h at 37°C as the cofactors/substrates are unstable at this temperature. Do not leave exposed to light for long periods.

7. Calculation of the rates as pmol per min per mass of sample or protein requires the determination of the apparent-extinction coefficient for the NO-mediated spectral shift, because this depends on the precise wavelengths and bandwidths of the filters used, the volume of the assay, and the shape of the wells used. This is determined by measuring the change in absorbance at 405–420 nm following oxidation of a known concentration of hemoglobin by NO (as the dissolved gas or from NO donors). The stoichiometry of the reaction is 1 mol of oxyhemoglobin monomer oxidized per mol of NO. Typically, the apparent-extinction coefficient is in the range 30,000–50,000 M^{-1}.
8. Where not specified, the reagents can be obtained from Sigma.

References

1. Knowles, R. G. and Moncada, S. (1994) Nitric oxide synthases in mammals. *Biochem. J.* **298,** 249–258.
2. Feelisch, M., Kubitzek, D., and Werringloer, J. (1996) The oxyhaemoglobin assay, in *Methods in Nitric Oxide Research* (Feelisch, M. and Stamler, J. S., eds.), Wiley, New York, pp. 472–473.
3. Gross, S. S., Jaffe, E. A., Levi, R., and Kilbourn, R. G. (1991) Cytokine-activated endothelial cells express an isotype of nitric oxide synthase which is tetrahydrobiopterin-dependent, calmodulin-dependent and inhibited by arginine analogs with a rank-order of potency characteristic of activated macrophages. *Biochem. Biophys. Res. Comm.* **178 (3),** 823–829
4. Stuehr, D. J. and Griffith, O. W. (1996) Purification, assay and properties of mammalian nitric oxide synthases in *Methods in Nitric Oxide Research* (Feelisch, M. and Stamler, J. S., eds.), Wiley, New York, p. 185.
5. Stuehr, D. J. and Griffith, O. W. (1996) Purification, assay and properties of mammalian nitric oxide synthases, in *Methods in Nitric Oxide Research* (Feelisch, M. and Stamler, J. S., eds.), Wiley, New York, pp. 183–184.
6. Charles, I. C., Scorer, C. A., Angeles Moro, M., Fernandez, C. Chubb, A., Dawson, J., Foxwell, N., Knowles, R. G., and Baylis, S. A. (1996) Expression of the three human NO synthase isozymes. *Methods Enzymol.* **268,** 449–460
7. Komori, Y., Hyun, J., Chiang, K., and Fukuto, J. M. (1995). The role of thiols in the apparent activation of rat brain nitric oxide synthase (NOS). *J. Biochem* **117,** 923–927
8. Hofmann, H. and Schmidt, H. H. H. W. (1995). Thiol dependence of nitric oxide synthase. *Biochemistry* **34 (41),** 13,443–13,452.

23

Use of Arginine Analogs as Inhibitors of Nitric Oxide Synthase in Rat-Aortic Rings

Rachel J. Russell

1. Introduction

Nitric oxide synthases (NOS) are expressed constitutively in vascular endothelial cells (eNOS; *1*) and can be induced in vascular smooth muscle following stimulation with cytokines and endotoxin (inducible (i) NOS; for review, *see* **ref. *2***). The discovery and characterization of inhibitors of NOS *(3,4)* led to the elucidation of the biological significance of NO synthesis within the vessel wall.

N^{G}Monomethyl-L-arginine (L-NMMA) inhibits the endothelium-dependent relaxation stimulated by acetylcholine (the Furchgott phenomenon) and causes an increase in basal tone in rings of rat aorta *(3)*. Thus demonstrating that the constitutive NOS has a physiological role in the vessel wall in both agonist-induced changes in vasodilator tone and in blood pressure control.

Stimulation of rat-aortic rings with lipopolysaccharide results in a gradual loss of tone, which is paralleled by a time-dependent expression of the iNOS. The fall in tone can be inhibited by L-NMMA and glucocorticoids *(5,6)* and is found to be largely independent of the presence of the endothelium [i.e., most of the induced NOS activity is present within the vascular smooth muscle *(7,6)*]. This induction of NO synthesis within the vessel wall contributes to the loss of tone.

Other L-arginine analogs have been shown to be inhibitors of NOS, e.g., N$^{\omega}$Nitro L-arginine methyl ester (L-NAME) and N$^{\omega}$-imino ethyl-L-ornithine (L-NIO), which exhibit qualitatively similar pharmacological responses in vitro. The advent of these inhibitors has led to the development of carefully designed assays that allow the constitutive NOS and iNOS to be studied

From: *Methods in Molecular Biology, Vol. 100. Nitric Oxide Protocols*
Edited by: Michael A. Titheradge © Humana Press Inc., Totowa, NJ

individually (as shown in this Chapter). Furthermore, such assays could be utilized for other vascular preparations in order to investigate any differences that may be exhibited within different vasculature beds.

2. Materials

1. Krebs bicarbonate Ringer solution *(8)*. Add 500 mL of Krebs concentrate (10 times concentrated, Sigma) to 4.5 L of distilled water (sterile and endotoxin tested). Dissolve 8.9 mg of indomethacin (Sigma) in 1 mL of 5% Na_2CO_3 (BDH) and add to Krebs giving a final concentration of 5 μ*M* indomethacin. Aerate the Krebs with 95% O_2:5% CO_2 before adding 12.5 mL of 1 *M* $CaCl_2$ (BDH) to give a final concentration of 2.5 m*M* $CaCl_2$ (*see* **Notes 1** and **2**). Make fresh each day.
2. Phenylephrine (Sigma). Make a 10 m*M* concentrated stock in 0.9% (w/v) saline, from which you can prepare dilutions ranging from 0.001 m*M* to 10 m*M*. The addition of 25 μL of 0.001 m*M* phenylephrine to the organ bath (followed by 75 μL of 0.001 m*M*, then 17.5 μL and 75 μL of 0.01 m*M*, etc.) will give a cumulative-dose response curve, in half-log units, ranging from 1 m*M* to 10 μ*M*. Prepare daily.
3. Acetylcholine bromide (Sigma). Make a 1 m*M* concentrated stock in 0.9% saline, from which you can prepare dilutions ranging from 0.01 m*M* to 1m*M*. The addition of 10 μL of 0.01 m*M* acetylcholine to the organ bath (followed by 10 μL of 0.01 m*M*, then 20, 40, 80 μL of 0.01 m*M*, etc.) will give a cumulative-dose response curve in the range of 5 m*M* to 2560 m*M*. Half-log unit increments are not sufficient to see the full-dose response curve. Prepare daily and keep on ice.
4. NOS inhibitors, e.g., L-NMMA, L-NAME, and L-NIO (Calbiochem; *see* **Note 3**). Make 100 m*M* concentrated stocks in 0.9% saline, from which you can prepare dilutions ranging from 0.1 m*M* 100 m*M*. The addition of 25 μL of 0.1 m*M* NOS inhibitor to the organ bath (followed by 75 μL of 0.1 m*M*, then 17.5 μL and 75 μL of 1 m*M*, etc.) will give a cumulative-dose response curve, in half-log units, ranging from 0.01 μ*M* to 300 μ*M*. Prepare daily and keep on ice. Concentrated stocks can be kept frozen for months.
5. Lipopolysaccharide W.S. typhosa 0901 (Difco). Make a 0.1 mg/mL concentrated stock in saline (*see* **Note 4**). The addition of 25 μL of the concentrated stock will give a final concentration in the organ bath of 100 ng/mL. Prepare daily.

3. Methods

3.1. Preparation of Tissue

1. Prepare and aerate 5 L of Krebs buffer and calibrate the isometric transducers (calibration will vary depending on the make of transducers used and should be explained in the appropriate manual).
2. Take three male Wistar rats (250–300 g) and kill them by a blow to the head (*see* **Note 5**), cutting the neck to allow drainage of blood. Quickly excise the thoracic aorta, i.e., the section from the heart to the diaphragm, taking care not to damage

the aorta. Wash in Krebs buffer and transport to the laboratory in fresh Krebs buffer. Note that one thoracic aorta should provide four aortic rings.

3. Trim the aorta of adhering fat and connective tissue, while bathed in Krebs buffer, and cut transversely into 4-mm rings (*see* **Note 6**). For endothelium-denuded rings, cut the aorta into 8-mm sections, rub the intimal surface with a pipe cleaner, and then trim them to 4-mm rings.
4. Mount the rings onto two L-shaped stainless-steel hooks (*see* **Note 6**); the upper hook should be connected to an isometric transducer and the lower hook to the base of a 25-mL organ bath containing Krebs buffer at 37°C gassed continuously with 95% O_2 5%:CO_2.
5. Set the rings at 2 g resting tension. Record the tension using isometric transducers connected to a chart recorder.
6. Leave the rings to equilibrate for 45 min, washing every 15 min during this period with fresh Krebs buffer. The rings should relax over this time course (*see* **Note 7**). Set all the rings to a 1 g tension, which will be used as the working baseline.

3.2. Inhibition of Constitutive NOS (eNOS) by L-Arginine Analogs

1. In endothelium-intact aortic rings (set at a 1 g baseline), obtain a cumulative-concentration contraction curve to phenylephrine (1–10,000 n*M*; *see* **Note 8a**). From this data, plot a sigmoidal curve and determine the ED_{10} and ED_{90} for phenylephrine. Note that once the ED_{10} and ED_{90} (n = 6) for phenylephrine have been determined, these concentrations may be used for all future experiments (as long as the rats are within the 250–300 g weight band and of the same source and strain).
2. Add phenylephrine, at a concentration shown above to give an ED_{90}, to fresh endothelium-intact aortic rings (set at a 1 g baseline; *see* **Note 9**).
3. Obtain a cumulative-concentration relaxation curve to acetylcholine (5–2560 nM) in each ring in order to assess the integrity of the endothelium (*see* **Note 8b**). Relaxation of greater than 60% is taken as an indicator of an intact endothelium.
4. Leave the rings to re-equilibrate for 45 min, washing every 15 min during this period with fresh Krebs buffer, to bring them back to baseline tension. For those tissues that do not quite return to the original 1 g baseline, bring them back using the chart-recorder pens.
5. In those rings that have been accessed to have an intact endothelium (> 60% relaxation to acetylcholine), contract them with phenylephrine (ED_{10}) before obtaining a cumulative-concentration contraction curve to the NOS inhibitors (0.01–300 μ*M*). Under these circumstances, inhibition of eNOS by an NOS inhibitor results in a loss of the vasodilator tone exerted by NO, i.e., resulting in a contraction (**Fig. 1**; *see* **Notes 8c**, **10**, and **11**).

3.3. Inhibition of iNOS by L-Arginine Analogs

1. In fresh endothelium-denuded aortic rings obtain a cumulative-concentration contraction curve to phenylephrine (1–10,000 n*M*). From this data, plot a sigmoidal curve and determine the ED_{90} for phenylephrine in each ring (*see* **Notes** ʼ and **12**).

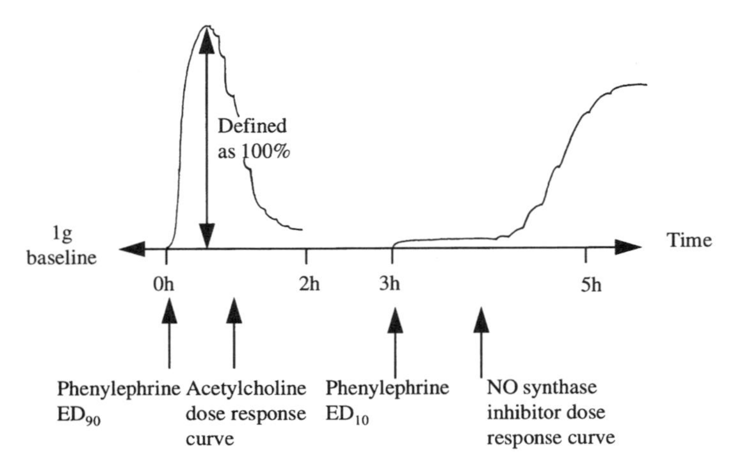

Fig. 1. Inhibition of constitutive NOS.

2. Leave the rings to re-equilibrate for 45 min, with washes every 15 min during this period with fresh Krebs buffer, to bring them back to baseline tension. For those tissues that do not quite return to the original 1 g baseline, bring them back using the chart-recorder pens.
3. Add lipopolysaccharide (100 ng/mL; *see* **Note 4**) to the organ baths 15 min prior to contracting the tissues submaximally with phenylephrine (ED_{90}), determined previously from the dose-response curve above.
4. Leave the rings for a 6 h period (*see* **Note 13**), during which induction of iNOS in the vessel wall results in a gradual time-dependent vasorelaxation, despite the continued presence of phenylephrine. Therefore, subsequent inhibition of the iNOS results in contraction of the ring.
5. In rings that show greater than 30% relaxation, 6 h after the addition of LPS, obtain a cumulative-concentration contraction curve to the NOS inhibitors (0.01–300 μ*M*; **Fig. 2**; *see* **Notes 8c** and **10**).

3.4. Data Analysis

3.4.1. Inhibition of the Constitutive NOS

The contraction induced by the NOS inhibitors is expressed as percentage contraction of that induced by the ED_{90} concentration of phenylephrine (*see* **Fig. 1**). Note that if the rings are showing spontaneous activity (apparent as oscillation of the pen recorder), take the measurement from the base of the oscillation.

3.4.2. Inhibition of the iNOS

The contraction induced by the NOS inhibitor is expressed as the percentage of the difference in tone of the phenylephrine-contracted tissue immediately

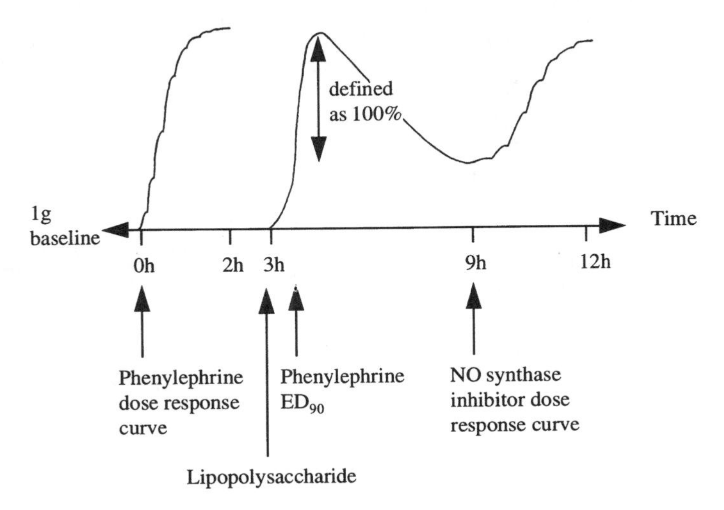

Fig. 2. Inhibition of iNOS.

before lipopolysaccharide addition and 6 h after the addition (not as a percentage contraction of that induced by the phenylephrine ED_{90}; *see* **Fig. 2**).

4. Notes

1. The addition of $CaCl_2$ to Krebs that has not been aerated with 95% O_2:5% CO_2, will result in the formation of a white precipitate of calcium phosphate. Further bubbling with 95% O_2:5% CO_2 should bring this back into solution.
2. Indomethacin does not dissolve in saline; dissolve in 1 mL of 5% Na_2CO_3 before adding to the Krebs buffer. The Na_2CO_3 has a shelf life of at least 6 mo.
3. A variety of NOS inhibitors, besides arginine analogs, are now commercially available. Some of them exhibit different mechanisms of action, have potential side effects, and show greater selectivity for one isoenzyme of NOS than another (*see* **Table 1**): hence, care must be taken to choose an appropriate NOS inhibitor for individual experiments.
4. Lipopolysaccharide is extremely toxic and should be handled with care, using gloves, face mask, and safety glasses.
5. Rats must be killed by a blow to the head, not by using anesthetic, as barbiturates are known to affect the way endothelium-intact aortic rings respond to acetylcholine and NOS inhibitors.
6. Care must be taken when excising the thoracic aorta and cleaning it of surrounding adventitia, as the endothelial cells are easily removed. Furthermore, care must be taken not to remove the endothelial cells when picking the rings up with the stainless-steel hooks. Removal of the endothelial cells is the most common reason for poor relaxation to acetylcholine.

Table 1
NOS Inhibitors Commercially Available

Compound	Mechanism of action	Points to be aware of
L-NA N^{ω}-nitro-L- arginine	Reversible and competitive inhibitor. Slow binding/slow reversal to nNOS and eNOS; rapidly reversible to iNOS.	Inhibits both NOS and NADPH-oxidation by nNOS. Selective (~10 fold) for nNOS and eNOS versus iNOS. Long half life *in vivo*.
L-NAME N^{ω}-nitro-L-arginine methyl ester		Active as L-NA after cleavage by esterases; the ester itself is inactive on the purified enzymes. Alkyl esters of L-arginine are also muscarinic antagonists and therefore poor choices as NOS inhibitors if muscarinic receptors are not blocked.
L-NMMA N^{G}-mono methyl-L-arginine	Reversible and competitive inhibitor.	Prevents NO formation, but does not inhibit the NADPH-oxidase reaction, therefore may lead to secondary release of superoxide by nNOS. Competes with L-arginine for entry into cells.
L-NIO N^{ω}-imino ethyl-L-ornithine (and L-NIL, N^{ω}-imino ethyl lysine)	Slow binding, competitive inhibitor of iNOS, rapid inhibitor of nNOS and eNOS.	Prevents NO formation, but does not inhibit the NADPH-oxidase reaction, therefore may lead to secondary release of superoxide by nNOS. Selective (~30-fold) for iNOS versus eNOS and nNOS. Also competes with L-arginine for entry into cells.
Aminoguanidine	Slow binding, competitive inhibitor of iNOS, rapid inhibitor of nNOS and eNOS.	Has many other effects besides inhibiting NOS including inhibiting aldose reductase, diamine oxidase (contributes to inactivation in vivo of

Table 1 *(continued)*

Compound	Mechanism of action	Points to be aware of
		histamine), formation of advanced glycosylation end products. Selective (~30-fold) for iNOS versus eNOS and nNOS.
Isothioureas (ITUs; e.g., S-ethyl isothiourea SEITU)	Reversible, competitive and very potent inhibitors.	Although ITUs are nano-molar potency inhibitors, they have difficulty entering cells, leading to lower than predicted potency in intact cells, tissues, and in vivo. Have well-documented pressor effects in vivo.
S-Methyl-thiocitrulline and S-Ethyl-thiocitrulline	Reversible, competitive and slow binding inhibitors to nNOS and eNOS.	Selective (~15–150 fold) for nNOS versus eNOS.

7. Failure of tissues to relax (and hence respond later) when initially set up at 2 g may be owing to a lack of $CaCl_2$ in the Krebs or the organ baths not being maintained at 37°C. You cannot correct for these problems; you have to start the experiment again with fresh tissues.
8. Care must be taken with all dose-response curves to ensure that a plateau is reached between each dose, otherwise misleading data will be obtained:
   a. For phenylephrine, a plateau is reached about 3–5 min after each addition.
   b. Acetylcholine is broken down very quickly and speed is of the essence here if a cumulative-dose response curve is to be obtained. Increasing concentrations must be added within a min (or as soon as a plateau is achieved) after each addition of acetylcholine.
   c. For NOS inhibitors, a plateau is not reached for at least 15 min after dosing.
9. The preliminary experiments to determine the phenylephrine-dose response curve are essential. Over-contracting with phenylephrine (> than ED_{90}), before the acetylcholine-dose response curve, will result in a subsequent poor relaxation, even if the endothelium is intact. Whereas, under-contraction with phenylephrine will result in the contractile tone not being maintained before the addition of acetylcholine. The latter can be corrected by the addition of more phenylephrine.
10. Some NOS inhibitors may have no effect against eNOS and cause further relaxation in tissues treated with lipopolysaccharide (instead of a contraction). This could be owing to a direct effect of the inhibitor (other than against NOS), a substrate effect, or a toxicity problem of either the inhibitor or the solvent in which it is made up.

11. Failure to give an ED_{10} of phenylephrine before the dose-response curve to NOS inhibitors will result in very little contractile response to these agents being apparent.
12. Removal of the endothelium results in a lot of spontaneous activity in rat-aortic rings. The tissues must be allowed to equilibrate and reach a natural baseline initially, before setting to 1 g tension to start experiments. Failure to do so results in poor contraction to phenylephrine. The initial phenylephrine dose-response curve tends to help stabilize the spontaneous contractions, making it easier to carry out the rest of the experiment.
13. Failure of tissues to relax 6 h after lipopolysaccharide is common, hence the necessity to express the results as a percentage of this fall in tone. Consistent failure in obtaining a 30% fall in tone could be owing to de-sensitization to lipopolysaccharide. This, in turn, may be owing to large quantities of endotoxin being found both stuck to the glassware and present in in-house distilled water (it is not removed by washing, only by heating at 180°C overnight or 220°C for 3 h, procedures impossible for organ baths). In this case, try using tissue-culture water (which has been tested for endotoxin) to make up the Krebs. Different types of lipopolysaccharide may also be tested.

References

1. Palmer, R. M. J., Ferrige, A. G., and Moncada, S. (1987) Nitric oxide release accounts for the biological activity of endothelium-derived relaxing factor. *Nature,* **327,** 524–526.
2. Stuehr, D. J. and Griffith, O. W. (1992) Mammalian nitric oxide synthases. *Adv. Enzymol. Rel. Areas Mol. Biol.* **65,** 287–346.
3. Palmer, R. M. J., Rees, D. D., Ashton, D. S., and Moncada, S. (1988) L-Arginine is the physiological precursor for the formation of nitric oxide in endothelium-dependent relaxation. *Biochem. Biophys. Res. Commun.* **153,** 1251–1256.
4. Rees, D. D., Palmer, R. M. J., Schulz, R., Hodson, H. F., and Moncada, S. (1990). Characterisation of three inhibitors of endothelial nitric oxide synthase *in vitro* and *in vivo. Br. J. Pharm.* **101,** 747–752.
5. Radomski, M. W., Palmer, R. M. J., and Moncada, S. (1990) Glucocorticoids inhibit the expression of an inducible, but not the constitutive, nitric oxide synthase in vascular endothelial cells . *Proc. Natl. Acad. Sci. USA* **87,** 10,043–10,047.
6. Rees, D. D., Cellek. S., Palmer, R. M. J., and Moncada, S. (1990) Dexamethasone prevents the induction by endotoxin of a nitric oxide synthase and the associated effects on vascular tone: an insight into endotoxin shock. *Biochem. Biophys. Res. Commun.* **173,** 541–547.
7. Knowles, R. G., Salter, M., Brooks, S. L., and Moncada, S. (1990) Anti-inflammatory glucocorticoids inhibit the induction of nitric oxide synthase in the lung, liver and aorta of the rat. *Biochem. Biophys. Res. Commun.* **172,** 1042–1048.
8. Krebs, H. A. and Henseleit, K. (1932) Urea formation in the animal body. *Z. Physiol. Chem.* **210,** 33–66.

24

Measurement of Biopterin and the Use of Inhibitors of Biopterin Biosynthesis

Kazuyuki Hatakeyama

1. Introduction

Tetrahydrobiopterin (BH_4) and related pteridines have received much attention since BH_4 was found to be an essential cofactor for nitric oxide synthases (NOS) *(1–3)*. As shown in **Fig. 1**, BH_4 is produced from guanosine 5′-triphosphate (GTP) by three enzymes. GTP cyclohydrolase I converts GTP to dihydroneopterin triphosphate, which is then converted to 6-pyruvoyl-tetrahydropterin by 6-pyruvoyltetrahydropterin synthase. This second unstable intermediate compound is converted to BH_4 through a two-step reduction of its side-chain carbonyl groups by sepiapterin reductase.

In most tissues, the activity of the enzyme catalyzing the second step, 6-pyruvoyltetrahydropterin reductase, is about 10-fold higher than that of GTP cyclohydrolase I, and the activity of the enzyme catalyzing the third and fourth steps, sepiapterin reductase, is 10-fold higher than that of 6-pyruvoyl-tetrahydropterin reductase. Thus, the efficiency of the conversion of each intermediate is high. The intermediates and their degradation products are either not detected or are quite low compared to the amount of BH_4.

However, in some types of human cells, a degradation product of the intermediate dihydroneopterin triphosphate accumulates. The enzyme activity of GTP cyclohydrolase I is induced by cytokines, along with that of the inducible (i) type of NOS. In murine cells, the levels of the induced activity are still lower than those of constitutively-expressed 6-pyruvoyltetrahydropterin synthase. But in human cells, including macrophages and the THP-1 myelomonocytoma-derived cell line, the induced activity of GTP cyclohydrolase I exceeds the activity of 6-pyruvoyltetrahydropterin synthase and,

From: *Methods in Molecular Biology, Vol. 100. Nitric Oxide Protocols*
Edited by: Michael A. Titheradge © Humana Press Inc., Totowa, NJ

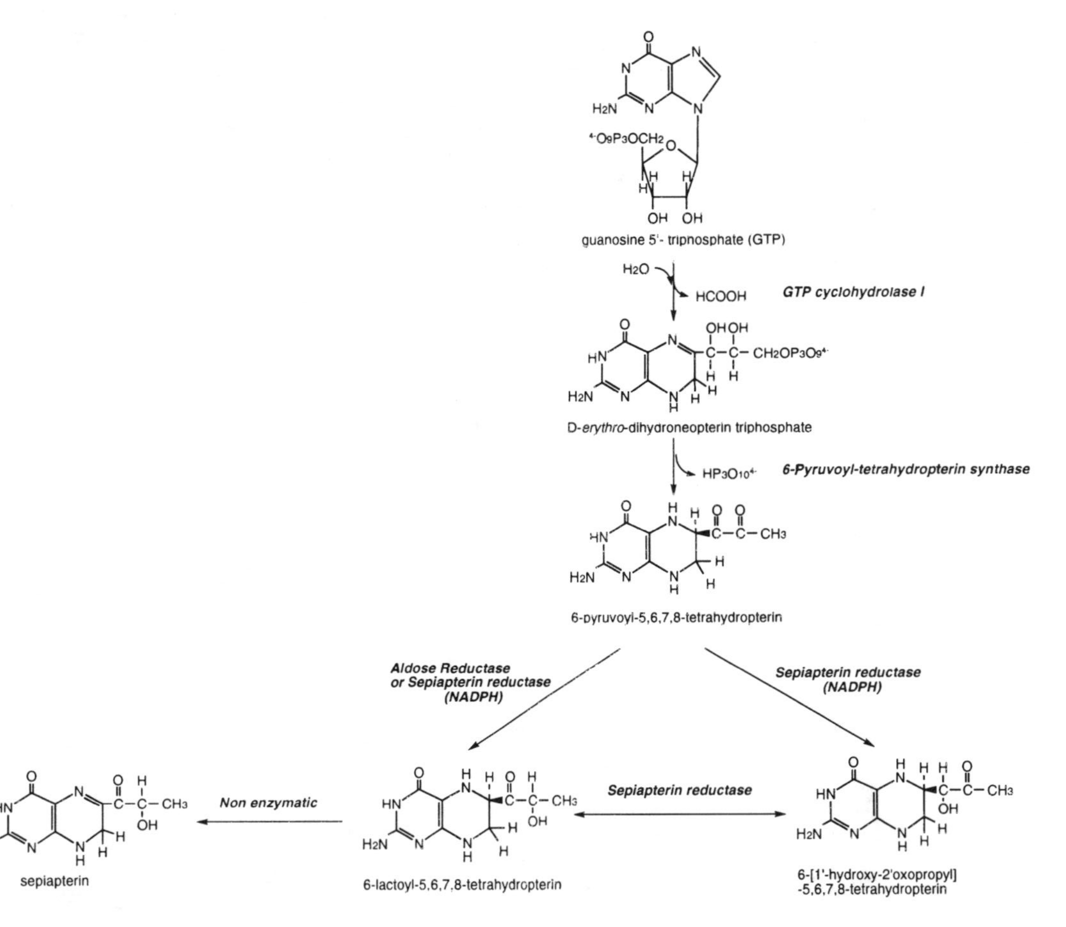
guanosine 5'- triphosphate (GTP)
H_2O
HCOOH
GTP cyclohydrolase I
D-*erythro*-dihydroneopterin triphosphate
$HP_3O_{10}^{4-}$
6-Pyruvoyl-tetrahydropterin synthase
6-pyruvoyl-5,6,7,8-tetrahydropterin
Aldose Reductase or Sepiapterin reductase (NADPH)
Sepiapterin reductase (NADPH)
Non enzymatic
Sepiapterin reductase
sepiapterin
6-lactoyl-5,6,7,8-tetrahydropterin
6-[1'-hydroxy-2'oxopropyl] -5,6,7,8-tetrahydropterin

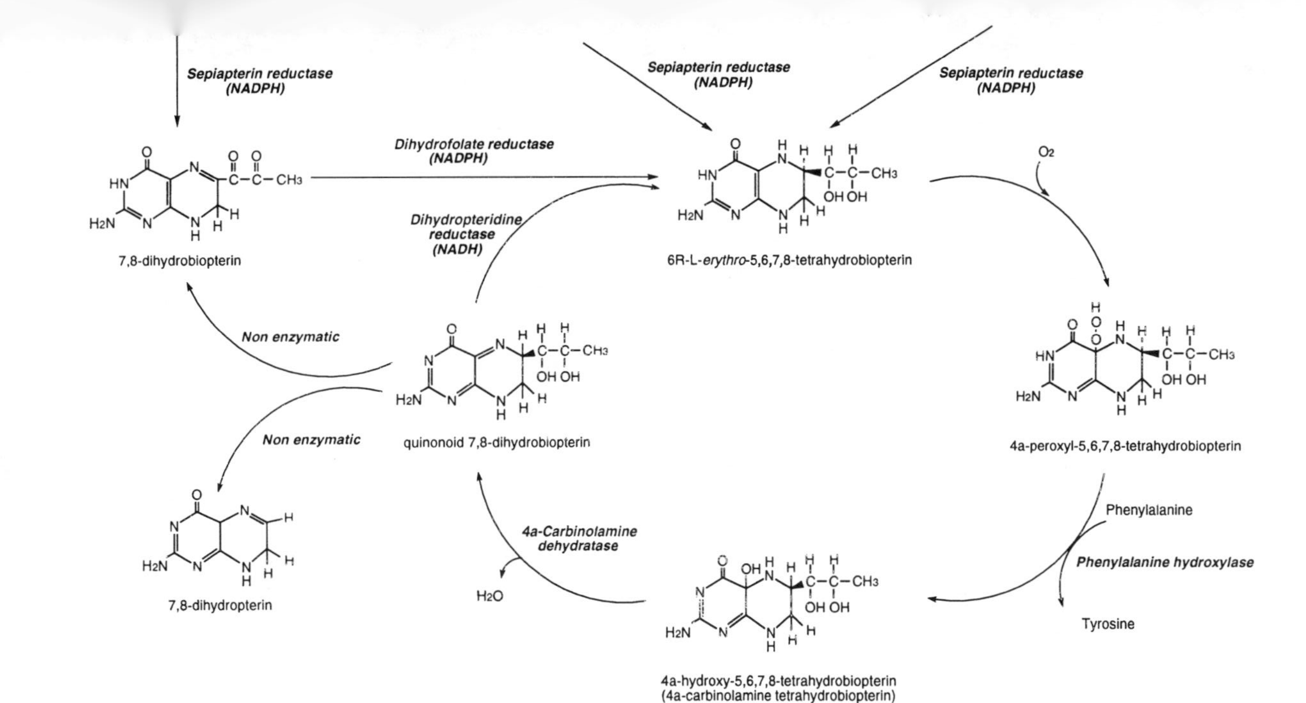

Fig. 1. Proposed scheme for BH_4 metabolism. The reactions of tyrosine hydroxylase, tryptophan hydroxylase, and NOS are not shown. The route for the conversion of 6-pyruvoyl-tetrahydropterin to BH_4 is controversial ***(5,45)***. Recent best-fit analysis of kinetic data for sepiapterin reductase (EC 1.1.1.153) ***(46)*** suggests that the following route is predominant: 6-pyruvoyl-tetrahydropterin is reduced at the 2'-oxo group to form 6-(1'-oxo-2'-hydroxypropyl)-tetrahydropterin (also called 6-lactoyl-tetrahydropterin) by sepiapterin reductase. Then, 6-(1'-oxo-2'-hydroxypropyl)-tetrahydropterin is isomerized by sepiapterin reductase to 6-(1'-hydroxy-2'-oxopropyl)-tetrahydropterin, which is reduced again at the 2'-oxo group by sepiapterin reductase. The conversion of 6-pyruvoyl-tetrahydropterin to 6-(1'-oxo-2'-hydroxypropyl)-tetrahydropterin is also catalyzed by aldose reductase (EC 1.1.1.21) ***(47,48)***. Regarding the salvage pathway from sepiapterin to BH_4, see the text.

hence, dihydroneopterin triphosphate accumulates much more than does BH_4 *(4)*. In these cells, dihydroneopterin triphosphate is converted to neopterin by being oxidized and dephosphorylated. Thus, the levels of neopterin, as well as of biopterin, reflect the activity of GTP cyclohydrolase I in such cells. The increased urinary excretion of neopterin has been frequently reported in association with a variety of stimulated immune activities *(5)*. No specific function of neopterin or dihydroneopterin triphosphate has been defined. Furthermore, these pteridines do not serve as cofactors for NOS *(6)*.

In addition to the *de novo* biosynthetic pathway of BH_4, there is a recycling pathway that regenerates BH_4 from 4a-hydroxytetrahydrobiopterin, an oxidized compound produced in hydroxylation reactions catalyzed by phenylalanine hydroxylase (**Fig. 1**). The role of this recycling pathway in NO production is not defined, but should be minor, because BH_4 is tightly bound to NOS and does not appear to be released with every enzymatic reaction as in the case of the aromatic amino-acid hydroxylases *(7)*.

There also is a salvage pathway that is important for experimental manipulations. BH_4 is generated by dihydrofolate reductase (EC 1.5.1.3) from a dihydro form of biopterin, 7,8-dihydrobiopterin (BH_2). The K_m of chicken-liver dihydrofolate reductase for BH_2 was first determined to be 2.7 μM, which was comparable to that for dihydrofolate, 0.6 μM *(8)*. However, the K_m and/or V_{max} for BH_2 and dihydrofolate of dihydrofolate reductases purified from various tissues were then reported to be: *rabbit brain*, 17 μM and 0.74 μmol/min/mg for BH_2 and 0.9 μM and 7.9 μmol/min/mg for dihydrofolate *(9)*; *beef liver*, 0.63 μmol/min/mg for BH_2 and 23.5 μmol/min/mg for dihydrofolate *(10)*; *rat liver*, 43 μM for BH_2 and 0.1 μM for dihydrofolate *(11)*. Although dihydrofolate reductase is thus highly specific for dihydrofolate, the reaction involving BH_2 was initially thought to be the final step of the *de novo* biosynthetic pathway for BH_4. This idea was later disproven by experiments using methotrexate, a specific inhibitor of dihydrofolate reductase *(12)*. Nevertheless, the possibility remains that dihydrofolate reductase has a role in the generation of BH_4, because BH_2, along with 7,8-dihydropterin, is nonenzymatically generated in vitro by auto-oxidation of BH_4 through an unstable intermediate quinonoid form of BH_2 *(13)*. It is not known how much BH_2 is generated in vivo. Thus, the physiological role of the dihydrofolate-reductase reaction is not yet fully defined. However, this enzyme, present in most cells, has become useful in experimentation because it provides a way to supplement BH_4 under the conditions where the *de novo* biosynthetic pathway is blocked. In addition to BH_2, sepiapterin, a pterin found in insects and not present in mammals, is commonly used for this purpose. In the cell, sepiapterin (6-1'-oxo-2'-hydroxy-propyl dihydropterin) is converted to BH_2 by sepiapterin reductase, which is then converted to BH_4 by dihydrofolate reductase (**Fig. 1**). Because dihydrofolate reductase does not act on sepiapterin at neutral pH (S. Katoh, personal communication), it is unlikely that sepiapterin is first reduced by dihydrofolate reductase to 6-lactoyl-tetrahydropterin (6-1'-oxo-2'-hydroxy-propyltetrahydropterin) and then converted to BH_4 by sepiapterin reductase.

In addition to sepiapterin, BH_4 itself can be used. However, BH_4 is not suitable for long incubations because it is unstable. For example, the half-life is less than 30 min in common culture media (M. Hoshiga and K. Hatakeyama, unpublished observation). BH_4 is nonenzymatically oxidized to BH_2 and/or dihydropterin ***(13)***. Even when BH_4 is supplied to cells, the salvage pathway is thought to play a role, because part of BH_4 might be regenerated in the cells by dihydrofolate reductase from the BH_2, which is generated from the oxidation of BH_4 in the media.

Two HPLC methods have been developed for measuring biopterin. One is to measure biopterin with a fluorescence detector after oxidizing samples with iodine in an acid or alkaline solution ***(14)***; the other is to directly measure BH_4 with an electrochemical detector ***(15–17)***. We describe the former method originally developed by Fukushima and Nixon ***(14)***, with some modifications by Bräutigam et al. ***(18)*** and ourselves, which enable a differential determination of BH_4 through the selective oxidation of the reduced forms of biopterin. This method is based on the fact that BH_4, BH_2, and quinonoid BH_2 are oxidized to biopterin with iodine in acid solution, whereas biopterin is stable under these conditions. Therefore, the amount of biopterin obtained by oxidation in acid represents total biopterin which is the sum of BH_4, BH_2, quinonoid BH_2, and biopterin. In contrast, in alkaline solution BH_4 is converted to pterin with iodine, whereas BH_2 is oxidized to biopterin. Biopterin is unaffected by the alkaline conditions. Hence, alkali oxidation measures the sum of BH_2 and biopterin. Therefore the difference between the values obtained in acid and alkali oxidations is a measure of BH_4 and quinonoid BH_2. In other words, the amount of base-labile biopterin serves as a measure of BH_4 and quinonoid BH_2.

2. Materials

1. Phosphate-buffered saline (PBS).
2. 0.2% Trypsin, 1 m*M* ethylenediaminetetraacetic acid (EDTA).
3. Extraction solution: 0.2 *M* Perchloric acid containing 10 m*M* dithiothreitol.
4. Acidic-iodine solution: 2.5% (w/v) Iodine and 10% (w/v) potassium iodide in 0.2*M* perchloric acid.
5. Alkaline iodine solution: 0.9% (w/v) Iodine and 1.8% (w/v) potassium iodide in 0.1 *M* sodium hydroxide.
6. Ascorbate solution: 2% (w/v) Ascorbate (should be freshly prepared).
7. 2 *M* Perchloric acid.
8. 1 *M* Ammonium hydroxide.
9. Glacial and 1 *M* acetic acid.

10. Dowex 50 [H^+] resin (200–400 mesh [X 12]) and Dowex 1 [OH^-] resin (200–400 mesh [X 12]).
11. The apparatus required for running high-pressure liquid chromatography (HPLC) includes an autoinjector, a pump, and a continuous flow fluorescence detector (*see* **Note 1**).
12. Column: A Partisil 10 ODS reversed-phase column (4.6 × 250 mm) (Whatman) (*see* **Note 2**).
13. Solvents: 50 m*M* Sodium acetate containing 0.1 m*M* EDTA and 5% (v/v) methanol. Reagent-grade deionized water should be used. For example, a Millipore laboratory system yields water of sufficient quality. The stock solution of 0.5 *M* sodium acetate and 1 m*M* EDTA is adjusted to pH 5.0 with sodium hydoxide and filtered through a 0.2 µm pore-sized membrane. The working solution is first degassed thoroughly and then methanol is added, to a final concentration of 5% (v/v).

3. Method

1. Remove the medium from cells growing in monolayers in 90-mm tissue culture dishes by gentle aspiration (*see* **Notes 3** and **4**). Wash the monolayers once with PBS (*see* **Note 5**).
2. Add ice-cold 0.2% trypsin, 1 m*M* EDTA solution and immediately aspirate it. Incubate the dishes for 1–3 min to allow the cells to detach from the dishes.
3. Add 5 mL of PBS and transfer the cells into a culture tube.
4. Centrifuge the cell suspension at 1000*g* for 2 min at 4°C and remove the last traces of PBS from the cell pellets.
5. After tapping the tube, add 1 mL of 0.2 *M* perchloric acid containing 10 m*M* dithiothreitol (DTT) and resuspend the cell pellets by pipetting gently (*see* **Note 6**). Incubate the tubes on ice for 20 min in the dark (*see* **Note 7**).
6. Transfer the solution to microcentrifuge tubes and collect the supernatant by centrifugation at 12,000*g* for 3 min at 4°C.
7. For acid oxidation, add 80 µL of acidic-iodine solution to 480 µL of the supernatant. For alkali oxidation, add 50 µL of 5 *M* sodium hydroxide and then 200 µL of alkaline-iodine solution to another 450 µL of the supernatant. Incubate both samples for 1 h at room temperature in the dark.
8. Add 100 µL of 2% ascorbic acid to the acidic mixture. Add 75 µL of 2% ascorbic acid to the alkaline mixture, which is acidified with 100 µL of 2 *M* perchloric acid beforehand (*see* **Note 8**).
9. Centrifuge each mixture at 12,000*g* for 5 min at room temperature (*see* **Note 9**) and recover the supernatant. Reserve the precipitates for protein determination.
10. Apply the supernatant fluid to a Dowex 50 [H^+] column (200–400 mesh [X 12]) (0.2 mL resin) packed in a Pasteur pipet equilibrated with water. Wash the column with 5 mL of water and elute biopterin with 2.8 mL of 1 *M* ammonium hydroxide (*see* **Notes 10** and **11**).
11. Apply the eluate to a Dowex 1 [OH^-] column (200–400 mesh [X 12]) (0.1 mL) packed in a 1-mL plastic pipet. Wash the column with 5 mL of water and elute pteridines with 2 mL of 1 *M* acetic acid (*see* **Note 12**).

12. Apply the acidified sample to a Whatman ODS column (4 × 250 mm) equilibrated with 50 m*M* sodium acetate (pH 5) containing 0.1 m*M* EDTA and 5% methanol at a flow rate of 0.8 mL/min. Monitor the column effluents fluorometrically using a continuous-flow fluorometer with an excitation-wave length of 350 and an emission-wavelength of 440 nm (*see* **Notes 13** and **14**).
13. Determine the biopterin peak area using an integrator and calculate the amount of biopterin eluted on the basis of the peak area of the standard biopterin (0.5 or 5 pmol) injected.
14. The amount of biopterin obtained from acid oxidation represents the sum of biopterin, BH_2, quinonoid BH_2, and BH_4. Estimate the amount of BH_4 and quinonoid BH_2 from the difference between the amount of biopterin recovered from acid oxidation and that from alkaline oxidation (*see* **Notes 15–17**).

4. Notes

1. Choice of fluorometer is an important consideration. A fluorometer with high sensitivity is needed to detect biopterin from tissues and cultured cells where the content of pterins is low.
2. Other ODS columns can be used, such as μBondapak 10 ODS reversed-phase column (Waters) ***(14)*** or Lichrosorb RP-18 reversed-phase column (7-μm particle size, Merck) ***(19)***.
3. When biopterin levels in the medium are measured, remove aliquots before aspirating the medium and treat as described in **step 7**.
4. Cells growing in suspension are treated from **step 5** after washing twice with PBS.
5. Serum contains trace amounts of oxidized biopterin (e.g., the amount in fetal calf serum (FCS) is 0.2 μ*M*) ***(18)*** and therefore it should be removed from the cells for accurate measurement of intracellular biopterin.
6. This step is to extract BH_4 and other biopterins from the cells. DTT is included to prevent the oxidation of BH_4 and other reduced biopterins ***(18)***.
7. Biopterin is sensitive to light.
8. At this stage, the yellow color of iodine should be visible, which indicates that iodine is not totally consumed and, hence, adequate iodine is available to oxidize reductants in the samples.
9. Biopterin in its fully oxidized form is stable at room temperature.
10. Several fluorescent compounds such as isoxanthopterin, lumazine, 6-hydroxylumazine, 7-hydroxylumazine, riboflavin, flavin mononucleotide (FMN), and flavin adenine (FAD) elute from the column with water. With ammonium hydoxide, neopterin and pterin are eluted in addition to biopterin ***(18)***.
11. This step may not be necessary for samples from cell culture. In rat hepatocytes, rat smooth-muscle cells and rat PC-12-pheochromocytoma cells, no contaminating peak elutes close to the biopterin peak.
12. This step is to concentrate the sample and accordingly may be omitted when biopterin content of the samples is high enough to be detected. In this case, samples obtained at **step 10** can be applied to the HPLC column at **step 12** after being acidified with a one fourteenth volume of glacial-acetic acid.

13. A typical chromatographic profile is shown in **Fig. 2**.
14. The column is regenerated by washing with H_2O and then absolute methanol.
15. BH_2 and biopterin are alkaline-stable and are recovered as biopterin in both acid and alkaline oxidations. BH_4 and quinonoid BH_2 are alkaline-unstable and are converted to pterin by losing its side chain *(14)*. Pterin is stable in alkali. Pterin recovered from acid oxidation is derived from endogenous pterin and the tetrahydro form of folate *(14)*.
16. Even in alkaline oxidation, 20–30% of BH_4 is recovered as biopterin, whereas 70–80% of BH_4 is recovered as pterin under the conditions.
17. Neopterin is recovered in a nearly 100% yield from neopterin and dihydroneopterin in acid oxidation. However, in alkaline oxidation only 65% of dihydroneopterin is recovered as neopterin *(14)*.
18. The use of inhibitors for BH_4 biosynthetic enzymes in cell-culture experiments: Specific-enzyme inhibitors of GTP cyclohydrolase I, sepiapterin reductase, and dihydrofolate reductase have been used in cell culture or in vivo to inhibit the generation of BH_4. No inhibitors for 6-pyruvoyltetrahydropterin synthase and 4a-carbinolamine dehydratase have been reported which are useful in cell culture or in vivo experiments.
    a. The inhibitor used for GTP cyclohydrolase I is 2,4-diamino-6-hydroxy-pyrimidine (DAHP) *(20)*. A median effective concentration of DAHP against rat GTP cyclohydrolase I (when assayed as described in Chapter 25, this volume) is 0.7 m*M* (T. Yoneyama and K. Hatakeyama, unpublished observation). In cell-culture experiments, DAHP has been used at concentrations of 1–20 m*M* for various cells, including murine fibroblasts *(19)*, murine-peritoneal macrophages *(21)*, rat vascular smooth-muscle cells *(22–24)*, rat glomerular-mesangial cells *(25)*, rat-glial cells *(26)*, canine arterial-endothelial cells *(27)*, porcine aortic-endothelial cells *(28)*, human umbilical-vein endothelial cells *(29)*, human thyrocytes *(30)*, murine-erythroleukemia cells *(31)*, and murine RAW-264 macrophages *(32)*. In DAHP-treated cells, BH_4 can be repleted by addition of sepiapterin, BH_2, or BH_4 (*see* Introduction).
    b. Several compounds have been reported to inhibit sepiapterin reductase, which thereby inhibits the BH_4 *de novo* biosynthetic pathway. Several catecholamines and indolamines have been found to inhibit the enzyme activity in vitro *(33)*. Among these, the most effective is *N*-acetylserotonin (NAS), the K_i value of which against purified rat sepiapterin reductase is 0.2 μ*M* *(33)*. NAS has been used for treatment of rat aortic smooth-muscle cells *(22)* and rat glomerular-mesangial cells *(25)* at concentrations of 0.1 and 1 mM, respectively, and results in 60 and 90% inhibition of nitrite production. However, NAS has little effect on BH_4 levels in murine RAW-264 macrophages even at a concentration of 5 m*M* *(32)*. Thus, the effect of NAS on BH_4 levels varies according to cell type.

       Furthermore, various inhibitors of general aldo-keto reductases have been examined for their effects on sepiapterin reductase activity by the same investigators. They found that ethacrynic acid, rutin, and dicumarol effectively

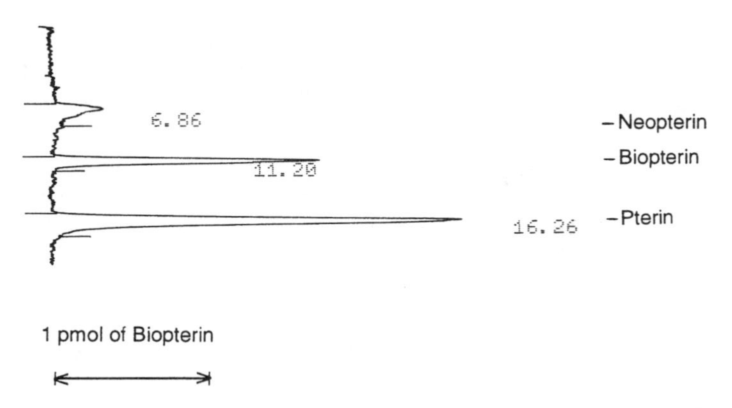

Fig. 2. A chromatographic profile of biopterin and pterin from rat hepatocytes. Isolation and culture of rat hepatocytes were performed as described *(49)*. Hepatocytes (5×10^6) were plated onto a 90-mm gelatin-coated Petrie dish and incubated for 24 h at 37°C. Cells were treated as described in the Methods Section. An injected sample was 100 μL from a 660-μL solution obtained at **step 8** for acid oxidation, which thus did not pass through Dowex columns.

inhibit the enzyme activity, the K_i values of which are estimated, using sepiapterin as substrate, to be 30, 60, and 0.6 μ*M*, respectively *(34)*. Ethacrynic acid and dicumarol have been used for treating murine-erythroleukemia cells at a concentration of 0.5 m*M* and found to decrease the cellular BH_4 to undetectable levels *(31)*. Phenprocoumon (4-hydroxy-3-(1-phenylpropyl)coumarin), a derivative of dicumarol, at a concentration of 50 μ*M* has also been shown to effectively block the cytokine-induced NO production in murine-peritoneal macrophages *(32)*.

Based on the fact that NAS and *N*-acetyldopamine inhibit the enzyme activity (**33**), Smith et al. found, by molecular modeling, additional sepiapterin-reductase inhibitors that show higher-competitive affinity *(35)*. *N*-chloroacetyldopamine, *N*-chloroacetylserotonin and *N*-methoxyacetylserotonin inhibit the activity of partially-purified bovine-sepiapterin reductase with K_i values of 14, 6, and 8 n*M*, respectively. In MOLT-4 T-cell leukemia and MCF-7 breast adenocarcinoma in culture, 0.1 m*M* *N*-methoxyacetylserotonin depletes BH_4 content by 97 and 50%, respectively, with no growth inhibition, whereas *N*-chloroacetylserotonin and *N*-chloroacetyldopamine similarly depletes BH_4 content, but simultaneously inhibits cell growth completely. Because BH_4 is depleted by both the former and the latter two compounds, the latter compounds are cytotoxic through some mechanism other than BH_4 depletion. Furthermore, BH_4 depletion and cell-growth inhibition of murine-erythroleukemia cells by *N*-chloroacetyldopamine, dicumarol, or ethacrynic acid are not reversed by addition of BH_4 or BH_4-related pterins, suggesting that these inhibitors have side effects on those cells apart from the inhibitory effects on sepiapterin reductase *(31)*. Alternatively, the activity of dihydrofolate reductase in those cells might not be enough to convert BH_2 to BH_4. Therefore it is recommended to perform control experiments to examine the effect of BH_4 supplementation on the observed effects of sepiapterin reductase inhibitors on the cells and, in some cases, to determine the extent of conversion of BH_2 to BH_4 in the cells.

c. Methotrexate is used to block the salvage pathway of BH_4 (**Fig. 1**) via inhibition of dihydrofolate reductase *(22,23,27)*. Because dihydrofolate reductase is not involved in *de novo* BH_4 biosynthesis, methotrexate alone has little effect on biopterin levels or on NO production *(22,23,25)*. But when sepiapterin or BH_2 is used as an external biopterin source under conditions where *de novo* BH_4 biosynthesis is blocked, methotrexate effectively blocks the rise of intracellular BH_4 levels. Methotrexate and aminopterin are highly specific inhibitors of dihydrofolate reductase, whereas these compounds are less potent inhibitors of dihydropteridine reductase. The K_i for methotrexate of dihydrofolate reductase is 3.4 p*M* (human recombinant) *(36)*, 7.3 p*M* (W1-L2 human-lymphoblastoid cells) *(37)*, and 5.3 p*M* (L1210 mouse-leukemia cells) *(37)*, whereas the K_i for methotrexate of dihydropteridine reductase is 38 μ*M* (sheep liver) *(38)*; the K_i for aminopterin of dihydrofolate reductase is 37 p*M* (*Escherichia coli*) *(39)*, 3.0 p*M* (W1-L2 human-lymphoblastoid cells) *(37)*, and 2.2 p*M* (L1210 mouse-leukemia cells) *(37)*, whereas the K_i for aminopterin of dihydropteridine reductase (sheep and bovine liver) is 20 μ*M* *(40,41)*. Thus, for example, methotrexate at a concentration of 10 μ*M* is inferred to inhibit only dihydrofolate reductase but not dihydropteridine reductase. Although a role for dihydropteridine reductase in NO production has yet to be determined, it should be recognized that at higher concentrations methotrexate and aminopterin inhibit dihydropteridine reductase activity as well.

d. Here are some considerations for the use of sepiapterin, BH_2, and BH_4 for replenishing BH_4 in cells. It has been observed that when cytokine-stimulated cells are treated with a combination of methotrexate and sepiapterin, NO production is lower than that in cells treated with methotrexate alone, in spite of the increase in intracellular biopterin *(19,22)*. This phenomenon may be explained by the possible inhibition of GTP cyclohydrolase I by accumulated BH_2 in the presence of sepiapterin and methotrexate; BH_2 accumulates because sepiapterin is first converted to BH_2, but BH_2 is not further converted to BH_4 by dihydrofolate reductase because the enzyme is inhibited by methotrexate. In the absence of GTP cyclohydrolase I feedback regulatory protein (GFRP), which mediates the feedback inhibition by BH_4 *(42)*, GTP cyclohydrolase I is not inhibited by BH_2 or sepiapterin even at a concentration of 100 μ*M*, the concentration used for the above experiments. In the presence of GFRP, the median-inhibitory concentrations of BH_2 and sepiapterin are 4 and 18 μ*M*, respectively, while that of BH_4 is 2 μ*M* *(42)*. Thus, in the cells that contain GFRP, BH_2 and sepiapterin might inhibit the *de novo*

BH_4 biosynthesis through GFRP. An alternative explanation of the phenomenon is our hypothesis that BH_2 directly inhibits NOS. Consistent with this hypothesis is the observation that BH_2 does not support NOS activity, although it binds to the enzyme with an affinity constant of 2.2 μM, a value 10-fold higher than that of BH_4 (0.23 μM) *(43)*. In fact, under conditions where BH_2 concentration exceeds BH_4 concentration by more than 10 μM, BH_2 inhibits the activity of NOS *(43)*. In murine-alveolar macrophages, it has been observed that sepiapterin or even BH_4 alone inhibits the NO production stimulated by interferon-γ *(44)*. This observation may be explained by the hypothesis that the activity of dihydrofolate reductase in these cells is too low to accumulate BH_2. The reason why BH_4 as well as sepiapterin inhibits NO production is that BH_4 is unstable in neutral solution at 37°C and is oxidized to BH_2 and/or dihydropterin in phosphate and Tris buffers *(13)*; the half life of BH_4 in Hank's balanced salt solution and in RPMI medium containing 10% fetal bovine serum (FBS) is 20–30 min (M. Hoshiga and K. Hatakeyama, unpublished observation). In fact, in the presence of methotrexate, BH_4 was observed to be ineffective in supporting NO production *(22)*. Thus the effects of sepiapterin, BH_2 and BH_4 on NO production vary, depending on the levels of sepiapterin reductase and dihydrofolate reductase in the cells. It is therefore recommended that the effect of varying concentrations of these pteridines be determined for the cells of interest.

Acknowledgment

I thank T. Yoneyama for discussion and preparation of figures and T. R. Billiar, S. M. Morris, Jr., and R. Shapiro, for helpful reviews of the manuscript.

References

1. Tayeh, M. A. and Marletta, M. A. (1989) Macrophage oxidation of L-arginine to nitric oxide, nitrite, and nitrate. *J. Biol. Chem.* **264,** 19,654–19,658.
2. Kwon, N. S., Nathan, C. F., and Stuehr, D. J. (1989) Reduced biopterin as a cofactor in the generation of nitrogen oxides by murine macrophages. *J. Biol. Chem.* **264,** 20,496–20,501.
3. Mayer, B., John, M., and Bohme, E. (1990) Purification of a Ca^{2+}/calmodulin-dependent nitric oxide synthase from porcine cerebellum. *FEBS Lett.* **277,** 215–219.
4. Werner, E. R., Werner-Felmayer, G., Fuchs, D., Hausen, A., Reibnegger, G., Yim, J. J., Pfleiderer, W., and Wachter, H. (1990) Tetrahydrobiopterin biosynthetic activities in human macrophages, fibroblasts, THP-1, and T 24 cells. *J. Biol. Chem.* **265,** 3189–3192.
5. Nichol, C. A., Smith, G. K., and Duch, D. S. (1985) Biosynthesis and metabolism of tetrahydrobiopterin and molybdopterin. *Ann. Rev. Biochem.* **54,** 729–764.
6. Kwon, N. S., Nathan, C. F., and Stuehr, D. J. (1989) Reduced biopterin as a cofactor in the generation of nitrogen oxides by murine macrophages. *J. Biol. Chem.* **264,** 20,496–20,501.

7. Giovanelli, J., Campos, K. L., and Kaufman, S. (1991) Tetrahydrobiopterin, a cofactor for rat cerebellar nitric oxide synthase, does not function as a reactant in the oxygenation of arginine. *Proc. Natl. Acad. Sci. USA* **88,** 7091–7095.
8. Kaufman, S. (1967) Metabolism of the phenylalanine hydroxylation cofactor. *J. Biol. Chem.* **242,** 3934–3943.
9. Abelson, H. T., Spector, R., Gorka, C., and Fosburg, M. (1978) Kinetics of tetrahydrobiopterin synthesis by rabbit brain diydrofolate reductase. *Biochem. J.* **171,** 267–268.
10. Reinhard, J. F., Chao, J. Y., Smith, G. K., Duch, D. S., and Nichol, C. A. (1984) A sensitive high-performance liquid chromatographic-fluorometric assay for dihydrofolate reductase in adult rat brain, using 7,8-dihydrobiopterin as substrate. *Anal. Biochem.* **140,** 548–552.
11. Stone, K. J. (1976) The role of tetrahydrofolate dehydrogenase in the hepatic supply of tetrahydropterin in rats. *Biochem. J.* **157,** 105–109.
12. Nichol, C. A., Lee, C. L., Edelstein, M. P., Chao, J. Y., and Duch, D. S. (1983) Biosynthesis of tetrahydrobiopterin by *de novo* and salvage pathways in adrenal medulla extracts, mammalian cell cultures, and rat brain *in vivo*. *Proc. Natl. Acad. Sci. USA* **80,** 1546–1550.
13. Davis, M. D., Kaufman, S., and Milstien, S. (1988) The auto-oxidation of tetrahydrobiopterin. *Eur. J. Biochem.* **173,** 345–351.
14. Fukushima, T. and Nixon, J. C. (1980) Analysis of reduced forms of biopterin in biological tissue and fluids. *Anal. Biochem.* **102,** 176–188.
15. Bräutigam, M., Dreesen, R., and Herken, H. (1982) Determination of reduced biopterins by high pressure liquid chromatography and subsequent electrochemical detection. *Hoppe-Seyler's Z. Physiol. Chem.* **363,** 341–343.
16. Lunte, C. E. and Kissinger, P. T. (1983) Determination of pterins in biological samples by liquid chromatography/electrochemistry with a dual-electrode detector. *Anal. Chem.* **55,** 1458–1462.
17. Smith, G. K. and Nichol, C. A. (1986) Synthesis, utilization, and structure of the tetrahydropterin intermediates in the bovine adrenal medullary de novo biosynthesis of tetrahydrobiopterin. *J. Biol. Chem.* **261,** 2725–2737.
18. Bräutigam, M., Dreesen, R., and Herken, H. (1984) Tetrahydrobiopterin and total biopterin content of neuroblastoma (N1E-115, N2A) and pheochromocytoma (PC-12) clones and the dependence of catecholamine synthesis on tetrahydrobiopterin concentration in PC 12 cells. *J. Neurochem.* **42,** 390–396.
19. Werner-Felmayer, G., Werner, E. R., Fuchs, D., Hausen, A., Reibnegger, G., and Wachter, H. (1990) Tetrahydrobiopterin-dependent formation of nitrite and nitrate in murine fibroblasts. *J. Exp. Med.* **172,** 1599–1607.
20. Gál, E. M., Nelson, J. M., and Sherman, A. D. (1978) Biopterin III. Purification and characterization of enzymes involved in the cerebral synthesis of 7,8-dihydrobiopterin. *Neurochem. Res.* **3,** 69–88.
21. Schoeden, G., Schneemann, M., Hofer, S., Guerrero, L., Blau, N., and Schaffner, A. (1993) Regulation of the L-arginine-dependent and tetrahydrobiopterin-dependent biosynthesis of nitric oxide in murine macrophages. *Eur. J. Biochem.* **213,** 833–839.

22. Gross, S. and Levi, R. (1992) Tetrahydrobiopterin synthesis; An absolute requirement for cytokine-induced nitric oxide generation by vascular smooth muscle. *J. Biol. Chem.* **267,** 25,722–25,729.
23. Nakayama, D. K., Geller, D. A., Di Silvio, M., Bloomgarden, G., Davies, P., Pitt, B. R., Hatakeyama, K., Kagamiyama, H., Simmons, R. L., and Billiar, T. R. (1994) Tetrahydrobiopterin synthesis and inducible nitric oxide production in pulmonary artery smooth muscle. *Am. J. Phys.* **266,** L455–L460.
24. Scott-Burden, T., Elizondo, E., Ge, T., Boulanger, C. M., and Vanhoutte, P. M. (1994) Simultaneous activation of adenyl cyclase and protein kinase C induces production of nitric oxide by vascular muscle cells. *J. Pharmacol. Exp. Ther.* **46,** 274–282.
25. Mühl, H. and Pfeilschifter, J. (1994) Tetrahydrobiopterin is a limiting factor of nitric oxide generation in interleukin 1b-stimulated rat glomerular mesangial cells. *Kidney Int.* **46,** 1302–1306.
26. Sakai, N., Kaufman, S., and Milstien, S. (1995) Parallel induction of nitric oxide and tetrahydrobiopterin synthesis by cytokines in rat glial cells. *J. Neurochem.* **65,** 895–902.
27. Cosentino, F. and Katušić, Z. S. (1995) Tetrahydrobiopterin and dysfunction of endothelial nitric oxide synthase in coronary arteries. *Circulation* **91,** 139–144.
28. Schmidt, K., Werner, E. R., Mayer, B., Wachter, H., and Kukovetz, W. R. (1992) Tetrahydrobiopterin-dependent formation of endothelium-derived relaxing factor (nitric oxide) in aortic endothelial cells. *Biochem. J.* **281,** 297–300.
29. Werner-Felmayer, G., Werner, E. R., Fuchs, D., Hausen, A., Reibnegger, G., Schmidt, K., Weiss, G., and Wachter, H. (1993) Pteridine biosynthesis in human endothelial cells. Impact on nitric oxide-mediated formation of cyclic GMP. *J. Biol. Chem.* **268,** 1842–1846.
30. Kasai, K., Hattori, Y., Nakanishi, N., Manaka, K., Banba, N., Motohashi, S., and Shimad, S. (1995) Regulation of inducible nitric oxide production by cytokines in human thyrocytes in culture. *Endocrinology* **136,** 4261–4270.
31. Zhuo, S., Fan, S., and Kaufman, S. (1996) Effects of depletion of intracellular tetrahydrobiopterin in murine erythroleukemia cells. *Exp. Cell Res.* **222,** 163–170.
32. Sakai, N., Kaufman, S., and Milstien, S. (1993) Tetrahydrobiopterin is required for cytokine-induced nitric oxide production in a murine macrophage cell line (RAW 264). *Mol. Pharmacol.* **43,** 6–10.
33. Katoh, S., Sueoka, T., and Yamada, S. (1982) Direct inhibition of brain sepiapterin reductase by a catecholamines and indoleamine. *Biochem. Biophys. Res. Commun.* **105,** 75–81.
34. Sueoka, T. and Katoh, S. (1985) Carbonyl reductase activity of sepiapterin reductase from rat erythrocytes. *Biochim. Biophys. Acta* **843,** 193–198.
35. Smith, G. K., Duch, D. S., Edelstein, M. P., and Bigham, E. C. (1992) New inhibitors of sepiapterin reductase; Lack of an effect of intracellular tetrahydrobiopterin depletion upon *in vitro* proliferation of two human cell lines. *J. Biol. Chem.* **267,** 5599–5607.

36. Appleman, J. R., Prendergast, N., Delcamp, T. J., Freisheim, J. H., and Blakley, R. L. (1988) Kinetics of the formation and isomerization of methorexate complexes of recombinant human dihydrofolate reductase. *J. Biol. Chem.* **263,** 10,304–10,313.
37. Jackson, R. C., Niethammer, D., and Hart, L. I. (1977) Reactivation of dihydrofolate reductase inhibited by methotrexate or aminopterin. *Arch. Bichem. Biophys.* **182,** 646–656.
38. Craine, J. E., Hall, E. S., and Kaufman, S. (1972) The isolation and characterization of dihydropteridine reductase from sheep liver. *J. Biol. Chem.* **247,** 6082–6091.
39. Baccanari, D. P. and Joyner, S. S. (1981) Dihydrofolate reductase hysteresis and its effect on inhibitor binding analyses. *Biochemistry* **20,** 1710–1716.
40. Cheema, S., Soldin, S. J., Knapp, A., Hofman, T., and Scrimgeour, K. G. (1973) Properties of purified quinonoid dihydropterin reductase. *Can. J. Biochem.* **51,** 1229–1239.
41. Aksnes, A. and Ljones, T. (1980) Steady state kinetics of dihydropteridine reductase: initial velocity and inhibition studies. *Arch. Bichem. Biophys.* **202,** 342–347.
42. Harada, T., Kagamiyama, H., and Hatakeyama, K. (1993) Feedback regulation mechanisms for the control of GTP cyclohydrolase I activity. *Science* **260,** 1507–1510.
43. Klatt, P., Schmid, M., Leopold, E., Schmidt, K., Werner, E. R., and Mayer, B. (1994) The pteridine binding site of brain nitric oxide synthase; tetrahydrobiopterin binding kinetics, specificity, and allosteric interaction with the substrate domain. *J. Biol. Chem.* **269,** 13,861–13,866.
44. Jorens, P. G., van Overveld, F. J., Bult, H., Vermeire, P. A., and Herman, A. G. (1992) Pterins inhibit nitric oxide synthase activity i rat alveolar macrophages. *Br. J. Pharmacol.* **107,** 1088–1091.
45. Kaufman, S. (1993) New tetrahydrobiopterin-dependent systems. *Ann. Rev. Nutr.* **13,** 261–286.
46. Sueoka, T., Hikita, H., and Katoh, S. (1990) Best-fit analysis of kinetic scheme for the stepwise reduction of the "diketo" group of 6-pyruvoyl tetrahydropterin by sepiapterin reductase, in *Enzymology and Molecular Biology of Carbonyl Metanolism 3,* (Weiner, H., Wermuth, B., and Crabb, D. W., eds.), Plenum, New York, pp. 229–239.
47. Milstien, S. and Kaufman, S. (1989) The biosynthesis of tetrahydrobiopterin in rat brain; purification and characterization of 6-pyruvoyl tetrahydropterin (2'-oxo)reductase. *J. Biol. Chem.* **264,** 8066–8073.
48. Steinerstauch, P., Wermuth, B., Leimbacher, W., and Curtius, H-Ch. (1989) Human liver 6-pyruvoyl tetrahydropterin reductase is biochemically and immunologically indistinguishable from aldose reductase. *Biochem. Biophys. Res. Commun.* **164,** 1130–1136.
49. Geller, D. A., Nussler, A. K., Di Silvio, M., Lowenstein, C. J., Shapiro, R. A., Wang, S. C., Simmons, R. L., and Billiar, T. R. (1993) Cytokines, endotoxin, and glucocorticoids regulate the expression of inducible nitric oxide synthase in hepatocytes. *Proc. Natl. Acad. Sci. USA* **90,** 522–526.

25

A Sensitive Assay for the Enzymatic Activity of GTP Cyclohydrolase I

Kazuyuki Hatakeyama and Toshie Yoneyama

1. Introduction

In addition to its well-known cofactor roles for aromatic amino-acid hydroxylases, an essential role of tetrahydrobiopterin (BH_4) for nitric oxide synthases (NOS) has been recently established *(1–3)*. All three isoforms of NOS contain high-affinity binding sites for BH_4. This cofactor is enzymatically synthesized in the cell from guanosine 5′-triphosphate (GTP) by the sequential actions of three distinct enzymes (*see* Chapter 24, **Fig. 1**, this volume). The first and rate-limiting enzyme in the biosynthetic pathway of BH_4 is GTP cyclohydrolase I (EC 3.5.4.16). Cellular expression of this enzyme is subject to varied and multi-level control that is cell-type specific. For example, the activity of GTP cyclohydrolase I in immune cells is normally either low or undetectable, but it is coinduced with the inducible NOS (NOS 2) in response to immunoreactive cytokines and bacterial lipopolysaccaride *(4–6)*. Hepatic GTP cyclohydrolase I is constitutively expressed and is regulated by BH_4 and phenylalanine through interaction with a specific GTP cyclohydrolase I feedback regulatory protein (GFRP) *(7)*. Although the role of this regulator protein in nonhepatic tissues is unclear, an interesting possibility is that it may be involved in regulating the production of NO by NOS. On the other hand, the two enzymes catalyzing the subsequent steps to BH_4 are constitutively expressed in cells and there is little evidence suggesting regulation of the activities of these enzymes.

GTP cyclohydrolase I is an allosteric enzyme composed of 10 identical subunits *(8)*. Its native-molecular weight is estimated to be 300 kDa by gel filtration and its subunit-molecular weight is 30 kDa by sodium dodecyl

From: *Methods in Molecular Biology, Vol. 100. Nitric Oxide Protocols*
Edited by: Michael A. Titheradge © Humana Press Inc., Totowa, NJ

sulfate-polyacrylamide gel electrophoresis (SDS-PAGE). The enzyme shows positive cooperativity to the substrate GTP. The $K_{0.5}$ value of the enzyme is 30 μM at a KCl concentration of 0.1 M, and the V_{max} value is 2700 nmol/h/mg protein. The $K_{0.5}$ value changes depending on the concentration of KCl, but the V_{max} value does not. GTP cyclohydrolase I hydrolyzes GTP to produce formic acid and dihydroneopterin triphosphate. Theoretically, dihydroneopterin triphosphate, formic acid, or related derivatives can be measured to monitor enzyme activity. The assay methods so far reported for measuring the activity of GTP cyclohydrolase I are classified into two groups according to the products measured. We discuss methods that are relevant for assay using crude extracts, which contain small amounts of GTP cyclohydrolase I.

Dihydroneopterin triphosphate can be measured directly, or with higher sensitivity as either its oxidized derivative, neopterin triphosphate, or its oxidized and dephosphorylated derivative, neopterin, which are strong fluorescence emitters. These compounds can be measured by using either a spectrofluorometer ***(9,10)*** or a high-pressure liquid chromatography (HPLC) system equipped with a continuous-flow fluorometer ***(11,12)***. The former method needs relatively large amounts of sample and, moreover, provides only fluorescent information on total products because there is no chromatographic separation involved. Accordingly, upon performing this method, one should include appropriate control experiments to ascertain that the fluorescence obtained really reflects the conversion of GTP to dihydroneopterin triphosphate. In contrast, the method using HPLC not only needs less sample, but also assures identification of the products through its separation step. HPLC methods suitable for the assays using crude extracts are those that detect neopterin ***(11)*** or neopterin triphosphate ***(12)*** because of their strong fluorescent properties and selectivity. Those two compounds show similar fluorescent intensity. Neopterin is obtained by oxidation and dephosphorylation of dihydroneopterin triphosphate. Neopterin triphosphate is obtained by only one step oxidation of dihydroneopterin triphosphate, but there is a disadvantage that the enzyme activity may be underestimated because dihydroneopterin triphosphate can be dephosphorylated during the enzyme reaction, especially in crude extracts. The resulting neopterin, mono- or di-phosphate forms of neopterin are not recovered as neopterin triphosphate.

Generally speaking, contaminant peaks that elute close to the peak to be measured often make the quantitation difficult. It may be necessary to consider another chromatographic system for neopterin triphosphate in order to obtain better separation from potential contaminating peaks if the method described below has problems in separation. Neopterin and neopterin triphosphate are different in charge, and hence the chromatographic systems separating neopterin or neopterin triphosphate are quite different and the accompanying contaminating peaks also would elute differently.

Formate release can be measured by using [8-^{14}C]GTP, which is labeled at the carbon atom released as formate during the enzymatic reaction. This was shown by using a method developed to estimate the amount of formate after converting it to carbon dioxide ***(13,14)***. Two more convenient methods also were developed to directly measure formate after isolating it from GTP with different procedures, based on the facts that charcoal does not adsorb formate but does adsorb GTP ***(15)*** and that ethyl acetate extracts formate but not GTP from the reaction mixture ***(16)***. These procedures are simple but less specific in comparison with HPLC methods. Better confirmation can be obtained by including several negative-control experiments, such as those in which the reaction is performed with cell lysates from control cells or with inhibitors of GTP cyclohydrolase I, but uncertainties in quantifying-enzymatic activity still may remain.

We describe here a neopterin method ***(8)*** slightly modified from the method originally reported by Viveros et al. ***(11)***, which was developed using the chromatographic system described by Fukushima and Nixon ***(17)***.

2. Materials

1. Phosphate-buffered saline (PBS).
2. 0.2% Trypsin, 1 m*M* ethylenediaminetetraacetic acid (EDTA).
3. 10X Reaction mixture: 500 m*M* Tris-HCl buffer, pH 7.5, containing 1 *M* KCl, 10 m*M* EDTA (*see* **Notes 1–3**).
4. 1X Homogenization buffer: 1X Reaction mixture containing 1 m*M* dithiothreitol (DTT) and several protease inhibitors (0.2 m*M* phenylmethylsulfonyl fluoride (PMSF), 1 μ*M* leupeptin, 1 μ*M* pepstatin) (*see* **Note 4**).
5. PMSF stock solution: Dissolve PMSF in isopropanol at a concentration of 34 mg/mL (0.2 *M*) and store either at –20°C or at room temperature for 1 mo.
6. Leupeptin stock solution: Dissolve leupeptin in water at a concentration of 1 mg/mL (2 m*M*) and store at –20°C.
7. Pepstatin-stock solution: Dissolve pepstatin in absolute methanol at a concentration of 1.4 mg/mL (2 m*M*) and store at –20°C.
8. Spin column: A NICK spin column (Pharmacia).
9. 100 m*M* GTP.
10. Oxidizing solution: 0.5% Iodine and 1.0% potassium iodide in 1 *M* HCl.
11. Ascorbate solution: 2% (w/v) Ascorbate (should be freshly prepared).
12. Alkaline phosphatase: 0.1 *M* Tris-HCl (pH 8.0 at 0.5 *M*) containing 30 unit/mL alkaline phosphatase (bovine intestinal mucosa, type VII-S, Sigma), 1.0 m*M* $MgCl_2$, and 0.1 m*M* $ZnCl_2$ (*see* **Note 5**). The enzyme solution can be stored at 4°C for several mo.
13. Stopping and protein-precipitating solution: 1 *M* Acetic acid.

14. The apparatus required for running HPLC include an autoinjector, a pump, and a continuous flow fluorescence detector (*see* **Note 6**).
15. Column: A Partisil 10 ODS reversed-phase column (4.6 × 250 mm) (Whatman) or a Cosmosil 10 C_{18} reversed-phase column (4.6 × 250 mm) (Nacalai Tesque, Japan) (*see* **Note 7**).
16. Solvents: 50 m*M* Sodium acetate containing 0.1 m*M* EDTA. Reagent-grade deionized water should be used. For example, a Millipore laboratory system yields water of sufficient quality. The stock solution of 0.5 *M* sodium acetate and 1 m*M* EDTA is adjusted to pH 5.0 with sodium hydroxide and filtered through a 0.2 µm pore-sized membrane. The working solution is first degassed thoroughly and then adjusted to contain 1–5% methanol. The percentage of methanol should be determined for each column lot.

3. Methods

3.1. Preparation of Cell Extracts

1. Detach cultured cells from culture dishes by trypsinization (0.2% trypsin/1 m*M* EDTA) (*see* **Notes 8** and **9**).
2. Suspend the cells in PBS for washing.
3. Recover cell pellets by centrifugation.
4. Sonicate the cell pellets in 600 µL of the homogenization buffer (*see* **Notes 10** and **11**).
5. Centrifuge the homogenate at 12,000*g* for 5 min at 4°C and recover the supernatant (*see* **Note 12**).
6. Centrifuge the supernatant through a spin column (Sephadex G-50) equilibrated with 1X reaction mixture (*see* **Note 13**).

3.2. Assay of Enzyme Activity

1. At 0°C, make the reaction mixture. To a microcentrifuge tube, add in the following order (*see* **Note 14**):
   5–490 µL Crude extracts in 1X reaction mixture
   490–5 µL 1X Reaction mixture (to give a final volume of 500 µL)
   5 µL 100 m*M* GTP
2. Incubate at 37°C for 1 h in the dark.
3. Put the reaction tube on ice.
4. Add 50 µL of the oxidizing solution and incubate in the dark at room temperature for 1 h.
5. Add 50 µL of the ascorbate solution and mix well (*see* **Note 15**).
6. Add 50 µL of 2 *M* sodium hydoxide and then 100 µL of the alkaline-phosphatase solution. Incubate at 37°C for 1 h.
7. Put the reaction tube on ice and add 100 µL of 1 *M* acetic acid.
8. Remove precipitates by centrifugation for 5 min in a microcentrifuge at room temperature.
9. Take a sample of the supernatant.

10. Inject the sample into the column manually or by using automatic sample injector (*see* **Note 16**). The flow rate is 0.8 mL/min.
11. Monitor the column effluents fluorometrically using a continuous-flow fluorometer with an excitation-wavelength set at 350 nm and an emission-wave length at 440 nm (*see* **Notes 17** and **18**).
12. Determine the neopterin-peak area using an integrator and calculate the amount of neopterin eluted on the basis of the peak area of the standard neopterin (0.5 or 5 pmol) injected (*see* **Note 19**).

4. Notes

1. EDTA is included to inhibit the enzyme activity of 6-pyruvoyltetrahydropterin synthase, which further converts dihydroneopterin triphosphate to 6-pyruvoyltetrahydropterin *(1,18,19)*. EDTA also inhibits the enzyme activities of some phosphatases, which are able to degrade GTP to guanosine 5′-diphosphate (GDP), guanosine 5′-monophosphate (GMP), and guanosine. These guanine compounds are not substrates for GTP cyclohydrolase I.
2. Phosphate buffer can be included at less than 20 m*M*. Higher concentrations inhibit the enzyme activity of alkaline phosphatase used in a later step.
3. GTP cyclohydrolase I is active between pH 7.0 and 10.0 *(8)*.
4. When a stock solution of PMSF is added to the working solution, it should be done while stirring the solution to prevent PMSF becoming crystallized in the surface of the solution. Because PMSF is unstable in aqueous solutions, a solution containing the inhibitor should be used immediately.
5. $ZnCl_2$ is needed for alkaline-phosphatase activity. Zn^{2+} is contained in the active center of the enzyme. $MgCl_2$ enhances the enzyme activity.
6. An adequate fluorometer is important for successful measurement. A fluorometer with the highest sensitivity is needed to detect the enzyme activities of tissues and cultured cells. At least one picomole of neopterin should be detectable.
7. Other octadecyl silane (ODS) columns can be used such as μBondapak 10 ODS reversed-phase column (Waters) or Lichrosorb 7 ODS reversed-phase column (Merck).
8. Tissue samples should be immediately homogenized or frozen for future use to prevent protein degradation.
9. Scraping cells from a culture dish with a rubber policeman causes significant loss of enzyme activities from the samples, probably owing to the rupture of the cells during the procedure (T. Yoneyama and K. Hatakeyama, unpublished observation for rat hepatocytes).
10. The volume of the homogenization buffer, which is usually around 2–10 volumes of a wet weight of the sample, is determined by the amount of GTP cyclohydrolase I activity in the sample.
11. Sonication makes the solution nonviscous in addition to lysing the cells.
12. The crude extracts can be stored at –70°C. After thawing the crude extracts, it is centrifuged to remove insoluble materials.

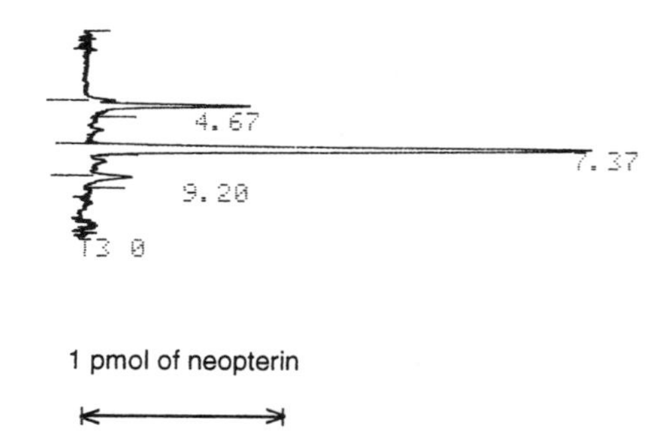

Fig. 1. A chromatographic profile of neopterin. Isolation and cell culture of rat hepatocytes were performed as described (20). Hepatocytes (5×10^6) were plated onto a 90-mm gelatin-coated Petrie dish and incubated for 2 h at 37°C. Crude extracts were obtained as described in the Appendix. Two-hundred microliters of the crude extracts was passed through a NICK spin column. Fifty microliters of the eluate was used for GTP cyclohydrolase I assay. After performing the enzyme assay as described in the Methods Section, 30 µL from 850 µL of the supernatant obtained at the **step 9** was injected into a Partisil 10 ODS reversed-phase column. Peak at 7.37 min, neopterin.

13. With this step, some unidentified compounds that, at least in rat liver, inhibit the enzyme activity can be removed. This step also allows the removal of endogenous neopterin, which can cause an overestimation of the enzyme activity. Furthermore, this step allows the removal of BH_4 and phenylalanine, which affect the GTP cyclohydrolase I activity through GFRP *(7)*.
14. GTP cyclohydrolase I shows positive cooperativity against the substrate GTP *(8)*. The GTP concentration that produces the half maximal velocity is 30 µ*M* at a KCl concentration of 0.1 *M*. As the concentration of KCl becomes higher, the K_m also become higher. However, an increase in KCl concentration produces little change in the V_{max} value of the enzyme. At a KCl concentration of 0.1 *M*, nearly maximal enzyme activity is obtained when the GTP concentration ≥ 0.2 m*M*. Under the current conditions described using a GTP concentration of 1 m*M*, we observed linearity of the enzyme reaction at least until 50% of GTP is consumed.
15. The appearance of the solution should change from yellow to transparent, if there is sufficient ascorbate to reduce the iodine.
16. The volume of the sample can be varied up to 500 µL, depending on the enzyme activities contained in the sample.
17. The chromatographic run can be done in a room where temperature is constant, but in order to obtain reproducible results, use of an incubator to keep the column temperature constant is recommended.
18. A typical chromatographic profile is shown in **Fig. 1**.
19. The column is regenerated by washing with H_2O and then absolute methanol.

Acknowledgment

We thank T. R. Billiar, S. M. Morris, Jr., and R. Shapiro, for helpful reviews of the manuscript.

References

1. Nichol, C. A., Smith, G. K., and Duch, D. S. (1985) Biosynthesis and metabolism of tetrahydrobiopterin and molybdopterin. *Ann. Rev. Biochem.* **54,** 729–764.
2. Marletta, M. (1993) Nitric oxide synthase structure and metabolism. *J. Biol. Chem.* **268,** 12,231–12,234.
3. Kaufman, S. (1993) New tetrahydrobiopterin-dependent systems. *Ann. Rev. Nutr.* **13,** 261–286.
4. Morris, S. M., Jr. and Billiar, T. R. (1994) New insights into the regulation of inducible nitric oxide synthesis. *Am. J. Physiol.* **266,** E829–E839.
5. Werner, E. R., Werner-Felmayer, G., and Wachter, H. (1993) Tetrahydrobiopterin and cytokines. *Proc. Soc. Exp. Biol. Med.* **203,** 1–12.
6. Nathan, C. and Xie, Q-w. (1994) Regulation of biosynthesis of nitric oxide. *J. Biol. Chem.* **268,** 13,725–13,728.
7. Harada, T., Kagamiyama, H., and Hatakeyama, K. (1993) Feedback regulation mechanisms for the control of GTP cyclohydrolase I activity. *Science* **260,** 1507–1510.
8. Hatakeyama, K., Harada, T., Suzuki, S., Watanabe, K., and Kagamiyama, H. (1989) Purification and characterization of rat liver GTP cyclohydrolase I; cooperative binding of GTP to the enzyme. *J. Biol. Chem.* **264,** 21,660–21,664.
9. Yim, J. J. and Brown, G. M. (1976) Characterization of guanosine triphosphate cyclohydrolase I purified from *Escherichia coli*. *J. Biol. Chem.* **251,** 5087–5094.
10. Fukushima, K., Richter, W. E., Jr., and Shiota, T. (1977) Partial purification of 6-(D-erythro-1',2',3'-trihydroxypropyl)-7,8-dihydropterin triphosphate synthetase from chicken liver. *J. Biol. Chem.* **252,** 5750–5755.
11. Viveros, O. H., Lee, C.-L., Abou-Donia, M. M., Nixon, J. C., and Nichol, C. A. (1981) Biopterin cofactor biosynthesis: Independent regulation of GTP cyclohydrolase I in adrenal medulla and cortex. *Science* **213,** 349–350.
12. Blau, N. and Niederwieser, A. (1983) Guanosine triphosphate cyclohydrolase I assay in human and rat liver using high-performance liquid chromatography of neopterin phosphates and guanine nucleotides. *Anal. Biochem.* **128,** 446–452.
13. Levenberg, B. and Kaczmarek, D. K. (1966) Enzymic release of carbon atom 8 from guanosine triphosphate, an early reaction in the conversion of purines to pteridines. *Biochim. Biophys. Acta.* **117,** 272–275.
14. Burg, A. W. and Brown, G. M. (1966) The biosynthesis of folic acid; VI. Enzymatic conversion of carbon atom 8 of guanosine triphosphate to formic acid. *Biochim. Biophys. Acta.* **117,** 275–278.
15. Burg, A. W. and Brown, G. M. (1968) The biosynthesis of folic acid; VIII. Purification and properties of the enzyme catalyzes the production of formate from carbon atom 8 of guanosine triphosphate. *J. Biol. Chem.* **243,** 2349–2358.

16. Guroff, G. and Strenkoki, C. A. (1966) Biosynthesis of pteridines and of phenylalanine hydroxylase cofactor in cell-free extracts of Pseudomonas species (ATCC 11299a). *J. Biol. Chem.* **241,** 2220–2227.
17. Fukushima, T. and Nixon, J. C. (1980) Analysis of reduced forms of biopterin in biological tissue and fluids. *Anal. Biochem.* **102,** 176–188.
18. Takikawa, S., Curtius, H.-Ch., Redweik, U., Leimbacher, W., and Ghisla, S. (1986) Biosynthesis of tetrahydrobiopterin; Purification and characterization of 6-pyruvoyl-tetrahydropterin synthase from human liver. *Eur. J. Biochem.* **161,** 295–302.
19. Inoue, Y., Kawasaki, Y., Harada, T., Hatakeyama, K., and Kagamiyama, H. (1991) Purification and cDNA cloning of rat 6-pyruvoyl-tetrahydropterin synthase. *J. Biol. Chem.* **266,** 20,791–20,796.

26

Simultaneous Measurement of Mitochondrial Function and NO

Ignacio Lizasoain and María A. Moro

1. Introduction

There is increasing evidence to support the concept that defects in mitochondrial-energy metabolism may underlie the pathology of several disorders such as neurodegenerative diseases. The exact mechanisms involved in the pathogenesis of these disorders are not known, but it is thought that there may be excessive production of oxygen-derived free radicals *(1–2)*. Nitric oxide (NO) is a free radical that plays important roles in diverse physiological processes, such as vasodilatation, inhibition of platelet aggregation, and neurotransmission. However, it can also be toxic under certain conditions and it has been postulated that the mechanisms of cell damage by NO include an inhibition of a number of cellular processes, one of which is mitochondrial respiration *(3–5)*.

Because of these points, the simultaneous study of the mitochondrial-respiratory chain and the NO concentration is important to determine the mechanisms involved in many pathological diseases. The method we present here allows the study of the effect of NO on mitochondrial function, concomitantly with the determination of the actual concentration of NO that exerts such effect, regardless of the system used to release this molecule. Either isolated mitochondria, submitochondrial particles, or intact cells can be used as biological systems for the simultaneous measurement of oxygen consumption and NO concentrations.

From: *Methods in Molecular Biology, Vol. 100. Nitric Oxide Protocols*
Edited by: Michael A. Titheradge © Humana Press Inc., Totowa, NJ

2. Materials

2.1. Preparation of Rat-Brain Mitochondria

1. Isolation buffer: 0.15 *M* KCl, 20 m*M* potassium phosphate, pH 7.6, at 4°C. Store at 4°C for 1 mo.
2. Ficoll: 10% solution w/v in isolation buffer (Sigma).
3. Dounce homogenizer (0.1 mm clearance).
4. Teflon/glass homogenizer (0.25 mm clearance).
5. Beckman Ultracentrifuge and SW50 rotor or similar.
6. Sorvall Centrifuge and SS-34 rotor or similar.

2.2. Preparation of the Oxygen Electrode

1. Sodium dithionite ($Na_2S_2O_4$) (Sigma).
2. Distilled water.
3. Clark-type oxygen micro-electrode (YSI Inc., Yellow Springs, OH or similar), consisting of a thermostated chamber with an inbuilt oxygen electrode.
4. Incubation buffer: 50 m*M* potassium phosphate, 100 μ*M* ethylene glycol-bis(β-aminoethyl ether)-N,N,N′,N′-tetraacetic acid (EGTA), adjusted to pH 7.2 with KOH. Store at 4°C for 1 mo.
5. NADH-disodium salt (50 μ*M* or 1 m*M* final concn.). Stock solution of 5 m*M* or 100 m*M*. Store at –20°C.
6. Succinic acid-disodium salt (5 m*M* final concn.). Stock solution of 500 m*M*. Store at –20°C.
7. L-ascorbic (5 m*M* final concn.) plus N,N,N′,N′-tetramethyl-*p*-phenylenediamine (TMPD; 0.5 m*M* final concn.). Stock solutions of 500 m*M* and 50 m*M*, respectively. Store at –20°C.
8. Glutamate/malate (2.5 m*M* each final concn.). Stock solutions of 250 m*M*. Store at –20°C.

2.3. Preparation of the NO Electrode

1. Pure NO gas in deoxygenated water (*see* **Note 1**).
2. 3-morpholinosydnonimine N-ethylcarbamide (SIN-1; 500 μM; Alexis Corporation; protect from light) plus superoxide dismutase (SOD; 25-800 units/mL, Sigma) are used as the NO-releasing system (*see* **Note 2**). Dissolve stocks in water. Make fresh as required.
3. Clark-type NO-sensitive electrode (Diamond General Development Corporation, MI, or similar) (*see* **Note 3**).
4. Place the tip of the NO electrode into the oxygen-electrode chamber through a gas-tight stopper in the top of the chamber. Alternatively, the chamber might be left open when the NO concentrations are sufficiently high, as those produced when a NO donor is used.

3. Methods

3.1. Preparation of Rat-Brain Mitochondria and Submitochondrial Particles

The experimental approach for isolating mitochondria is basically the same irrespective of the source of tissue. However, the yield of mitochondria and the ease with which pure preparations may be obtained depends very much on the type of tissue and, of course, the amount of tissue available. In this Chapter, we describe an experimental protocol for preparing rat brain mitochondria by a modification of the method of Partridge et al. *(6)*. (For isolation of mitochondria from different types of cells, *see* **ref.** *7*.) If intact cells are to be used, *see* **ref.** *8*.

1. Kill five adult female Wistar rats by decapitation. Remove all brain tissue rostral to the cerebellum and place it into ice-cold isolation buffer. With small scissors, mince the tissue into small cubes and wash in ice-cold isolation buffer to remove as much blood as possible; the final washing medium should be free of blood.
2. Transfer the minced brain tissue to a pre-cooled Dounce homogenizer, add 50 mL of cold isolation buffer, and homogenize the tissue using 10–12 up and down strokes. Centrifuge the homogenate for 3 min at 1300*g* at 4°C (Sorvall centrifuge with SS-34 rotor or similar). Set aside the supernatant at 4°C and resuspend the pellet in 30 mL of cold isolation buffer using 4–6 passes in the Dounce homogenizer. Centrifuge the resulting suspension for 3 min at 1300*g* at 4°C. Combine the resulting supernatant with the previously set-aside supernatant.
3. Centrifuge the supernatant for 10 min at 17,000*g* at 4°C (Sorvall centrifuge with SS-34 rotor or similar). Resuspend the pellet in 20 mL of cold isolation buffer using four up and down strokes in a tight-fitting Teflon/glass homogenizer. It is important not to process too much tissue at one time.
4. Layer the resulting suspension over 14 mL of 10% Ficoll solution (w/v in isolation buffer). Centrifuge it for 45 min at 100,000*g* at 4°C (Beckman Ultracentrifuge or similar).
5. Discard the upper phases and gently resuspend the brownish mitochondrial pellet in 10 mL of isolation buffer and centrifuge at 9800*g* for 10 min at 4°C. The mitochondria form a soft brown pellet. If a dark-brown button is found, this should be discarded as it consists of pelleted red-blood cells.
6. Resuspend the resulting pellet in a small volume of incubation buffer. Store the suspension in aliquots at –70°C until studied or used in the preparation of submitochondrial particles (SMP).
7. Subject a freshly prepared suspension of brain mitochondria to three cycles of freeze–thaw. Centrifuge it for 10 min at 20,000*g* at 4°C to obtain a high yield of SMP. Pool and store SMP from multiple rats in aliquots at –70°C until studied (*see* **Note 4**).

3.2. Preparation of the Oxygen Electrode

3.2.1. Calibration of the Oxygen Electrode

1. Pipet 2 mL of distilled water into the electrode chamber and add a few crystals of sodium dithionite ($Na_2S_2O_4$). The oxygen concentration rapidly falls to zero. The output of the electrode to the chart recorder also falls and the position of the pen on the chart recorder is set to zero.
2. Wash out the electrode chamber several times with distilled water to remove sodium dithionite. Pipet 2 mL of distilled water, previously equilibrated with air at 37°C into the electrode chamber, itself thermostatted at 37°C.
3. Set the chart recorder to 90% of its deflection by suitable choice of sensitivity. This level now corresponds to 260 μ*M* dissolved oxygen. This provides two known oxygen concentrations, i.e., 0 and 260 μ*M*, from which other concentrations can be determined.

3.2.2. Measurements of the Respiration Rate

1. Pipet 2 mL of incubation buffer into the electrode chamber, which is thermostatted at 37°C with continuous stirring.
2. Add mitochondria or SMP to give a final concentration of 0.5 mg of protein/mL and measure the background rate for 1–2 min.
3. To quantify complex I-, III-, and IV-dependent respiration, add glutamate/malate to a final concentration of 2.5 m*M* if mitochondria are used or NADH to a final concentration of 50 μ*M* if SMPs are used (*see* **Note 5**). Measure the rate of oxygen uptake for 1–2 min.
4. To quantify complex II-, III-, and IV-dependent respiration, add succinate to a final concentration of 5 m*M*. Measure the rate of oxygen uptake for 1–2 min.
5. To quantify complex IV-dependent respiration, add ascorbate (final concentration of 5 m*M*) plus TMPD (final concentration of 0.5 m*M*). Measure the rate of oxygen uptake for 1–2 min.
6. Typical values for SMPs are given in **Table 1**.
7. Expose SMPs to different compounds, e.g., SIN-1 (final concentration of 500 μ*M*) ± SOD (final concentration of 400 units/mL), then add substrates and measure the rate of oxygen uptake for 1–2 min (*see* **Table 1**).

3.3. Preparation of the NO Electrode

3.3.1. Calibration of the NO Electrode

1. Pipet 2 mL of incubation buffer into the same electrode chamber (the NO probe is located inside the chamber). Add anaerobic solutions of pure NO gas in deoxygenated water using a microsyringe (to a maximal final concentration of 1–10 μ*M* (*see* **Note 1**).
2. Wash out the electrode chamber several times with distilled water between measurements.

Table 1
Effects of SIN-1 ± SOD on SMP Respiration from NADH, Succinate, and TMPD/Ascorbate

Respiration rate (μ*M* of O_2/min) from substrate[a]			
	NADH	Succinate	TMPD/Ascorbate
Control	11.9 ± 0.5	9.2 ± 0.9	19.5 ± 1.5
SIN-1	10.8 ± 0.8[ns]	8.9 ± 0.4[ns]	20.2 ± 1.0[ns]
SIN-1 + SOD	6.0 ± 1.1[b]	3.5 ± 0.5[b]	8.5 ± 0.3[b]

[a]The concentration used were: SIN-1, 500 μ*M*; SOD, 400 units/mL; NADH, 50 μ*M*; succinate, 5 m*M*; TMPD/Ascorbate, 0.5 m*M*/5 m*M*. The data are means ± SEM from 3–6 determinations.

[b]Significantly different from control, $P < 0.05$ (Student's unpaired *t*-test); ns, no significant difference from control. (From Lizasoain et al., **ref.** *5*.)

3. Calculate standard curves from the peak height of the response using the anaerobic solutions of pure NO gas. The minimum detection limit for NO is approx 0.1 μ*M*. Take into account that low NO concentrations are more effective at low-oxygen tension (*see* **Note 6**).

3.3.2. Measurements of NO Release

1. Pipet 2 mL of incubation buffer into the oxygen-electrode chamber thermostatted at 37°C with continuous stirring.
2. Add mitochondria or SMP to give a final concentration of 0.5 mg of protein/mL and measure the background rate for 1–2 min.
3. Add different substrates for mitochondrial respiration. Check that the NO release is not modified.
4. Expose SMPs to different compounds, e.g., SIN-1 (final concentration of 500 μ*M*) ± SOD (final concentration of 25–800 units/mL) (*see* **Note 2**), then add substrates and measure the release of NO for 1–2 min. (Typical values of inhibition of NADH respiration and production of NO in these conditions are shown in **Fig. 1**.)

4. Notes

1. To prepare pure NO gas in deoxygenated water, fill a gas bulb with NO (British Oxygen or similar supplier) from a cylinder and seal with silicone-rubber injection septa. Remove with a syringe 10–1000 μL and inject into another gas bulb filled with water, which has been deoxygenated by gassing with He for 1 h, to give stock solutions of NO of 0.01-1.0 m*M*. These solutions are stable at least for 24 h ***(9)***.
2. SOD is added to remove superoxide, which is formed from SIN-1 concomitantly with the release of NO ***(10)***.
3. Clark-type NO-sensitive electrodes are highly selective for NO and do not detect nitrite, nitrate, peroxynitrite, nitroxyl, superoxide, or hydrogen peroxide ***(11)***.

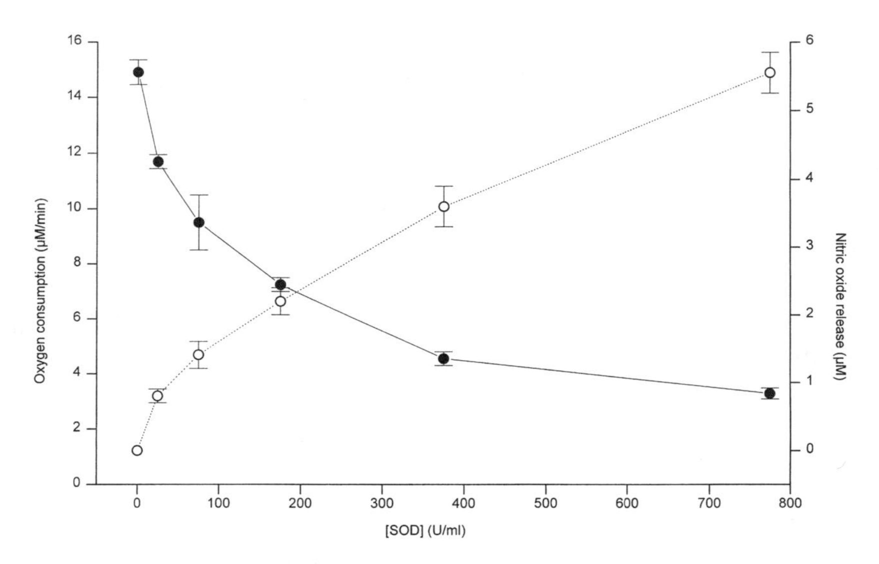

Fig. 1. Production of NO and inhibition of NADH respiration by SIN-1 + SOD. SIN-1 was present at 500 µ*M* and SOD at 0–800 units/mL as shown. ∘, NO concentrations; •, respiration rates. (Used with permission from **ref. *5*.**)

4. It is often useful to separate the mitochondrial compartments on a smaller scale or separate fractions so that segments of the metabolic chain may be studied without interference. The preparation of SMP can be produced by different methods: sonication, detergents, cycles of freeze-thaw, etc. (*see* **ref. *12***).
5. Glutamate/malate is used for mitochondria because NADH is unable to cross the intact mitochondrial inner membrane. To test the reversibility of effects on respiration, add NADH to a final concentration of 1 m*M* from a stock solution of 100 m*M* to sustain oxygen consumption for more than 10 min.
6. It is known that there is competition between oxygen and NO for cytochrome oxidase and therefore the percentage inhibition of the respiratory chain increases as the oxygen tension decreases ***(13)***. Then low NO concentrations will be more effective at low-oxygen tension.

References

1. Halliwell, B. (1992) Reactive oxygen species and the central nervous system. *J. Neurochem.* **59,** 1609–1623.
2. Olanow, C. W. (1993) A radical hypothesis for neurodegeneration. *Trends Neurosci.* **16,** 439–444.
3. Cleeter, M. W. J., Cooper, J. M., Darley-Usmar, V. M., Moncada, S., and Schapira, A. H. V. (1994) Reversible inhibition of cytochrome c oxidase, the terminal enzyme of the mitochondrial respiratory chain, by nitric oxide. Implications for neurodegenerative diseases. *FEBS Lett.* **345,** 50–54.

4. Schweizer, M. and Richter, C. (1994) Nitric oxide potently and reversibly deenergizes mitochondria at low oxygen tension. *Biochem. Biophys. Res. Commun.* **204,** 169–175.
5. Lizasoain, I., Moro, M. A., Knowles, R. G., Darley-Usmar, V. M., and Moncada, S. (1996) Nitric oxide and peroxynitrite exert distinct effects on mitochondrial respiration which are differentially blocked by glutathione or glucose. *Biochem. J.* **314,** 877–880.
6. Partridge, R. S., Monroe, S. M., Parks, J. K., Johnson, K., Parker, W. D., Eaton, G. R., and Eaton, S. S. (1994) Spin trapping of azidyl and hydroxyl radicals in azide-inhibited rat brain submitochondrial particles. *Arch. Biochem. Biophys.* **310,** 210–217.
7. Rickwood, D., Wilson, M. T., and Darley-Usmar, V. M. (1987) Isolation and characteristics of intact mitochondria, in *Mitochondria: a practical approach.* (Darley-Usmar, V. M., Rickwood, D., and Wilson, M. T., eds.), IRL Press Limited, Oxford, England, pp. 1–16.
8. Brown, G. C., Bolaños, J. P., Heales, S. J. R., and Clark, J. B. (1995) Nitric oxide produced by activated astrocytes rapidly and reversibly inhibits cellular respiration. *Neurosci. Lett.* **193,** 201–204.
9. Palmer, R. M. J., Ferrige, A. G., and Moncada, S. (1987) Nitric oxide release accounts for the biological activity of endothelium-derived relaxing factor. *Nature (London*), **327,** 524–526.
10. Feelisch, M., Ostrowski, J., and Noack, E. (1989) On the mechanism of NO release from sydnonimines. *J. Cardiovasc. Pharm.* **14,** 13–22.
11. Schmidt, K., Klatt, P., and Mayer, B. (1994) Reaction of peroxynitrite with oxyhaemoglobin: interference with photometrical determination of nitric oxide. *Biochem. J.* **301,** 645–647.
12. Ragan, C. I., Wilson, M. T., Darley-Usmar, V. M., and Lowe, P. N. (1987) Subfractionation of mitochondria and isolation of the proteins of oxidative phosphorylation, in *Mitochondria: a practical approach*, (Darley-Usmar, V. M., Rickwood, D., and Wilson, M. T., eds.), IRL Press Limited, Oxford, England, pp. 79–112.
13. Torres, J., Darley-Usmar, V. M., and Wilson, M. T. (1995) Inhibition of cytochrome c oxidase turnover by nitric oxide: mechanism and implications for control of respiration. *Biochem. J.* **312,** 169–173.

27

Release of NO from Donor Compounds: A Mathematical Model for Calculation of NO Concentrations in the Presence of Oxygen

Kurt Schmidt, Wolfgang Desch, Peter Klatt, Walter R. Kukovetz, and Bernd Mayer

1. Introduction

For studies on the biological role of nitric oxide (NO), especially long-term effects of NO on transcriptional or translational regulation of protein expression, it is desired to expose cells to a well-defined NO concentration over a certain period of time. Application of NO gas or NO solutions does not meet this goal, as it gives rise to relatively high initial-NO concentrations followed by a rapid decline owing to autoxidation. This problem can be avoided using NO-donor compounds, which should be stable in solutions of high or low pH and decompose in a first-order reaction to release stoichiometric amounts of NO after dilution in physiological buffers. Though decomposition of most of the currently available NO donors is not that simple, nucleophilic complexes of NO with amines (NONOates) appear to meet these criteria and release NO in a first-order reaction at physiological pH *(1)*. However, owing to the complex third-order kinetics of NO autoxidation, i.e., the fact that the rate of NO autoxidation increases with increasing concentrations of NO *(2)*, it is difficult to predict the actual NO concentration at a given time point even in this ideal situation.

In this study, we present a mathematical model for the simulation of concentration-time profiles of NO released from ideal-donor compounds, e.g., NONOates, in physiological buffers. The model is based on a system of two

From: *Methods in Molecular Biology, Vol. 100. Nitric Oxide Protocols*
Edited by: Michael A. Titheradge © Humana Press Inc., Totowa, NJ

differential equations describing the first-order decomposition of the NO donor in association with the overall third-order reaction of NO with oxygen. Though there is no closed formula to express the solution of this equation system, the solution can be computed by any standard numerical-equation solver or simulation software, provided that the following input parameters are known:

1. Initial concentration of the donor;
2. Decomposition rate constant of the donor;
3. Stoichiometry of NO release;
4. O_2 concentration of the buffer; and
5. Rate constant of NO autoxidation.

For standard experimental conditions, most of these parameters are published, making a determination of these values unnecessary in most cases (*see* **Notes 1–3**). In addition, we provide graphs that are based on dimensionless variables and allow a rough but very simple estimation of the NO concentration for a wide variety of experimental situations without using sophisticated mathematics software.

1.1. Calculation of NO Concentrations by Numerical Simulation

Upon decomposition of donor compounds in aerobic solutions, the actual concentration of NO depends on the decomposition rate of the donor and the autoxidation rate of NO. The following system of differential equations describes the concentration of NO [$c_{NO}(t)$] and the concentration of the donor [$c_D(t)$] at time t.

$$\frac{d}{dt} c_D(t) = -k_1 c_D(t) \quad (1)$$

$$\frac{d}{dt} c_{NO}(t) = k_1 c_D(t)\, e_{NO} - k_2 o_2 c_{NO}(t)^2 \quad (2)$$

$$c_D(0) = c_0,\ c_{NO}(0) = 0 \quad (3)$$

Equation 1 describes the decomposition of the donor following first order kinetics with a rate constant of k_1. The first term in **Eq. 2** models the production of NO, assuming that each mol of donor yields e_{NO} mol of NO. The second term in **Eq. 2** describes the autoxidation of NO as a third-order reaction depending on $c_{NO}(t)^2$ and the oxygen concentration o_2. If the latter is not limiting, we may assume that o_2 remains constant. The rate constant for the autoxidation reaction is denoted by k_2. Condition **(3)** gives the initial state of the solution with c_0 denoting the initial concentration of the donor.

Using this equation system, the solution $c_{NO}(t)$ can be easily computed by any standard numerical-differential equation solver for each given set of parameters and initial conditions the solution (*see* **Subheading 3.1.**).

1.2. Estimation of NO Concentrations by a Graphical Method

Numerical solution of differential equations is a prerequisite for the exact modeling of NO concentration-time profiles. Because appropriate computer software may not be always available, we provide a graphical tool to roughly estimate $c_{NO}(t)$. According to **Eqs. 1** and **2**, the concentration of NO depends on five parameters ($k_1, k_2, c_0, o_2, e_{NO}$) and the solution $c_{NO}(t)$ must, therefore, be recomputed for each given set of parameter. To reduce the number of parameters, we rescaled the model to dimensionless variables defined as follows:

$$\tau = k_1 t$$

$$u_D(\tau) = \frac{1}{c_0} c_D(\tau/k_1)$$

$$u_{NO}(\tau) = \frac{1}{c_0 e_{NO}} c_{NO}(\tau/k_1).$$

Introducing the new parameter

$$\alpha = \frac{k_2 o_2 c_0 e_{NO}}{k_1}$$

the model equations read now

$$\frac{d}{d\tau} u_D(\tau) = -u_D(\tau) \tag{4}$$

$$\frac{d}{d\tau} u_{NO}(\tau) = u_D(\tau) - \alpha u_{NO}(\tau)^2 \tag{5}$$

$$u_D(0) = 1,\ u_{NO}(0) = 0. \tag{6}$$

The graphical output of this equation system is a plot of u_{NO} ($= c_{NO}\, c_0^{-1}\, e_{NO}^{-1}$) vs τ ($= k_1 t$) for a given value of α. In contrast to the plot c_{NO} vs t which describes c_{NO} for only one single-experimental condition, plots of u_{NO} vs τ allow the estimation of c_{NO} for a wide variety of desired experimental situations (*see* **Subheading 3.2.**).

2. Materials

The differential equations presented were solved using a second-order Runge–Kutta algorithm from the commercially available software packages MATLAB (Version 4.0. The MathWorks, Inc., Natick, MA) or Mathematica (Version 2.2.2., Wolfram Research, Inc., Champaign, IL).

3. Methods

3.1. Numerical Simulation

The following protocol describes the procedure for calculating $c_{NO}(t)$ with the commercially available software package Mathematica. As an example, the concentration-time profile of NO released by decomposition of 8.3 μ*M* diethylamine/NO complex (DEA/NO) at pH 7.4 and 37°C is simulated for a time range 0–1200 s. The following abbreviations and units were used: k1, rate constant for decomposition of the donor (s^{-1}); k2, rate constant for NO autoxidation (M^{-2} s^{-1}); c0, initial concentration of the donor (*M*); eNO, mol NO released per mol donor; o2, concentration of oxygen (*M*); t, time (s); cd[t], concentration of the donor at time t (*M*); cno[t], concentration of NO at time t (M). Note that user-defined parameters and variables must be written in lower case, whereas the initial letters of system-defined commands and functions must be written in upper case.

1. Start the program.
2. Define the rate constant for decomposition of the donor as follows: k1 = 5.2 10^-3 (*see* **Note 1**). Press the enter key.
3. Define the stoichiometry of NO release (mol NO released per mol donor) as follows: eNO = 1 (*see* **Note 1**). Press the enter key.
4. Define the rate constant for NO autoxidation as follows: k2 = 13.6 10^6 (*see* **Note 2**). Press the enter key.
5. Define the oxygen concentration of the buffer as follows: o2 = 1.85 10^-4 (*see* **Note 3**). Press the enter key.
6. Define the initial concentration of the donor as follows: c0 = 8.3 10^-6. Press the enter key.
7. Define the equations and the command for numerical differentiation as follows: NDSolve[{cd'[t]==-k1 cd[t], cd[0]==c0, cno'[t]==k1 cd[t] eNO – k2 o2 cno[t]^2, cno[0]==0}, {cd,cno},{t,0,1200}]. Press the enter key. As output, interpolating functions describing cd (concentration of the donor) and cno (concentration of NO) for the time range 0–1200 s will be obtained.
8. Define the plot command as follows: Plot[Evaluate[cno[t] /. %],{t,0,1200}]. Press the enter key. As output, an NO concentration-time profile in the time range 0–1200 s will be obtained (**Fig. 1**, dotted line).

To validate this numerical simulation, we monitored NO release from 8.3 μ*M* DEA/NO in 100 m*M* phosphate buffer (pH 7.4, 37°C) with a Clark-type NO-sensitive electrode (for details, *see* Chapter 11). As evident from **Fig. 1**, simulated (dotted line) and experimentally observed data (solid line) were identical, demonstrating that **Eqs. 1** and **2** exactly describe release and inactivation of NO in aerobic solution.

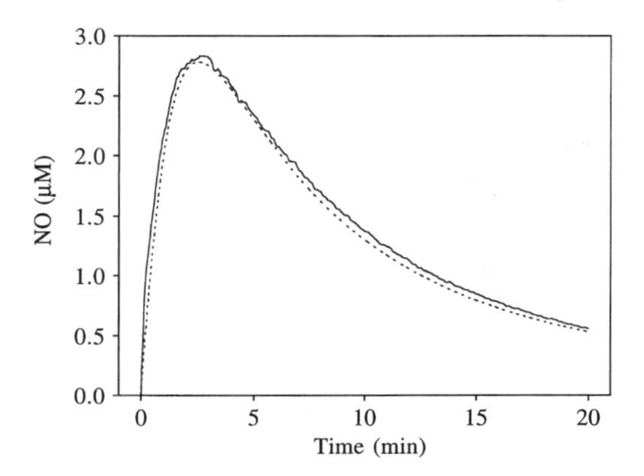

Fig. 1. Concentration–time profiles of NO release from DEA/NO. The simulated curve (dotted line) and a representative trace obtained with a Clark-type NO-sensitive electrode (solid line) are shown for decomposition of 8.3 μ*M* DEA/NO in 100 m*M* phosphate buffer, pH 7.4, at 37°C. Simulation was performed by numerical differentiation of **Eqs. 1** and **2** with the following set of parameters: $k_1 = 5.2 \times 10^{-3}\ s^{-1}$, $e_{NO} = 1$, $k_2 = 13.6 \times 10^6\ M^{-2}\ s^{-1}$, and $o_2 = 1.85 \times 10^{-4}\ M$.

3.2. Graphical Estimation

If an appropriate computer software is not available, the plots presented in **Figs. 2** and **3** can be used for the graphical estimation of $c_{NO}(t)$. The data were obtained by numerical differentiation of **Eqs. 4** and **5** using α values in the range 0.1–10^5. The following examples explain how to use these plots to estimate the actual concentration of NO.

Example 1: We determine the NO concentration 5 min after application of 8.3 μ*M* DEA/NO at 37°C and pH 7.4.

(1) Calculation of α and τ:

$$\alpha = \frac{k_2\, o_2\, c_0\, e_{NO}}{k_1} = \frac{(13.6 \times 10^6\ M^{-2}s^{-1})(1.85 \times 10^{-4}\ M)(8.3 \times 10^{-6}\ M)(1)}{5.2 \times 10^{-3}\ s^{-1}} = 4.0$$

$$\tau = k_1\, t = (5.2 \times 10^{-3}\ s^{-1})(300\ s) = 1.6.$$

(2) Graphical estimation of u_{NO} at $\tau = 1.6$ by interpolation between the curves for $\alpha = 2$ and $\alpha = 5$ in **Fig. 2:**

$$u_{NO} \sim 0.28.$$

(3) Calculation of c_{NO}:

$$c_{NO} = u_{NO}\, c_0\, e_{NO} = (0.28)(8.3 \times 10^{-6}\ M)(1) = 2.3 \times 10^{-6}\ M.$$

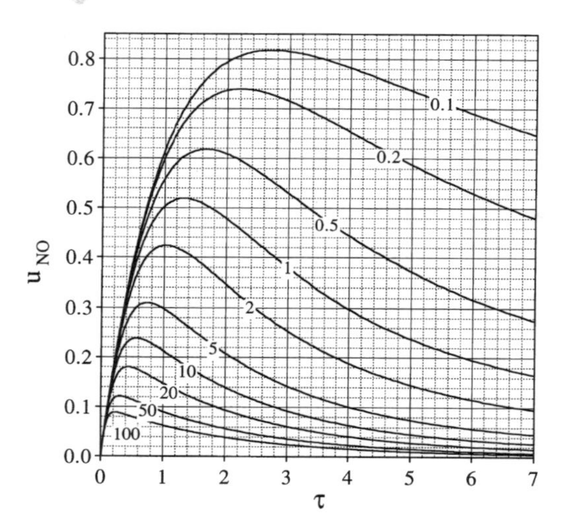

Fig. 2. Graphical plots estimation of NO when α is in the range 0.1–100. Data were calculated by numerical differentiation of **Eqs. 4** and **5** using the indicated α values. The use of these plots is demonstrated in Example 1.

Thus, the estimated NO concentration is ~2.3 μM and is identical with the actually measured NO concentration shown in **Fig. 1**.

Example 2: We determine the NO concentration 150 min after application of 100 μM Spermine/NO at 37°C and pH 7.4 using published decomposition parameters *(1)*.

(1) Calculation of α and τ

$$\alpha = \frac{(13.6 \times 10^6\, M^{-2}\mathrm{s}^{-1})(1.85 \times 10^{-4}\, M)(1.0 \times 10^{-4}\, M)(2)}{0.3 \times 10^{-3}\, \mathrm{s}^{-1}} = 1677$$

$$\tau = (0.3 \times 10^{-3}\, \mathrm{s}^{-1})(9000\, \mathrm{s}) = 2.7.$$

(2) Graphical estimation of log $u_{NO}0$ at τ = 2.7 by interpolation between the curves for α = 1000 and α = 2000 in **Fig. 3**, and calculation of u_{NO}:

$$\log (u_{NO}) \sim -2.15$$

$$u_{NO} \sim 0.0071.$$

(3) Calculation of c_{NO}:

$$c_{NO} = u_{NO} \times c_0 \times e_{NO} = (0.0071)(1.0 \times 10^{-4}\, M)(2) = 1.42 \times 10^{-6}\, M.$$

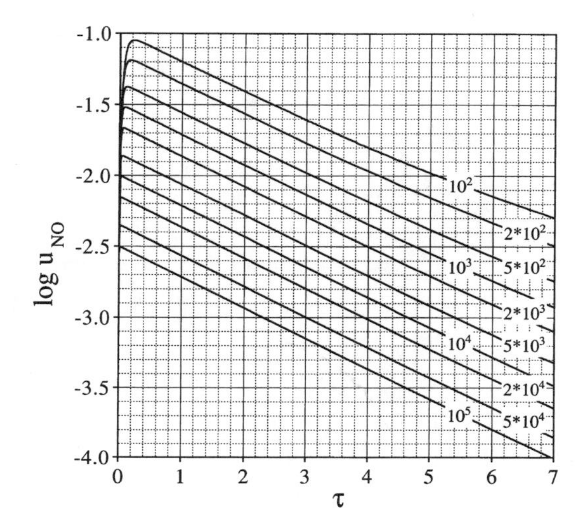

Fig. 3. Graphical plots estimation of NO when α is in the range 100–100,000. Data were calculated by numerical differentiation of **Eqs. 4** and **5** using the indicated α values. The use of these plots is demonstrated in Example 2.

The estimate of 1.4 μ*M* for the NO concentration is in good agreement with the value of 1.5 μ*M* experimentally obtained after addition of Spermine/NO to cultured-endothelial cells *(3)* (*see* **Note 4**).

4. Notes

1. Among available NO donors, NONOates may be the best choice to achieve a predictable release of NO, as the rates of decomposition are mainly determined by pH and temperature but not by the presence of reducing agents *(4)*. For decomposition of DEA/NO at pH 7.4 and 37°C, we measured a rate constant of $5.2 \times 10^{-3}\ s^{-1}$ which is almost identical to that found by Maragos et al. *(1)*. However, the published e_{NO} values may be less reliable. For the decomposition of DEA/NO at 37°C and pH 7.4, an e_{NO} value of 1.5 was described by Maragos et al. *(1)*, whereas we determined a value close to 1.0. Interestingly, other batches of DEA/NO yielded ***eNO*** values up to 2.0, indicating that the stoichiometry of NO release may be affected by yet-unknown factors. It is recommended, therefore, to check each batch of NONOate for the stoichiometry of nitrite release.
2. A determination of NO autoxidation is not necessary, as the rate is independent of pH *(5–7)* and only slightly affected by temperature *(5,6)*. The rate constants obtained at 25°C and 37°C (9.2 and $13.6 \times 10^6\ M^{-2}\ s^{-1}$, respectively) agree well with kinetic data published previously under different experimental conditions *(5–9)*, and thus may be used for calculating NO concentrations in aerobic solutions.

3. Similarly, the determination of O_2 is not necessary as in aerobic, physiological buffers the concentration will be close to 200 μM. Although this value slightly varies depending on the salt concentration and the temperature of the buffer (under our conditions we measured 210 and 185 μM at 25°C and 37°C, respectively), the differences are marginal and do not considerably affect the results.
4. Because the validity of the computer model is limited to conditions in which autoxidation is the predominant inactivation reaction, actual NO concentrations will be overestimated in systems containing efficient scavengers such as heme proteins or superoxide. Thus, the predictions of the model may be correct for aerobic-buffer solutions, but could yield false results when NO donors are added to cells or tissues. Intracellular heme could promote a significant decrease in the exogenous-NO concentration, and superoxide produced by the cells could lead to rapid conversion of NO to peroxynitrite. However, neither of the competing reactions appears to contribute appreciably to NO consumption in cell-culture experiments, as revealed by the excellent accordance of calculated and measured concentrations of NO 2.5 h after addition of a donor compound to cultured-endothelial cells (Example 2). However, to account for superoxide release that could be more pronounced in other experimental situations, it is recommended that the experiments be carried out in the presence of sufficient superoxide dismutase (1000 U/mL) to outcompete the peroxynitrite reaction.

References

1. Maragos, C. M., Morley, D., Wink, D. A., Dunams, T. M., Saavedra, J. E., Hoffman, A., Bove, A. A., Isaac, L., Hrabie, J. A., and Keefer, L. K. (1991) Complexes of NO with nucleophiles as agents for the controlled biological release of nitric oxide. Vasorelaxant Effects. *J. Med. Chem.* **34,** 3242–3247.
2. Ford, P. C., Wink, D. A., and Stanbury, D. M. (1993) Autoxidation kinetics of aqueous nitric oxide. *FEBS Lett.* **326,** 1–3.
3. Brunner, F., Stessel, H., and Kukovetz, W. R. (1995) Novel guanylyl cyclase inhibitor, ODQ reveals role of nitric oxide, but not of cyclic GMP in endothelin-1 secretion. *FEBS Lett.* **376,** 262–266.
4. Morley, D. and Keefer, L. K. (1993) Nitric oxide/nucleophile complexes: a unique class of nitric oxide-based vasodilators. *J. Cardiovasc. Pharmacol.* **22,** S3–S9.
5. Wink, D. A., Darbyshire, J. F., Nims, R. W., Saavedra, J. E., and Ford, P. C. (1993) Reactions of the bioregulatory agent nitric oxide in oxygenated aqueous media: Determination of the kinetics for oxidation and nitrosation by intermediates generated in the NO/O_2 reaction. *Chem. Res. Toxicol.* **6,** 23–27.
6. Lewis, R. S. and Deen, W. M. (1994) Kinetics of the reaction of nitric oxide with oxygen in aqueous solutions. *Chem. Res. Toxicol.* **7,** 568–574.
7. Goldstein, S. and Czapski, G. (1995) Kinetics of nitric oxide autoxidation in aqueous solution in the absence and presence of various reductants. The nature of the oxidizing intermediates. *J. Am. Chem. Soc.* **117,** 12,078–12,084.

8. Mayer, B., Klatt, P., Werner, E. R., and Schmidt, K. (1995) Kinetics and mechanism of tetrahydrobiopterin-induced oxidation of nitric oxide. *J. Biol. Chem.* **270,** 655–659.
9. Kharitonov, V. G., Sundquist, A. R., and Sharma, V. S. (1994) Kinetics of nitric oxide autoxidation in aqueous solution. *J. Biol. Chem.* **269,** 5881–5883.

Barry W. Allen

28

The Determination of Nitrotyrosine Residues in Proteins

Andrew J. Gow, Molly McClelland, Sarah E. Garner, Stuart Malcolm, and Harry Ischiropoulos

1. Introduction

Nitration of the ortho position of tyrosine results in the formation of 3-nitrotyrosine. Nitration of tyrosine residues in proteins using tetranitromethane has been used extensively to investigate the role of tyrosine residues in the function of many proteins *(1)*. However, the existence of tyrosine-nitrated proteins in vivo was not investigated until the discovery of a potential nitrating agent. It was shown that the major protein modification following the reaction of peroxynitrite with proteins is 3-nitrotyrosine *(2)*. Peroxynitrite is an oxidant formed by the near diffusion-limited reaction between two free radicals, nitric oxide (NO) and superoxide *(3)*. Inflammatory cells, endothelium, and, potentially, other cells generate peroxynitrite upon stimulation of NO and superoxide production *(4–6)*. Therefore, we and others have explored the possibility of using 3-nitrotyrosine as a marker for peroxynitrite-mediated oxidative stress. Published data has provided evidence that 3-nitrotyrosine is formed in a variety of human diseases and animal models of disease *(7–17)*. Endogenous-tyrosine nitration is almost certainly derived via enzymatically produced NO although NO itself is not a nitrating agent *(7–9)*. Chemically, protein-tyrosine residues can be nitrated by tetranitromethane *(1)*, nitric acid plus sulfuric acid *(18)*, nitrogen dioxide *(19)*, the acidification of nitrite *(20)*, and the reaction of nitrite with hypochlorous acid *(21)*. However, under pathophysiological conditions, it appears that peroxynitrite is the proximal species for the formation of protein nitrotyrosine in vivo *(22)* and that CO_2 is a catalyst for peroxynitrite-mediated nitration of tyrosine residues *(22–24)* (*see* **Fig. 1**).

From: *Methods in Molecular Biology, Vol. 100. Nitric Oxide Protocols*
Edited by: Michael A. Titheradge © Humana Press Inc., Totowa, NJ

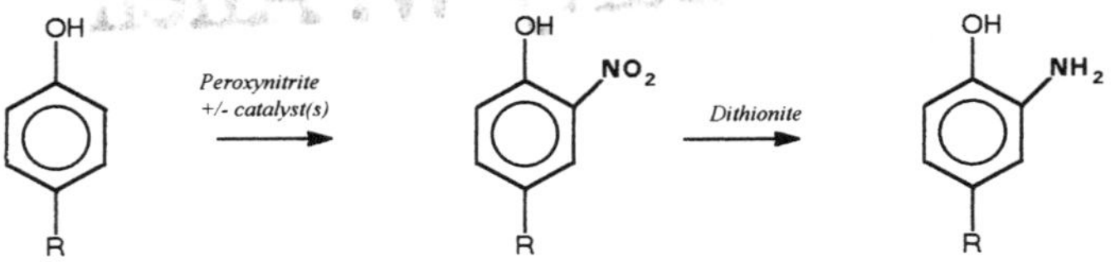

Fig. 1. Chemical structures: Tyrosine, 3-nitrotyrosine, and 3-aminotyrosine.

Several methods for the detection of nitrotyrosine have been developed and utilized *(7–17)*. One of the most recent methods is the use of polyclonal and monoclonal antinitrotyrosine antibodies *(7–17)*. These antinitrotyrosine antibodies showed high affinity and specificity for protein nitrotyrosine *(7,17)*. In this chapter, we describe methodologies for immunohistochemical detection of tyrosine-nitrated proteins and a solid-phase immunoradiochemical assay for quantification of protein nitrotyrosine in tissue homogenates. The immunohistochemical procedure was developed for using the purified monoclonal antinitrotyrosine antibody clone 1A6. This protocol has been tested mostly in rat tissues such as lungs, aorta, and brain that have been fixed in paraformaldehyde, cryoprotected, and frozen. (For tissue fixation, cryoprotection, freezing, and tissue cutting we recommend the protocols described in detail in **ref.** ***25***.)

2. Materials

1. Antibody buffer (AB) for immunoradiochemical assay: Add 0.5 g of Carnation milk powder to 100 mL of Tris-buffered saline containing 0.005% Tween-20 (TTBS), and gently stir until dissolved.
2. Block solution for immunoradiochemical assay: Add 5 g of nonfat, dried-milk powder (Carnation) to 100 mL of TTBS, and gently stir until dissolved.
3. Block solution for immunohistochemistry: Add 0.4 g of fatty acid free-bovine serum albumin (BSA) and 1 g of goat serum to 10 mL of 0.1 *M* phosphate-buffered saline (PBS), pH 7.2, containing 0.3% Triton X100.
4. Anti-nitrotyrosine antibody solution: Both affinity-purified polyclonal and monoclonal anti-nitrotyrosine antibodies (Upstate Biotechonology Inc., Lake Placid, NY) are added to Block solution for immunohistochemistry at 1:50, 1:100, 1:200, 1:500, and 1:1000 dilutions, or to AB for immunoradiochemical assay at 2 μg/mL.
5. Secondary-antibody solution: Rat plasma absorbed anti-mouse Cy3-conjugated affiniPure IgG (Jackson Immunoresearch Lab., West Grove, PA) is added to Block solution at a dilution of 1:250 for immunohistochemistry. Sheep anti-mouse ^{125}I-IgG (Amersham Arlington Heights, IL) is added to AB, at a concentration of 0.1–0.2 mCi/mL for immunoradiochemical assay.

6. 10 mM Nitrotyrosine solution: Dissolve 226 mg of 3-nitrotyrosine (Aldrich Chemical Co.) in 100 mL of PBS, pH 6.0, and then raise the pH to 7.2–7.4 by dropwise addition of 5 M NaOH.
7. 2 M Dithionite solution: Place 348 mg of sodium dithionite (sodium hydrosulfite) crystals in a 1.5-mL microcentrifuge tube and flush with N_2. Add 1 mL of degassed PBS, pH 9.0. Dithionite solution should be made fresh before each use.
8. Peroxynitrite: Peroxynitrite can be synthesized as described in this volume, *see* Chapter 20; or in **ref. *2***. It is also available from Alexis Corp., San Diego, CA, or Upstate Biotechonology Inc.
9. Nitrated-protein standard solution: Dissolve 2–4 mg of fatty acid free-BSA (which contains 20 tyrosine residues) in 1 mL of 100 mM phosphate buffer pH 7.4 containing freshly prepared 50 mM bicarbonate and react with 1–2 mM peroxynitrite as follows: Add 2–10 μL of peroxynitrite to the side of the tube right above the protein solution and vortex immediately. To measure the concentration of nitrotyrosine, an aliquot of the reacted protein is made alkaline with the addition of 10 M NaOH and a second aliquot is made acidic by the addition of 12 M HCl. These protein solutions are scanned from 300 nm to 500 nm. The absorbance at 430 nm at acidic pH is subtracted from the absorbance at 430 nm at alkaline pH and the concentration of nitrotyrosine calculated using an extinction coefficient of 4400 M^{-1} cm^{-1} (*see* **Note 1**).

3. Methods

3.1. Immunohistochemical Detection of Protein Nitrotyrosine

1. Wash the slides for 5 min in 0.1 M PBS pH 7.2.
2. Incubate twice for 3 min in borohydride (50 mg/100 mL) in 0.1 M PBS, pH 7.2. Borohydride will not reduce nitrotyrosine to aminotyrosine, but it will eliminate substantially background fluorescence (*see* **Note 2**).
3. Wash for 5 min in 0.1 M PBS, pH 7.2.
4. Place 100 μL, or more to completely cover the specimen, of Block solution on the slide.
5. Wash for 5 min in 0.1 M PBS, pH 7.2, containing 0.3% Triton X100.
6. Incubate for 2 h with the primary antibody diluted in Block solution by placing 100 μL or more on top of the specimen to completely cover it (*see* **Note 3**). All incubations should be performed in humidified chambers and during long incubation the tissues should be periodically checked to avoid dryness.
7. Wash for 5 min in 0.1 M PBS, pH 7.2, containing 0.3% Triton X100.
8. Incubate for 1 h with the secondary antibody diluted in Block solution by placing 100 μL or more on top of the specimen. Centrifuge the diluted solution for 3 min at 8000*g* and use the supernatant to cover the tissue (*see* **Note 4**).
9. Wash twice for 5 min in 0.1 M PBS, pH 7.2, containing 0.3% Triton X100.
10. Wash twice for 2 min in one part of 0.1 M PBS, pH 7.2, containing 0.3% Triton X100, four parts of 0.1 M PBS.

11. Wash twice for 2 min in 0.1 *M* PBS, pH 7.2.
12. Air dry for approx 20 min.
13. Mount and cover slip with an aqueous-based mounting medium like Mowoil.
14. Cover mounted slides with aluminum foil or place them in a box to avoid bleaching of the fluorescence.

3.2. Immunohistochemistry Controls

3.2.1. Tissue Staining with Antigen-Competed Primary Antibody

The protocol outlined in **Subheading 3.1.** should be followed using anti-nitrotyrosine that has first been competed with 3-nitrotyrosine.

1. Add anti-nitrotyrosine antibody to blocking solution containing 10 m*M* 3-nitrotyrosine solution to the same dilution as used in **step 6** of **Subheading 3.1.**
2. Incubate by gently rocking for 1 h at room temperature.
3. Use in **step 6** of **Subheading 3.1.** in place of primary-antibody solution.

3.2.2. Reduction of Nitrotyrosine to Aminotyrosine with Dithionite

Nitrotyrosine can be reduced to aminotyrosine with dithionite under alkaline conditions (*see* **Fig. 1**).

1. Place the slides in PBS that is made alkaline (pH 9.0) by the addition of sodium hydroxide.
2. Add a freshly prepared dithionite solution to a final concentration of 0.5 *M*.
3. Proceed from **step 1** of **Subheading 3.1.** using freshly reduced slides.

3.2.3. Tissue staining with Nonspecific Purified Mouse IgG

Repeat steps in **Subheading 3.1.**, using nonspecific mouse IgG in **step 6** instead of primary antibody.

3.2.4. Generation of Positive Slides by Peroxynitrite Treatment

1. Wash the slides for 5 min in 0.1 *M* PBS, pH 7.2.
2. 2 μL of peroxynitrite (100 m*M*) is placed directly above the tissue.
3. Incubate the slides for 5 min at room temperature.
4. Wash the slides for 5 min in 0.1 *M* PBS, pH 7.2.
5. Proceed from **step 2** of **Subheading 3.1.**

3.3. Solid-Phase Immunoradiochemical Assay for Nitrotyrosine Quantitation

1. Soak the nitrocellulose in Tris-buffered saline (TBS) pH 7.5.
2. Place the soaked nitrocellulose onto a 96-well Bio-Dot microfiltration unit (Bio-Rad, Hercules, CA) under vacuum.
3. Load samples and standards onto the nitrocellulose in a maximal-sample volume of 400 μL (*see* **Note 6**).

4. Wash the blot for 5 min with 0.005% TTBS.
5. Place the blot in blocking solution.
6. Incubate for 1 h at room temperature with gentle shaking.
7. Wash the blot once with TTBS for 5 min.
8. Place the blot in anti-nitrotyrosine solution.
9. Incubate the blot for 15 h at room temperature with gentle shaking. To preserve antibody the incubation can be performed in sealed bags on top of a rocking platform. If a larger volume of primary antibody is used, the blot should be covered to avoid evaporation during long incubations.
10. Wash the blot twice in TTBS.
11. Place the blot in the radiolabeled secondary-antibody solution (*see* **Note 7**).
12. Incubate for 3 h at room temperature with gentle shaking.
13. Wash the blot in TTBS once and three times in TBS.
14. Air dry the blot and measure the radioactivity of each sample by Beta scanning using an Ambis 400 imaging detector. The net counts of radioactivity (corrected for background counts from a sample blank) are obtained using the AMBIS image analysis software v4.1 and then plotted on a semi-logarithmic plot.
15. For sample blanks use three conditions:
    a. Fatty acid free-BSA that was not reacted with peroxynitrite.
    b. Samples reduced with dithionite; and
    c. Samples reacted with antigen-competed antibody.
16. The concentration of nitrotyrosine in each sample is determined from the linear portion of the sigmoidal curve from the semilog plot of net counts vs antigen concentration of the nitrated-BSA standard. Once the concentration of nitrotyrosine per protein spot is determined it can be plotted against the protein concentration per spot. The slope of the line from the linear regression analysis of this plot represents the concentration of nitrotyrosine per μg or mg of protein.

4. Notes

1. An appropriate negative control can be generated by the reverse order of addition as follows; the same concentration of peroxynitrite is added to the buffer, allowed to decompose for 5 min, followed by the addition of the protein. Chemically nitrated proteins are stable indefinitely. The nitro group is neither readily oxidized nor reduced to aminotyrosine. However, nitrotyrosine can be chemically reduced to colorless aminotyrosine by dithionite.
2. To eliminate background fluorescence, the reduction with borohydride is a critical step and must be performed with fresh solutions of borohydride. Red-blood cells show a relatively high intensity of intrinsic fluorescence in the region of rhodamine. Successful reduction with borohydride eliminates this fluorescence.
3. We recommend to initially stain tissues using a serial dilution of the primary antibody, i.e., 1:50, 1:100, 1:200, 1:500, and 1:1000 and the secondary antibody at 1:250 dilution in order to establish the best primary antibody concentration. The concentration of the secondary antibody can be also adjusted if needed to obtain the best signal-to-noise ratio (SNR).

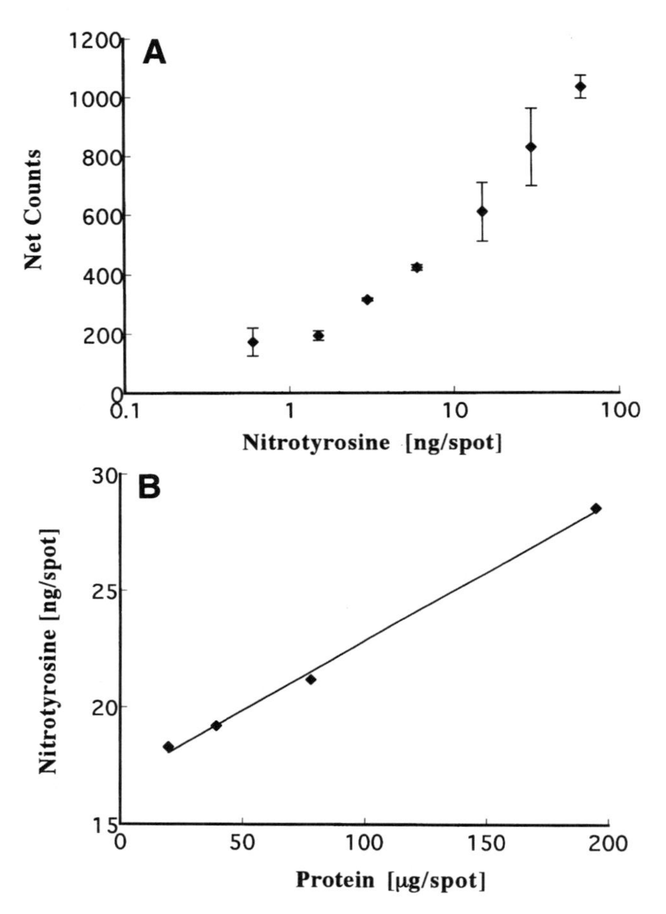

Fig. 2. **(A)** Representative plot of standard curve using the monoclonal antinitrotyrosine antibody and tyrosine-nitrated BSA. **(B)** Representative plot for nitrotyrosine analysis in endotoxin-treated rat-liver homogenate.

4. This method utilizes goat antimouse IgG that has been absorbed by incubation with rat plasma. Absorption of the secondary antibody with rat plasma eliminates nonspecific binding with rat tissue of the secondary antibody. In addition, nonspecific binding of the secondary antibody is also significantly reduced by blocking the tissues with normal goat serum.
5. The staining can be diffuse and not localized in some instances because nitrotyrosine is a small hapten and may be present in more than one protein and a relatively high concentration of the primary antibody may be required to observe a clear signal. However, this should not be confused with nonspecific staining before the staining is compared with all the negative controls.

6. Eight to twelve different concentrations of the peroxynitrite-modified BSA standard are loaded in duplicate for each blot. We usually run standards in the range of 1–100 ng nitrotyrosine per spot. The concentration of sample protein to achieve the best results must be determined experimentally. Initially we run eight different protein concentration in the range of 1 µg to 500 µg protein in 400 µL TBS. The ability of nitrocellulose to bind protein is approximately 100 $\mu g/cm^2$. For tissue homogenates, we dilute the homogenates first in 50 m*M* Tris buffer, pH 7.2, with 0.5 *M* PMSF, 1 µg/mL leupeptin, and 1 µg/mL aprotinin.

   Avoid storing tissue homogenates with azide. Oxidation of azide by heme proteins results in artifactual formation of nitrotyrosine. Tissue homogenates should be made in the presence of Butylated hydroxytoluene and stored in the presence of protease inhibitors at –80°C. Even under these conditions, loss of binding in the solid-phase radioimmunoassy has been found with prolonged storage time. We recommend to use fresh-tissue homogenates. Tissue homogenates with high content of lipids or membrane fractions can be solubilized in detergents such as Triton X100 or Brij and sodium dodecyl sulfate (SDS). However, the amount of detergent should be kept low, i.e., 0.1%, because at higher concentration they will interfere with protein binding to nitrocellulose. Protein binding to nitrocellulose can be visualized after loading with PonceauS staining.
7. The ^{125}I-radiolabeled-secondary antibody will decay with time and it should not be used beyond the manufacturers recommended time period.

References

1. Nielsen, A. T. (1995) in *Nitrocarbons*, VCH Publishers, New York, pp 40–41.
2. Ischiropoulos, H., Zhu, L., Chen, J., Tsai, J.-H. M., Martin, J. C., Smith, C. D., and Beckman, J. S. (1992) Peroxynitrite-mediated tyrosine nitration catalyzed by superoxide dismutase. *Arch. Biochem. Biophys.* **298,** 431–437.
3. Beckman, J. S., Chen, J., Ischiropoulos, H., and Crow, J. P. (1994) Examining the oxidative chemistry of peroxynitrite. *Methods Enzymol.* **233,** 229–240.
4. Ischiropoulos, H., Zhu, L., and Beckman, J. S. (1992) Peroxynitrite formation from macrophage-derived nitric oxide. *Arch. Biochem. Biophys.* **298,** 446–45.
5. Carreras, M. C., Pargament, G. A., Catz, S. D., Poderoso, J. J., and Boveris, A. (1994) Kinetics of nitric oxide and hydrogen peroxide production and formation of peroxynitrite during the respiratory burst of human neutrophils. *FEBS Lett.* **341,** 65–68.
6. Kooy, N. W. and Royall, J. A. (1994) Agonist-induced peroxynitrite production from endothelial cells. *Arch. Biochem. Biophys.* **310,** 352–359.
7. Beckman, J. S., Ye, Y.-Z., Anderson, P. G., Chen, J, Accavitti, M. A., Tarpey, M. M., and White, C. R. (1994) Extensive nitration of protein tyrosine in human atherosclerosis detected by immunohistochemistry. *Biol. Chem. Hoppe-Seyler* **375,** 81–88.
8. Haddad, I. Y., Pataki, G., Hu, P., Galliani, C., Beckman, J. S., and Matalon S. (1994) Quantitation of nitrotyrosine levels in lung sections of patients and animals with acute lung injury. *J. Clin. Invest.* **94,** 2407–2413.

9. Kooy, N. W., Royall, J. A., Ye, Y.-Z., Kelly, D. R., and Beckman, J. S. (1995) Evidence for in vivo peroxynitrite production in human acute lung injury. *Am. J. Resp. Crit. Care Med.* **151,** 1250–1254.
10. Kaur, H. and Halliwell, B. (1994) Evidence for nitric oxide-mediated oxidative damage in chronic inflammation. Nitrotyrosine in serum and synovial fluid from rheumatoid patients. *FEBS Lett.* **350,** 9–12.
11. Ischiropoulos, H., Al-Mehdi, A. B., and Fisher, A. B. (1995) Reactive species in rat lung injury: contribution of peroxynitrite. *Am. J. Physiol.* L185–L164.
12. Szabo, C., Salzman, A. L., and Ischiropoulos, H. (1995) Endotoxin triggers the expression of an inducible isoform of nitric oxide synthase and the formation of peroxynitrite in the rat aorta *in vivo*. *FEBS Lett.* **363,** 235–238.
13. Wizemann, T. M., Gardner, C. R., Laskin, J. D., Quinones, S., Durham, K. D., Golle, N. L., Ohnishi, S. T., and Laskin, D. L. (1994) Production of nitric oxide and peroxynitrite in the lung during acute endotoxemia. *J. Leuk. Biol.* **56,** 759–768.
14. Miller. M. J. S., Thompson, J. H., Zhang, X.-J., Saodowska-Krowicka, H., Kakkis, J. L., Munshi, U. K., Sandoval, M., Rossi, J. L., Eloby-Childress, S., Beckman, J. S., Ye, Y. Z., Rodi, C. P., Manning, P. T., Currie, M. G., and Clark, D. A. (1995) Role of inducible nitric oxide synthase expression and peroxynitrite formation in the guinea pig ileitis. *Gastroenterology* **109,** 1475–1483.
15. Schulz, J. B., Matthews, R. T., Jenkins, B. G., Ferrante, R. J., Siwek, D., Henshaw, D. R., Cipolloni, P. B., Mecocci, P., Kowall, N. W., Rosen, B. R., and Beal, M. F. (1995) Blockade of neuronal nitric oxide synthase protects against excitotoxicity *in vivo*. *J. Neurosci.* **15,** 8419–8429.
16. Basarga, O., Michaels, F. H., Zheng. Y. M., Borboski, L. E., Spitsin, S. V., Fu, Z. F., Tawadros, R., and Koprowski, H. (1995) Activation of the inducible form o nitric oxide synthase in the brains of patients with multiple sclerosis. *Proc. Natl Acad. Sci USA* **92,** 12,041–12,045.
17. Ischiropoulos, H., Beers, M. F., Ohnishi, S. T., Fisher, D., Garner, S. E., an Thom, S. R. (1996) Nitric oxide production and perivascular tyrosine nitration i brain following carbon monoxide poisoning in the rat. *J. Clin. Invest.* **97** 2260–2267.
18. Olah, G. A., Malhotra, R., and Narang, S. C. (1989) in *Nitration, Methods an Mechanisms* Organic Nitro Chemistry Series, VCH Publishers, New York pp. 10–116.
19. Knowles, M. E., McWeeny, D. J., Couchman, L., and Thorogood, M. (1974 Interaction of nitrite with proteins at gastric pH. *Nature* **247,** 288–289.
20. Prutz, W. A., Monig, H., Butler, J., and E. J. Land. (1985) Reactions of nitroge dioxide in aqueous model systems: Oxidation of tyrosine units in peptides ar proteins. *Arch. Biochem. Biophys.* **243,** 125–134.
21. Eiserich, J. P., Cross, C. E., Jones, D., Halliwell, B., and Van der Vliet, A. (199 Formation of nitrating and chlorinating species from the reaction of nitrite wi hypochlorous acid. *J. Biol. Chem.* **271,** 19,199–19,208.

22. Gow, A., D. Duran, Thom, S. R., and Ischiropoulos, H. (1996) Carbon dioxide catalyzed protein tyrosine nitration by peroxynitrite. *Arch. Biochem. Biophys.* **333,** 42–48.
23. Denicola, A. Trujillo, M., Freeman, B. A., and Radi, R. (1996) Peroxynitrite reaction with carbon dioxide/bicarbonate: Kinetics and influence on peroxynitrite-mediated oxidation reactions. *Arch. Biochem. Biophys.* **333,** 49–54.
24. Uppu, R. M., Squadrito, G. L., and Pryor, W. A. (1996) Acceleration of peroxynitrite oxidations by carbon dioxide. *Archives Biochem. Biophys.* **327,** 335–343.
25. Harlow, E. and Lane, D. (1988) *Antibodies: A Laboratory Manual.* Cold Spring Harbor Laboratory, pp. 553–612.

29

Measurement of DNA Damage Using the Comet Assay

Steven Thomas, Michael H. L. Green, Jillian E. Lowe, and Irene C. Green

1. Introduction

Nitric oxide (NO) has the potential to damage DNA directly, and also to react with O_2 to form the nitrosating species N_2O_3 or to react with O_2^- to form the powerful oxidizing agent peroxynitrite (*see* Chapter 20, this volume). The DNA-damaging activity of NO has been reviewed by Tannenbaum et al. *(1)*.

The comet assay (single-cell gel electrophoresis), a sensitive method for detecting DNA-strand breaks in eukaryotic cells, was first described by Ostling and Johanson *(2)* and has been developed by Singh et al. *(3)* and Olive et al. *(4)*. The assay has been reviewed by McKelvey-Martin et al. *(5)* and by Fairbairn et al. *(6)*. The protocol in this chapter is based on the method by Singh with slight modifications. A detailed description of the Comet assay has recently been published *(7)*.

A single-cell suspension of the cell culture or tissue is embedded in agarose on a frosted microscope slide (*see* **Fig. 1**). The cells are lysed in a high-salt solution that removes the cell contents except for nuclear material. The highly supercoiled DNA is then exposed to an alkaline buffer and commences to unwind from sites of strand breakage. An electric current is applied causing fragmented DNA to move towards the anode, giving the appearance of a "comet tail" while undamaged DNA remains in the nucleus. (*see* **Fig. 2**).

The comet assay can be used with almost any cell type. It does not require the use of growing cells, and can be used with material treated in vitro or in vivo. It has a specific advantage over other sensitive methods for detecting DNA damage in that it is a single-cell assay and can detect variation in the

From: *Methods in Molecular Biology, Vol. 100. Nitric Oxide Protocols*
Edited by: Michael A. Titheradge © Humana Press Inc., Totowa, NJ

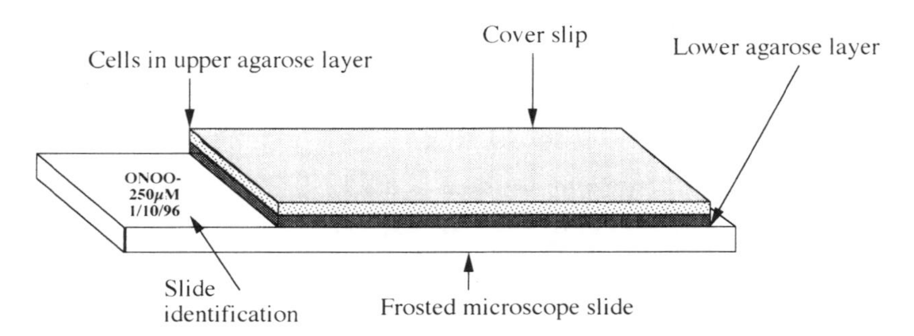

Fig. 1. Comet slide showing arrangement of the agarose layers.

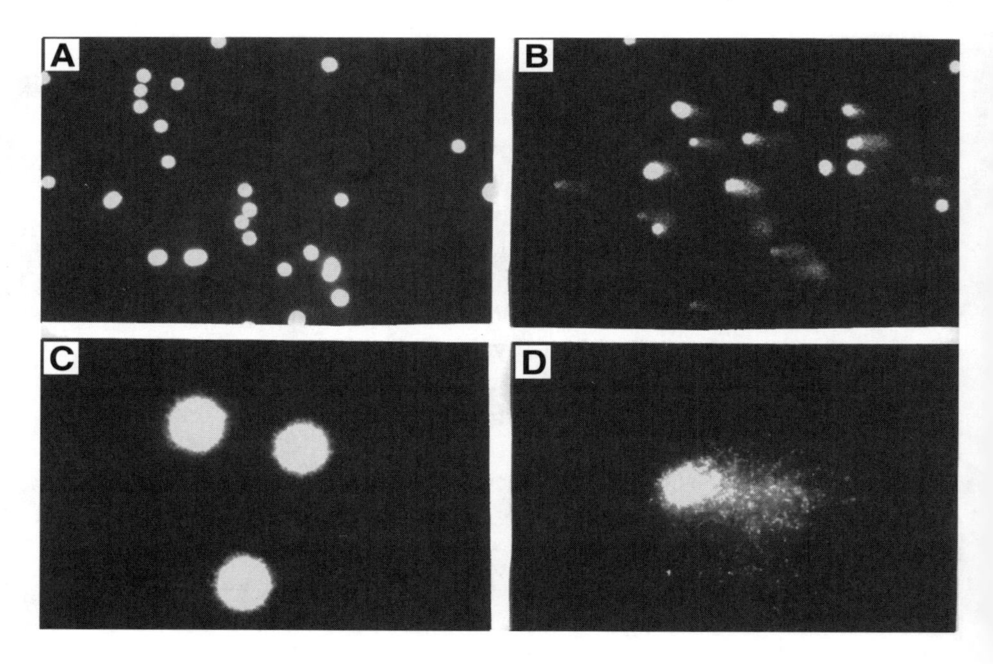

Fig. 2. Comets formed in HIT-T15 cells following treatment with peroxynitrite synthesized by the ozonation of sodium azide. This provides a solution of low ionic strength and free from hydrogen peroxide (*see* Chapter 20, this volume): **(A)** "Inactivated" control, ×10 magnification, **(B)** peroxynitrite, 250 μ*M*, ×10 magnification, **(C)** "inactivated" control, ×40 magnification, and **(D)** peroxynitrite, 250 μ*M*, ×40 magnification.

extent of damage both between cells of the same type and between different types of cell in mixed populations.

The comet assay can distinguish between breakage of the sugar-phosphate backbone and/or damage to DNA bases and using a time-course experiment can distinguish between immediate damage and strand breakage occurring as an intermediate stage of repair. It is also possible to detect specific types of DNA-damage using a purified DNA repair enzyme *(8)*.

2. Materials

2.1. Slide Preparation

1. Agarose: Sigma Type I, NuSieve low melting-point agarose (Flowgen).
2. Culture medium: As appropriate for the cell type. For our hamster cell line (HIT-T15) we use clear RPMI 1640 medium purchased as a sterile solution and stored at 4°C.
3. Serum: We use 5% fetal calf serum (FCS) for our hamster cell line.
4. Dulbecco's "A" phosphate buffered saline (PBS): 135 m*M* NaCl, 2.68 m*M* KCl, 8.1 m*M* Na_2HPO_4, and 1.47 m*M* KH_2PO_4 at pH 7.3. One tablet (Oxoid, Basingstoke, UK) is dissolved in 100 mL of double-deionized water and autoclaved at 115°C for 10 min. This solution can be stored at room temperature indefinitely.
5. Trypsin-EDTA and Hanks Balanced Salt Solution (HBSS): Purchased as sterile solutions and stored at 4°C.
6. Lysis solution: 2.5 *M* NaCl, 200 m*M* NaOH, 100 m*M* ethylenediaminetetraacetic acid, disodium salt (EDTA-Na_2), 10 m*M* Tris base in double-deionized water. Adjust carefully with 3 *M* NaOH to pH 10. Make up fresh once a week and store at room temperature. Place in refrigerator 1 h before use. Just before use add 1% (v/v) Triton X-100 and 10% (v/v) of DMSO (*see* **Note 1**).
7. Standard histology staining troughs and racks: cover with black electrical tape to exclude light.

2.2. Electrophoresis

1. Electrophoresis buffer: 300 m*M* NaOH, 1 m*M* EDTA-Na_2 in double-deionized water. Make up fresh each day using 500-mL bottles of autoclaved double-deionized water. Store some at 4°C and some at room temperature.
2. Neutralization buffer: 400 m*M* Tris base in double-deionized water. Adjust to pH 7.5 with concentrated HCl. Filter and sterilize. Store at room temperature until required.
3. Staining solution: 20 μg/mL ethidium bromide in double-deionized water. Store at 4°C until required.
4. Power pack: Capable of achieving 300 mA at 20 V. Because electrophoresis is carried out in 0.3 *M* NaOH, a substantial current is necessary to achieve the correct voltage.
5. Gel electrophoresis boxes: Commercially available boxes are satisfactory. The size of the box will determine both the size of the experiment and the size of the power pack required. Cover the boxes with black electrical tape to protect samples from light. We use workshop-made perspex boxes, 28 × 24 × 7.5 cm, which hold 1.5 L of electrophoresis buffer and 18 slides placed in two rows.

2.3. Scoring

1. Fluorescence microscope with a filter capable of detecting ethidium bromide or propidium iodide (*see* **Note 2**). We use a ×10 objective with a video camera for scoring and a ×40 objective for observing more closely.
2. Semiautomated image analysis system: This is desirable for objective analysis, although it is possible to analyze comets without software (see nonautomated scoring, below). A number of companies offer comet analysis software and hardware. We currently use the Casys systems of Synoptics (Cambridge, UK) and the Komet analysis software of kinetic imaging (Liverpool, UK). A sensitized or integrating monochrome camera is required (standard charge-coupled device (CCD) cameras are insufficiently sensitive to detect comets by fluorescence).

2.4. Safety

The protocol as described here should not be used for infectious material. Appropriate safety procedures should be used for handling lysis solution and electrophoresis buffer which are caustic, for ethidium bromide which is mutagenic, and for electrophoresis.

3. Methods

3.1. Slide Preparation

1. Prepare 3 mL of 0.6% regular agarose and 1.5 mL of 1.2% low-melting point (LMP) agarose in clear-culture medium (*see* **Note 3**). Microwave or heat to near boiling to dissolve the agar. Do not heat medium longer than is necessary to melt the agar.
2. Place the normal agarose into a water bath at 45°C and the LMP agarose into a 37°C water bath. Add serum (*see* **Note 3**) to the LMP agar after it has cooled.
3. Prelabel frosted slides (*see* **Note 4**) clearly for each treatment (in pencil). Warm the slides in an incubator (50°C). Place the warm slides on a metal baking tray on the bench (not over ice). Add 90 μL of regular agarose solution to the slide and immediately lower a 22 × 50 mm coverslip onto the slide guiding gently with a pipet tip, being careful not to create bubbles. Place the tray over ice for 10 min to allow the agar to solidify (*see* **Note 5**).
4. For our hamster-cell line, we incubate the cells in HBSS (minus Ca and Mg) for 5 min. Trypsinize the cells with trypsin-EDTA (*see* **Note 6**). Keep looking at the cells; do not leave for a fixed time. It is crucial to trypsinize for an absolute minimum of time if good controls are to be attained. Scoring comets in the presence of some clumped nuclei is more desirable than obtaining damaged controls or digested cells (*see* **Note 6**). Add a double volume of medium with serum and spin down the cells to remove trypsin (*see* **Note 7**). Resuspend the cells in complete medium at a density of approx 4×10^5/mL. At least 10-fold lower cell densities can be used if material is scarce.

5. Remove tray containing slides with regular agarose. Gently slide the coverslips off two slides. Add 85 µL of cell suspension and 85 µL of 1.2% LMP agarose into an Eppendorf in the 37°C waterbath and mix well. Add 75 µL of the cell suspension/agar mixture to each of the duplicate slides. Replace the coverslip as before. Allow to set on ice for a further 5 min.
6. Cells can be treated on the slide if they have not already been treated. Calculate the concentration of test agent assuming that 60 µL of medium containing this agent will be added to 165 µL of agar and cells. For example, remove the coverslips from duplicate slides. Place a drop of peroxynitrite solution onto the surface of a fresh coverslip. Add PBS to the surface of the top agar layer (to give a final volume of 60 µL when combined with peroxynitrite solution) and immediately drop the coverslip containing the peroxynitrite solution onto the surface of the top agar layer. Place the slides on damp laboratory tissue in a closed box. Incubate for a treatment time of e.g., 15 min at 37°C.
7. Add 1% (v/v) Triton X-100 and DMSO (10% v/v) to the cold-lysis solution immediately before use.
8. Gently remove the coverslips and place the slides on edge in a staining rack. Carefully immerse the rack into a staining trough containing cold (4°C), freshly made lysis solution. Protect from light and place in a 4°C refrigerator for a minimum of 1 h (*see* **Note 8**).

   Following lysis, all steps should be performed under dim yellow lights to prevent additional DNA damage.

3.2. Electrophoresis

1. Remove the rack of slides from the lysis solution and place the slides on the gel tray in the electrophoresis box. Slides should be placed close together and as close as possible to the anode. Blank slides should be placed in any empty spaces to prevent any movement and ensure consistent electrophoresis.
2. Gently add electrophoresis buffer to the gel box, just covering the slides (avoiding air bubbles). The electrophoresis buffer should be at a defined temperature (obtained by making it up with a mixture of distilled water stored at 4°C and at room temperature); we use 15°C. The volume of buffer chosen should completely, but only just, cover the slides. The shelf carrying the slides in the gel box must be completely horizontal, or the electrophoresis will not be the same between slides.
3. Place the gel elecrophoresis box at a defined temperature (we use 10°C) and allow to stand for exactly 40 min to allow unwinding of the DNA. A 4°C refrigerator may be preferred. Different times or temperatures may be used, but they must be exactly specified (*see* **Notes 9** and **10**).
4. Apply 20 V for 24 min and check the current. To obtain the same current in each experiment, the same voltage, volume of buffer, and gel box should be used (*see* **Note 11**).

5. Following electrophoresis, gently place the slides flat on a staining tray. Carefully rinse the slides with neutralization buffer. Drain for 5 min. Repeat twice.
6. Stain each slide with 60 μL of ethidium-bromide solution and add a fresh coverslip (*see* **Note 2**).
7. Store the slides at 4°C in a sealed box containing moist laboratory tissue. Slides are best scored immediately, but can still be scored after several days. Slides can also be dried and stored indefinitely (*see* **Note 12**).
8. Remove the slides from the box, wipe the base dry, and allow them to warm before placing them on the stage of a fluorescence microscope.
9. Locate fields containing comets. Have strict criteria setting out the grounds for rejecting a specific comet and always score unless a comet fails a specific criterion. Avoid areas close to the edge of the agar and comets that are grossly out of focus.

3.3. Scoring Procedures

1. When choosing a commercial comet-analysis package it is necessary to have a system which allows for:
   a. Consistency of measurement;
   b. Ability to recover from errors (e.g., pressing the wrong button); and
   c. Speed.

   A good system should be able to score a full set of slides in 1–2 h, depending on the experience of the operator.
2. Semiautomated scoring: A number of parameters are available to describe and quantitate damage:
   a. Threshold: An arbitrary brightness threshold is necessary to distinguish between a comet and its background. A robust definition is required for consistency within and between experiments.
   b. Comet length is the simplest parameter available to measure DNA damage and is capable of yielding entirely credible results (*see* **Note 13**).
   c. Tail moment is the product of the percentage of total DNA in the tail distribution and the displacement of the centers of mass of the head and tail *(4)* (*see* **Note 14**).
3. Nonautomated scoring: Satisfactory data can be obtained without image analysis software and may be appropriate where there is a proportion of highly damaged cells in an otherwise intact population. Comets with varying degrees of damage can be allocated an appropriate size class. Number the classes 0–4 and assign 100 comets from a treatment to the appropriate damage classes *(9,10)*.
4. Troubleshooting. (*See* **Notes 15** and **16.**)

4. Notes

1. The original procedure of Singh et al. *(3)* included sodium sarcosinate in the lysis solution. This is not required for efficient lysis and causes trouble with precipitation at 4°C. The original procedure also included a third agar layer above the cells. This does not seem to be necessary.

2. Other DNA staining procedures such as propidium iodide could be used with the appropriate filter for fluorescence microscopy. We have found ethidium bromide to be brighter than DAPI (diamino-2-phenylindole) and acridine orange to be subject to fading.
3. We prefer to use agar made up with same medium and serum in which we normally culture the cells.
4. Procedures exist for using plain slides on which a layer of agarose has been allowed to dry *(11)*.
5. Problems may be encountered with agar detaching from slides. If you encounter this problem:
   a. Ensure that the slides are hot when the bottom layer is added. The agar layer should not begin to set until it has come into intimate contact with the surface of the slide.
   b. Ensure that the agar has been left to set long enough and that the ice-tray is cold enough; and
   c. LMP agarose may deteriorate on the shelf.

   If all possibilities have been eliminated, and your agar is still detaching from the slide simply increase the concentration slightly.
6. When preparing a single-cell suspension, trypsin-EDTA appears to cause less damage than trypsin alone. A short treatment with a standard concentration of trypsin-EDTA is preferable to reducing the concentration and treating for a longer period. The aim is to minimize damage during treatment, not to obtain a pure single-cell suspension, especially when treating tissue such as rat or human islets of Langerhans. It is almost impossible to dissagregate an islet completely without destroying the outer cells. It is relatively easy to discount clumps of cells but difficult to allow for damaged cells. We have recently had promising results using a fractionating method to provide single-cell suspensions in islets, with either trypsin or dispase *(12)*. As a last resort the assay can be performed on cells that have been allowed to attach to the frosted microscope slide *(13)*. This allows the cells to repair transient damage owing to trypsinisation, but we find the comets hard to visualize because of their vicinity to the frosting on the slide.
7. Different centrifugation regimes for pelleting cells will be required depending on the cell type under investigation. We use 250*g* for 5 min for our hamster cell line.
8. The time that slides can be left in lysis solution is flexible, at least up to 6 h, provided that the pH of the lysis mix is no greater than 10. Earlier versions of the assay incorporated sodium sarcosinate in the lysis solution, which tended to precipitate out over longer periods. We do not recommend leaving slides in lysis solution overnight, although others have found this satisfactory.
9. Use of a shorter length of time in the electrophoresis buffer before applying the current will give some reduction in sensitivity but may avoid an unaceptable level of Comet formation on control slides. It is essential to use the same time and incubation temperature in all experiments.
10. Increasing the temperature of the electrophoresis buffer increases the sensitivity of the assay, but may also cause damage in controls. In our hands, a temperature

of 15°C and an unwinding time of 40 min appear to give maximal discrimination but are on the borderline of producing acceptable controls. Other workers might prefer some loss of sensitivity and a more robust assay.

11. By altering the voltage or time of electrophoresis, comets of any desired length can be obtained. Again, exactly the same voltage and time must be specified in all experiments.
12. Slides can be kept for several days and we have scored them successfully after they have been sent between laboratories by courier. A better procedure is to air dry the slides, which can be reconstituted days or months later (P. L. Olive, personal communication). To air dry the slides, remove the coverslips and leave at room temperature or in an incubator overnight. To reconstitute the slides, apply 100 μL of PBS and a cover slip. Allow the slide to sit for 1–2 h before staining and scoring. If this method is used the concentration of agar in the bottom layer should be increased to 1% and the volume doubled. Otherwise the reconstituted comets are too near the surface of the slide to be readily detectable.
13. Measurement of comet length has some limitations. Comet length increases linearly with damage only over a restricted range of doses. Further damage increases the proportion of DNA in the comet tail, but does not increase tail length. The comet length is also dependent on the threshold chosen.
14. The tail moment increases linearly over a wider range of doses than comet length, but fails when the center of the original nucleus is no longer the brightest part of the comet.
15. When the laboratory is warm and humid, excessive comet formation tends to occur in controls. This appears to be owing to condensation forming on the surface of the agar while the slides are held over ice, and causing the medium on the slides to become hypotonic. By removing coverslips from only one or two slides at a time, and leaving the agar surface exposed to the atmosphere for the minimum period, this problem can largely be overcome. Working in an air-conditioned room is an even more acceptable solution.
16. Excessive fluorescence background may be a problem of slide cleaning. The simplest solution is to use new slides for each experiment because it can be difficult to clean slides satisfactorily. If slides are to be re-used, the agar should be removed as soon as possible with very hot water and the slides kept wet. Slides should be soaked in Decon, then sonicated, then cleaned again.

Acknowledgments

S. Thomas is a Biotechnology and Biological Sciences Research Council CASE student, with Dr. R. Knowles, Glaxo Wellcome, UK. We thank the British Diabetic Association, the EC STEP Programme, and the British Council–German Academic Exchange Service for financial assistance during the development of this work.

References

1. Tannenbaum, S. R., Tamir, S., Derojas-Walker, T., and Wishnok, J. S. (1994) DNA damage and cytotoxicity caused by nitric oxide. *ACS Symp. Series* **553,** 120–135.
2. Ostling, O. and Johanson, K. J. (1984) Microelectrophoretic study of radiation-induced DNA damages in individual mammalian cells. *Biochem. Biophys. Res. Commun.* **123,** 291–298.
3. Singh, N. P., McCoy, M. T., Tice, R. R., and Schneider, E. L. (1988) A simple technique for quantitation of low level of DNA damage in individual cells. *Expl. Cell Res.* **175,** 184–191.
4. Olive, P. L., Banath, J. P., and Durand, R. E. (1990) Heterogeneity in radiation-induced DNA damage and repair in tumor and normal cells measured using the "Comet" assay. *Radiat. Res.* **122,** 86–94.
5. McKelvey-Martin, V. J., Green, M. H. L., Schmezer, P., Pool-Zobel, B. L., De Meo, M. P., and Collins, A. (1993) The single cell gell electrophoresis (SCGE) assay (Comet assay): A European review. *Mutation Res.* **288,** 47–63.
6. Fairbairn, D. W., Olive, P. L., and O'Neill, K. L. (1995) The comet assay: a comprehensive review. *Mutation Res.* **339,** 37–59.
7. Green, M. H. L., Lowe, J. E., Delaney, C. A., and Green, I. C. (1996) Use of the comet assay to detect nitric oxide-dependent DNA damage in mammalian cells, *Methods Enzymol.* **269,** 243–266.
8. Collins, A. R., Duthie, S. J., and Dobson, V. L. (1993) Direct enzymatic detection of endogenous oxidative base damage in human lymphocyte DNA. *Carcinogenesis* **14,** 1733–1735.
9. Anderson, D., Yu, T.-W., Phillips, B. J., and Schmezer, P. (1994) The effect of antioxidants and other modifying agents on oxygen-radical-generated DNA damage in human lymphocytes in the COMET assay. *Mutation Res.* **307,** 261–272.
10. Collins, A. R., Aiguo, M., and Duthie, S. J. (1995) The kinetics of repair of oxidative DNA damage (strand breaks and oxidised pyrimidines) in human cells. *Mutation Res.* **336,** 69–77.
11. Klaude, M., Eriksson, S., Nygren, J., and Ahnstrom, G. (1996) The Comet assay—mechanisms and technical considerations. *Mutation Res.* **363,** 89–96.
12. Kohnert, K.-D. and Hehmke, B. (1986) Preparation of suspensions of pancreatic islet cells: a comparison of methods. *J. Biochem. Biophys. Meth.* **12,** 81–88.
13. Singh, N. P., Tice, R. R., Stephens, R. E., and Schneider, E. L. (1991) A microgel electrophoresis technique for the direct quantitation of DNA damage and repair in individual fibroblasts cultured on microscope slides. *Mutation Res.* **252,** 289–296.

Barry W. Allen

30

Methods for the Study of NO-Induced Apoptosis in Cultured Cells

Anne C. Loweth and Noel G. Morgan

1. Introduction

In 1972, Kerr et al. *(1)* described a mechanism of cell death that was morphologically distinct from necrosis and they coined the term "apoptosis" [derived from the Greek *apo* (from) and *ptosis* (to fall)], to mean "falling away," as of leaves from a tree. Although descriptions of cell death consistent with what is now termed apoptosis have been made since 1885 (reviewed in **ref.** *2*), the importance of this concept has only gained widespread acceptance in recent years. Since 1972, apoptosis has been studied in many cell types and its importance in normal development, tissue mass homeostasis, programmed cell death, and disease is becoming increasingly evident. Unlike necrosis, or accidental cell death, which results from overwhelming damage to a cell making survival impossible, apoptosis is an active, regulated process undertaken by healthy cells in response to appropriate external and internal stimuli, leading to a controlled packaging of the cell contents and, ultimately, death.

Necrosis is characterized by cell swelling, organelle damage, and early cell lysis, causing release of the cell contents which, in vivo, invokes an inflammatory response prior to phagocytosis. The morphological changes associated with apoptosis are quite distinct from necrotic changes: the nucleus decreases in size and the chromatin condenses, becoming marginated at the periphery of the nucleus, and the DNA undergoes fragmentation. A well-characterized form of DNA cleavage in apoptosis is internucleosomal cleavage. Endogenous endonucleases cleave the DNA at linker regions between nucleosomes, which remain protected owing to their association with histone proteins. This hydrolysis results in oligonucleosomal fragmentation of the DNA, generating

From: *Methods in Molecular Biology, Vol. 100. Nitric Oxide Protocols*
Edited by: Michael A. Titheradge © Humana Press Inc., Totowa, NJ

discrete lengths that increase in size in increments of approx 180–200 base pairs. Although not evident in every instance of apoptotic-cell death, this type of fragmentation is extremely common in apoptosis and is widely used as a hallmark of apoptotic death, because the DNA forms a characteristic "ladder" pattern upon agarose-gel electrophoresis. In addition to this oligonucleosomal fragmentation, high molecular weight (50–300 kbp) fragmentation of DNA has also been identified in apoptotic cells *(3)*. The size of fragments generated corresponds well with that of the chromatin loop domains of the nuclear scaffold and may represent higher-order DNA degradation. In addition to endonucleases, proteases may also be involved in chromatin degradation (by proteolysis of histone, for example) resulting in chromatin unfolding *(4)*.

Cell detachment, either from neighboring cells or from the substratum of cultured cells, is an early event in apoptosis. Cells reduce in size, become rounded and lose surface structures such as microvilli. Internally, cytoskeletal filaments and ribosomes begin to aggregate and the smooth endoplasmic reticulum dilates. However, the plasma membrane retains its integrity and, at this stage, the cell is still able to exclude vital dyes such as Trypan Blue. As apoptosis progresses, plasma membrane blebbing is seen and the cell contents become packaged into membrane-bound apoptotic bodies that bud off from the cell and, in vivo, are phagocytosed by neighboring cells or macrophages. Further cell-surface alterations, including exposure of immature glycans and translocation of phosphatidylserine to the outer leaflet of the plasma membrane, are thought to label cells as "apoptotic" in order to direct phagocytosis by macrophages and neighboring cells *(5)*. In vitro, phagocytosis does not occur, and secondary necrosis is often seen as plasma membranes rupture and cell viability is lost. In vivo, this stage would occur in the phagosomes of the engulfing cell (reviewed in **ref.** ***6***). A cell may undergo the entire apoptotic process in as little as 35 min, rendering cell clearance by this mechanism an unobtrusive process, evident in perhaps only 1–4% of the cells of a tissue at any given time *(2)*. As cell contents are not lost during the process, apoptosis does not generate an inflammatory response under normal conditions in vivo.

2. Materials

2.1. Safety Regulations

Some of the reagents used in these methods are extremely toxic, particularly osmium tetroxide, ethidium bromide, and gluteraldehyde. It is essential to familiarize yourself with the requirements for the safe handling of these compounds and the necessary disposal routes before use. Information on safe handling and disposal should be sought from the product supplier.

2.2. Acridine Orange Staining

1. Phosphate-buffered saline (PBS): 137 m*M* NaCl, 10 m*M* Na_2HPO_4, 27 m*M* KCl, pH 7.4.
2. Acridine orange solution: 10 µg/mL solution of acridine orange in PBS. Store protected from light.

2.3. DNA Agarose-Gel Electrophoresis

1. Lysis buffer: 50 m*M* Tris-HCl, pH 8.0, 10 m*M* ethylenediaminetetraacetic acid (EDTA), 0.5% sodium lauroyl sarcosine. Autoclave at 121°C for 15 min before use. Supplement immediately before use with 0.5 mg/mL of proteinase K and store on ice.
2. RNase solution: Dissolve DNase-free RNase in molecular biology-grade water to give a 1 mg/mL concentration. Heat to 100°C for 10 min to ensure all DNase activity is destroyed *(7)*. Aliquot and store at –20°C.
3. Loading buffer: 1% low-melting temperature agarose, 10 m*M* EDTA, pH 8.0, 40% sucrose. Store in aliquots at –20°C.
4. 10x TPE buffer: Dissolve 142 g of Tris-HCl in 500 mL of distilled water, add 15.5 mL of 85% phosphoric acid and 40 mL of 0.5 *M* EDTA, pH 8.0. Make up to 1 L with distilled water.

2.4. Electronmicroscopy

1. 0.1 *M* Phosphate buffer, pH 7.2: Dissolve 1.71 g of sucrose in 36 mL of 0.2 *M* Na_2HPO_4 plus 14 mL of 0.2 *M* NaH_2PO_4. Make up to 100 mL with distilled water. Store at 4°C.
2. Fixative solution: 2.5% (v/v) gluteraldehyde in 0.1 *M* phosphate buffer, pH 7.2. Store at 4°C. CAUTION: gluteraldehyde is very toxic: wear gloves and work in a fume hood.
3. Osmium tetroxide: Make a 1% (w/v) solution of osmium tetroxide in 0.1 *M* phosphate buffer, pH 7.2. CAUTION: osmium is highly toxic: wear gloves and work in a fume hood.
4. Uranyl acetate: make up a 3% (w/v) solution in 70% ethanol.
5. Lead citrate: Dissolve 1.33 g of lead nitrate and 1.76 g of sodium citrate in 30 mL of distilled water. Shake vigourously for 1min then leave to stand, with intermittent shaking, for 30 min. Add 8 mL of 1 *M* NaOH and make up to 50 mL (pH 12 ± 0.1).

Most reagents are available from SIGMA Ltd, UK. Reagents for electronmicroscopy are available from Taab, UK and Agar Scientific, UK.

3. Methods

The identification of apoptotic cells can be based principally on visualization of the morphological changes occuring in the nucleus during apoptosis.

Specifically, the chromatin condensation and margination and the oligonucleosomal DNA fragmentation may be identified using a variety of techniques.

3.1. Acridine Orange Staining

Chromatin condensation may be visualized by staining the cellular DNA with the dye acridine orange, which fluoresces brightly under UV illumination after binding to chromatin.

1. Following culture of the cells with an apoptosis-inducing agent, harvest the cell samples by centrifugation (500*g* for 2 min). **Note:** if adherent cells are being used, it is important to collect detached cells as well as those still adhering, because detachment is often an early event in apoptosis. Adherent cells may be collected by first washing the monolayer with PBS, then incubating with trypsin-EDTA (0.05%, 0.02% (w/v), respectively) for 1 min. Aspirate the trypsin and incubate at 37°C for 3–5 min, until the cells become detached, and resuspend in culture medium.
2. Mix 10 µL of the cell suspension with an equal volume of 10 µg/mL acridine orange solution on a microscope slide and cover with a coverslip. Examine under a fluorescence microscope.
3. Count 100–200 intact cells per sample and note the number of these having apoptotic nuclei (*see* **Note 1**). Express the results as percentage apoptosis.

3.2. DNA Gel Electrophoresis

1. Pellet 10^6 cells by centrifugation and remove the supernatant, taking care to remove all traces of medium.
2. Resuspend the cells, on ice, in 20 µL of lysis buffer.
3. Use gentle pipetting and vortexing to ensure complete lysis of the sample, befor incubating on a dri-block for 1 h at 55°C.
4. Spin down the residue in a microfuge and add 5 µL of 1.0 mg/mL RNase A prio to incubation for a further 1 h at 55°C. Spin again briefly to collect the conden sate. The samples may now be stored at –20°C or prepared immediately for elec trophoresis, as follows.
5. Heat the lysate to 70°C and mix with 10 µL of loading buffer, pre-heate to 70°C.
6. Prepare a 2% agarose gel, by dissolving 0.5 g of agarose with heating (micro wave oven, full power for 35 s) in 25 mL of 1 X TPE buffer (approx 200 mL c 1 X TPE should be prepared). Cool to below 50°C before adding ethidium bro mide (*see* **Note 2**) at 0.5 µg/mL.
   (CAUTION: ethidium bromide is a carcinogen and mutagen. Wear gloves at a times and discard all contaminated debris and liquids via designated dispos routes.)
7. Pour into a Horizon 58 minigel apparatus, or equivalent, and leave to set fc 15 min.

8. Load the samples onto the dry gel and allow to set for 5 min. Pour on running buffer (1 X TPE) to cover the gel to a depth of 1–2 mm and electrophorese at a constant 40V for 2 h.
9. Visualize the gel on a transilluminator, under UV light (*see* **Note 3**).

3.3. Transmission Electronmicroscopy

1. Harvest at least 10^6 cells per sample and pellet by centrifugation. Remove the supernatant and add 1 mL of fixative (CAUTION: handle gluteraldehyde in a fumehood and wear gloves).
2. Leave on ice for at least 4 h (or overnight) to fix.
3. Wash the cells by adding 1 mL of phosphate buffer and leaving on ice for 30 min. Repeat twice.
4. Remove the final wash and cover the pellet with 1% osmium tetroxide (CAUTION: extremely toxic: wear gloves, work in fumehood, avoid contact and inhalation), and incubate on ice for 1 h.
5. Remove the osmium tetroxide and wash 3 times in distilled water.
6. Dehydrate the cell pellet by sequential immersion in 30, 60, 90, and 100% acetone at 15 min intervals. Change 100% acetone at least three times and ensure that the last change is in dry acetone. (Samples may be stored at 4°C in 60% acetone prior to further processing.)
7. Incubate the dehydrated samples in 50% Spurr's resin/acetone mixture for 1 h, with rotation.
8. Change to 100% resin and incubate the pellet overnight, with rotation.
9. Following three further changes of resin, embed the pellet by baking overnight in an oven at 60°C.
10. Cut 60–70 nm ultramicrotome sections, using a glass knife (these can be made on a LKB Knifemaker II) and a Reichert Ultramicrotome (*see* **Note 4**). Float the sections onto water and position onto grids. Air dry the grids on filter papers before proceeding.
11. Stain in uranyl acetate for 10 min (*see* **Note 5**). Wash thoroughly in distilled water (rinse 15 times in each of 3 fresh beakers of water).
12. Stain in lead citrate for 5 min. Wash thoroughly and leave grids to air dry on filter paper.
13. Examine the sections by transmission-electron microscopy. Score 100–200 cells per sample as apoptotic or nonapoptotic and express as percentage apoptotic (*see* **Note 6**).

4. Notes

1. Apoptotic nuclei appear brightly stained owing to the condensed chromatin that is usually fragmented and marginated, and frequently appears as a crescent or horseshoe shape (**Fig. 1**). Viable cells have barely discernible, translucent nuclei. Necrotic nuclei may contain condensed chromatin but, in contrast to apoptotic cells, this appears flocculated. The percentage of apoptotic cells should be established by counting a minimum of 100–200 cells per sample. In this method, it is

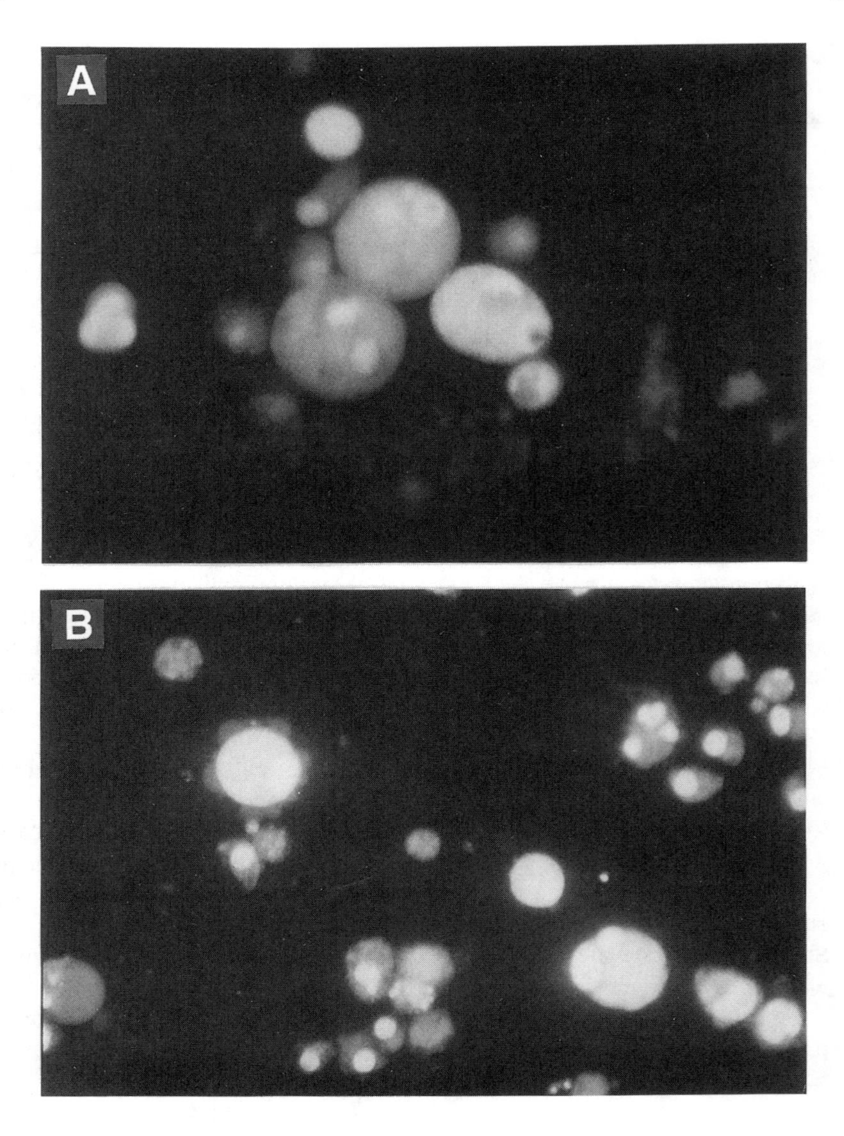

Fig. 1. Use of acridine orange staining to detect apoptotic nuclei. RINm5F cells were exposed to 1 m*M* GSNO for 24 h. Detached cells were harvested, resuspended in 20 μL of medium. 10 μL of the cell suspension were mixed with an equal volume of 10 μg/mL acridine orange in PBS and transferred to a microscope slide for examination under a fluorescence microscope. Apoptotic nuclei can be identified by the marginated, condensed (brightly stained) chromation **(A)**. The chromatin may also appear fragmented **(B)** and membrane blebbing is commonly observed **(B)**.

Fig. 2. Agarose gel electrophoresis of DNA from lysed cells. Following culture, with and without exposure to 1 m*M* GSNO for 24 h, 10^6 cells per condition were pelleted and resuspended, on ice, with 20 μL of lysis buffer. Following incubation of the lysate for 2 h at 55°C, 10 μL of loading buffer was added (at 70°C) prior to loading onto a 2% agarose gel (containing ethidium bromide). The gel was developed at 40V for 2 h, prior to visualizing under UV light. DNA from apoptotic cells is seen as a "ladder" of 180–200 bp increments, corresponding to oligonucleosomal fragments (lane 3). Viable, nonapoptotic cells contain only intact, high molecular-weight DNA (lane 2). A 100 bp DNA ladder is shown for comparison (lane 1).

essential to evaluate only intact cells since, in the case of fragmented cells it is not easy to distinguish those that have died via necrosis from those which have undergone secondary necrosis after initially entering apoptosis. In both cases, these cells will have lost their membrane integrity and will appear lysed. Because cells enter the apoptotic pathway at different times during an experiment, not all the cells examined at any one time point will be at the same stage of apoptosis and, therefore, only a proportion of the cells examined will display the nuclear features characteristic of apoptosis. In our studies, clonal pancreatic β-cells were exposed to chemical NO donors such as S-nitrosoglutathione (GSNO) or 3-morpholinosydnonimine (SIN-1), for 1–48 h, over a concentration range of 10–2000 μ*M*.

2. Because post-staining with ethidium bromide is inefficient owing to the density of the gel, it is recommended that the dye is incorporated into the gel prior to electrophoresis.
3. Using this method, DNA from apoptotic cells can be identified from the characteristic ladder pattern, whereas viable, nonapoptotic cells contain only intact, high molecular-weight DNA. Examples are shown in **Fig. 2**.
4. The cutting of ultramicrotome sections is technically difficult and will require considerable practice.
5. It is a good idea to place small drops of stain onto a block of wax in a Petri dish. Grids can then be inverted and floated in the stain.
6. Examples of electronmicrographs are given in **Fig. 3**. The granular appearance of the chromatin in **Fig. 3A** is typical of a nonapoptotic nucleus, whereas the dense, smooth-edged, and marginated chromatin in **Fig. 3B** is characteristic of an apoptotic nucleus.

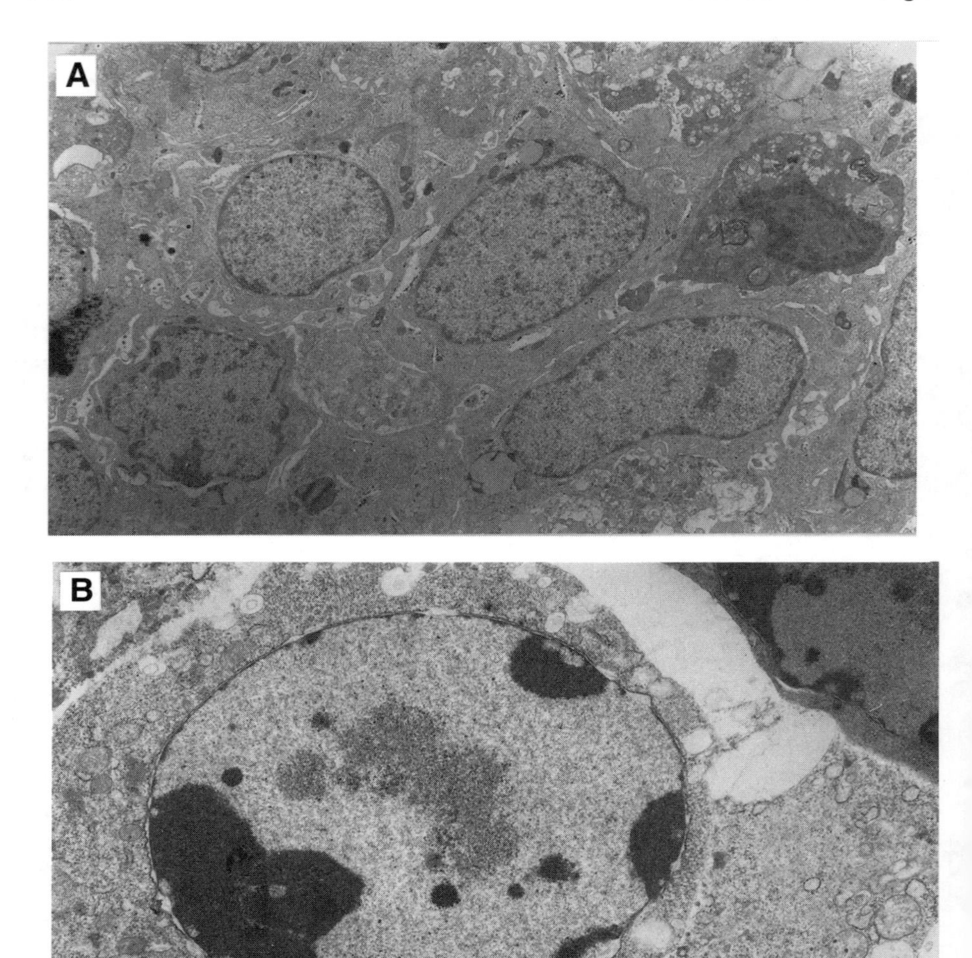

Fig. 3. Electronmicrographs of apoptotic and nonapoptotic cells. Following treatment, cells were harvested and fixed in 2.5% gluteraldehyde, post-fixed in osmium tretroxide then embedded in epoxy resin prior to staining in uranyl acetate and lead citrate. Ultramicrotome sections were examined using a Jeol100CXII transmission electronmicroscope. The chromatin of nonapoptotic cells **(A)** remains granular in appearance. Apoptotic nuclei **(B)** contain densely stained, marginated, and condensed chromatin. The examples shown are **(A)** untreated rat-islet cells and **(B)** a GSNO-treated RINm5F cell.

7. The techniques described in this chapter can provide data that, taken together, allow evaluation, both qualitatively and quantitatively, of the extent of apoptosis in a sample of cells, following any given treatment. It should be emphasized that it is preferable to use a combination of techniques to confirm apoptosis, rather than any one individually. Typically, we recommend acridine-orange staining and gel electrophoresis to establish the dynamics of apoptosis in response to a particular treatment before undertaking electronmicroscopy, because this latter procedure is relatively labor-intensive.
8. An increasing number of kits are now becoming available from commercial suppliers which purport to measure apoptosis by exploiting the ability of terminal deoxynucleotidyl transferase (TdT) to end label the oligonucleosomal fragments formed during apoptosis, allowing quantitation by fluorescent or colorimetric methods. Whereas these protocols do not normally claim to label only the double-stranded DNA nicks found on oligonuclesomal fragments (and therefore apoptotic cells) and all recommend the use of alternative techniques to confirm apoptosis, we would re-emphasize the potential limitations of these methods. We have evaluated one such cell-death detection ELISA kit, which depends upon selective binding of oligonucleosomal fragments to anti-DNA and anti-histone antibodies, followed by colorimetric detection. Our experience was that the extent of labeling obtained using the kit bore little relationship to the amount of DNA laddering observed by gel electrophoresis, under a variety of conditions. Indeed, high background levels were often obtained under circumstances when no DNA laddering could be seen and under conditions that were not associated with activation of apoptosis using more stringent criteria. We are confident that, under our experimental conditions, these discrepancies reflect nonspecific labeling of DNA by the kit reagents rather than differences in sensitivity between the techniques. On this basis, we strongly recommend that classical morphological criteria are applied initially to identify apoptotic cells after treatment with NO donors or exposure to conditions designed to elevate NO, and that less rigorous methods are only used once activation of apoptosis has been confirmed uequivocally. Some recently introduced methods (often marketed in kit form) can provide more rapid results but depend on indirect detection of DNA strand breaks and cannot be relied upon to distinguish apoptosis from necrosis. Use of these methods should only be accepted if the evidence for apoptosis is supported by more direct morphological criteria.

References

1. Kerr, J. F. R., Wyllie, A. H., and Currie, A. R. (1972) Apoptosis: a basic biological phenomenon with wide-ranging implications in tissue kinetics. *Br. J. Cancer* **26,** 239–257
2. Majno, G. and Joris, I. (1995) Apoptosis, oncosis and necrosis: An overview of cell death. *Am. J. Pathol.* **146,** 3–15

3. Walker, P. R., Smith, C., Youdale, T., Leblanc, J., Whitfield, J. F., and Sikorska, M. (1991) Topoisomerase II-reactive chemotherapeutic drugs induce apoptosis in thymocytes. *Cancer Res.* **51,** 1078–1085
4. Kokileva, L. (1994) Multi-step chromatin degradation in apoptosis and DNA breakdown in apoptosis. *Int. Arch. Allergy Immunol.* **105,** 339–343
5. Savill, J., Dransfield, I., Hogg, N., and Haslett, C. (1990) Vitronectin receptor-mediated phagocytosis of cells undergoing apoptosis. *Nature* **343,** 170–173
6. Arends, M. J. and Wyllie, A. H. (1991) Apoptosis: mechanisms and role in pathology. *Int. Rev. Exp. Path.* **32,** 223–255
7. Sambrook, J., Fritsch, E. F., and Maniatis, T. (1989) *Molecular Cloning: A Laboratory Manual,* (2nd ed.), Cold Spring Harbor Press, Cold Spring Harbor, NY.

Index